Economic Geography

This book series serves as a broad platform for scientific contributions in the field of Economic Geography and its sub-disciplines. *Economic Geography* wants to explore theoretical approaches and new perspectives and developments in the field of contemporary economic geography. The series welcomes proposals on the geography of economic systems and spaces, geographies of transnational investments and trade, globalization, urban economic geography, development geography, climate and environmental economic geography and other forms of spatial organization and distribution of economic activities or assets.

Some topics covered by the series are:

- Geography of innovation, knowledge and learning
- Geographies of retailing and consumption spaces
- Geographies of finance and money
- Neoliberal transformation, urban poverty and labor geography
- Value chain and global production networks
- Agro-food systems and food geographies
- Globalization, crisis and regional inequalities
- Regional growth and competitiveness
- Social and human capital, regional entrepreneurship
- Local and regional economic development, practice and policy
- New service economy and changing economic structures of metropolitan city regions
- Industrial clustering and agglomeration economies in manufacturing industry
- Geography of resources and goods
- Leisure and tourism geography

Publishing a broad portfolio of peer-reviewed scientific books *Economic Geography* contains research monographs, edited volumes, advanced and undergraduate level textbooks, as well as conference proceedings. The books can range from theoretical approaches to empirical studies and contain interdisciplinary approaches, case studies and best-practice assessments. Comparative studies between regions of all spatial scales are also welcome in this series. Economic Geography appeals to scientists, practitioners and students in the field.

If you are interested in contributing to this book series, please contact the Publisher.

More information about this series at http://www.springer.com/series/15653

Canfei He · Xiyan Mao

Environmental Economic Geography in China

 Springer

Canfei He
College of Urban and Environmental
Sciences
Peking University
Beijing, China

Xiyan Mao
School of Geography and Ocean Science
Nanjing University
Nanjing, China

ISSN 2520-1417   ISSN 2520-1425   (electronic)
Economic Geography
ISBN 978-981-15-8990-4   ISBN 978-981-15-8991-1   (eBook)
https://doi.org/10.1007/978-981-15-8991-1

This Springer imprint is published by the registered company Springer Nature Singapore Pte Ltd.
The registered company address is: 152 Beach Road, #21-01/04 Gateway East, Singapore 189721, Singapore

Foreword

In recent decades China has undergone a remarkable economic transformation. In 1980 it was only the 10th of economy in the world. At that time China's economy was a mere 8% of that of the world leader, the United States. According to the International Monetary Fund, in 2019 China was the second-largest economy, with an overall GDP that equalled two-thirds of that of the North American giant. At this rate of change, China is likely to become the largest global economy by 2030, if not before.

This breakneck and long-sustained speed of growth has made China the envy of the world. However, it has also come at a cost, as rapid development is—as in many fast-growing countries elsewhere—producing a number of important economic, social, and environmental problems. The first problem is related to the rise of interpersonal inequalities. China has witnessed a rapid rise of a large middle-class, which is, undoubtedly, good news. However, social polarization has also risen rapidly: while a small number of billionaires are amassing large fortunes, large swaths of the Chinese population are not party to the proceeds of rapid economic growth. Those being left behind are struggling to make ends meet, making economic growth far from inclusive.

Second, the country also experienced territorial divergence. This was especially the case during the 1990s and early 2000s. The concentration of economic activity along the coast and, in particular, in a number of large, dynamic, and increasingly prosperous cities has created large territorial imbalances, which could trigger—as we are seeing elsewhere in the world—economic and social tensions.

Third, economic agglomeration in and around large metropoles along the eastern seaboard has unclenched a number of environmental problems in a country that is already prone to natural environmental risks. The first stages of development were built around mass production and the building of large and often highly polluting plants. This contributed to a serious rise in harmful emissions that have become entrenched across many areas of the country and, unfortunately, are also becoming part of China's international trademark. Air pollution now represents a constant threat to the health of the population of many northern Chinese cities and a barrier for the expansion of existing or the inception of new economic activity. The massive pressure on water resources is also exacerbating the recurrent draught problems that have for long challenged many regions in the North of the country. Moreover, the rising demand in electricity and the need to resort to multiple ways of producing it—from nuclear plants to new dams, while still keeping a large dependence on coal-fired electricity production—is not without serious environmental consequences. China's hunger for electricity is making the transition to more environmentally friendly energy sources difficult.

Economic geography has confronted the challenges of rising interpersonal and territorial inequality profusely. The breadth and depth of research on those topics in China and elsewhere have allowed economic geographers not just to identify the causes behind the increasing social and territorial polarization, but also to influence the policy debate and to propose novel and workable policy solutions.

It is therefore all the more surprising that, as indicated by He and Mao in their book, economic geographers have shied away from delving in sufficient depth into the environmental consequences of the agglomeration of economic activities in an increasingly limited number of spaces. This implies that addressing the environmental problems that besiege China through an economic geography lens is now more needed than ever. Environmental Economic Geography in China by He and Mao fills this gap in our knowledge brilliantly. When economic geography is at the root of the problem, economic geography should also be part of the solution.

In Environmental Economic Geography in China, Canfei He and Xiyan Mao, two world-leading Chinese economic geographers, ably map the changing geographical location of pollution-intensive industries in China and match those changes to the evolution of pollution emissions and of waste across China. In doing so, they open up a fundamental series of questions about what are the environmental consequences of rapid industrialisation and economic growth, which should be at the heart of any analysis concerned with the environmental challenges that China now faces. This leads to an in-depth overview and examination of issues such as how industrial transition has affected the rise in environmental challenges, while, at the same time, enquiring about the extent to which changes in legislation can actually contribute to ease many of the environmental problems affecting China as a whole. But their work is not limited to this. It also covers a whole additional raft of vital environmental issues, such as the role of trade and of the regional division of labour for industrial pollution and how export upgrading can contribute to improvements in environmental performance across China.

He and Mao put together a refreshing and highly novel view of how economic geography and the concentration of economic activity over recent decades are at the root of the environmental challenges China now faces. They also indicate that economic geography can provide solutions to the environmental problems that ail China. From this perspective, Environmental Economic Geography in China represents an essential read for all those decision-makers, researchers, academics, and students that strive to reconcile the benefits of a rapid and still necessary economic development with a more harmonious, integrated, and, above all, environmentally sustainable expansion of economic activity across the country. Reading the book thus provides readers with an essential overview of how to improve the health and quality of life of current Chinese citizens by reducing emissions and making a better more sustainable use of energy and natural resources, while guaranteeing that the well-being and welfare of future generations is not jeopardized.

March 2020
Andrés Rodríguez-Pose
London School of Economics
London, England

Parts of this monograph have been published in the following journal articles but with significant modifications.

1. Chapter 4, modified article originally published in [Yang Xin and He Canfei (2015) Do polluting plants locate in the borders of jurisdictions: evidence from China? Habitat International, 50, pp. 140–148]. Published with kind permission. All Rights Reserved.
2. Chapter 6, modified article originally published in [He Canfei, Pan Fenghua and Yan Yan (2012) Is Economic Transition Harmful to China's Urban Environment? Evidence from Industrial Air Pollution in Chinese Cities, Urban Studies, 49(8), pp. 1767–1790]. Published with kind permission. All Rights Reserved.
3. Chapter 7, significantly modified article originally published in [He Canfei, Mao Xiyan and Zhu Xiaodong (2018) Industrial dynamics and environmental pollution in urban China, Journal of Cleaner Production, 195, pp. 1512–1522]. Published with kind permission. All Rights Reserved.
4. Chapter 8, significantly modified article originally published in [Mao Xiyan and He Canfei (2018) A trade-related pollution trap for economies in transition? Journal of Cleaner Production, 200, pp. 781–790]. Published with kind permission. All Rights Reserved.
5. Chapter 11, significantly modified article originally published in [Mao Xiyan and He Canfei (2017) Export upgrading and environmental performance: evidence from China, Geoforum, 86, pp. 150–159]. Published with kind permission. All Rights Reserved.

Acknowledgements

We acknowledge the financial support of the National Natural Science Foundation of China (No. 41731278, No. 41425001 and No. 41801104). The authors are responsible for all errors and interpretations.

Contents

List of Figures

List of Tables

Chapter 1
Introduction: When Economic Geography Meets the Environment

1.1 Responding to Environmental Challenges

The changing social-technical paradigms accelerate the pace of development but simultaneously disturb the balance of our planet at an unprecedented speed. The nature-society interactions face various dilemmas between development and conservation more than ever (Zimmerer 2010). The rapid processes of industrialisation, urbanization, and globalization account for an increasing proportion of the determinants of environmental changes such as climate change, environmental pollution, deforestation, loss of biodiversity, and depletion of natural resource. Environmental changes, in turn, challenge the sustainable development of human society and requires a wide range of positive feedbacks, including industrial transformation, economic transition, smart growth of cities, and global cooperation. In this regard, all United Nations Member States adopt the 2030 Agenda for Sustainable Development in 2015, which proposes 17 Sustainable Development Goals (SDGs) representing the urgent challenges that confront the human society as a whole. SDGs seek to coordinate the socio-economic development with eco-environmental conservation, and therefore require double efforts from both physical sciences and social sciences. The call of SDGs reiterates the interdisciplinary nature of environmental-related studies, hence attracting huge attention from scientists, including geographers.

Understanding and responding to environmental challenges, particularly the sustainability issues, requires geographical wisdom. Firstly, environmental challenges do not occur in a vacuum but depend on their spatial–temporal contexts. Just like the agricultural production in China, when the northern regions are suffering from the drought, the southern regions are struggling with the flood. That is to say, in one particular place, environmental challenges are less likely to be simply determined by one or a few factors, but instead depend on a wide array of complex and interactive factors, which is usually different from place to place. Thus, the causes of environmental problems, the formation of environmental problems, and the capabilities of responding to environmental problems tend to be place-specific. As a result, even the same environmental issues may manifest themselves in different ways across

© Springer Nature Singapore Pte Ltd. 2020
C. He and X. Mao, *Environmental Economic Geography in China*,
Economic Geography, https://doi.org/10.1007/978-981-15-8991-1_1

places, then challenging the places in different manners. To better understand how environmental issues come into being in one place and then affect it, a geographical perspective is essential. There is no one-size-fits-all problem.

Secondly, environmental challenges are also spatially interdependent. The socio-economic and eco-environmental systems operate at different scales, whose boundaries do not have to be consistent. In such a setting, the anthropogenic environmental changes can be transboundary. There are two channels for border-crossing. One is that the pollution emissions produced in one particular place can move transboundary, and thereby affect surrounding (or even more distant) places. The long-range transport of air pollutant is exactly the case. It is reported by scientific findings that the long-range transport is one of the dominant sources of the fine particulate matters ($PM_{2.5}$) concentration in Beijing (Yang et al. 2016; Li et al. 2017; Zhang et al. 2018a, b). There are also findings indicating that transboundary air pollution contributes 42% of the $PM_{2.5}$ concentration in central-east China, which is even traced to northern China (Shi et al. 2020).

Another typical case facing transboundary pollution is the river basin. A river basin is a physiographic unit drained by a river and its tributaries, usually covering several administrative units. Pollutants produced by the socio-economic activities in the upper reaches will be transported along the river, and thereby impact the environment in the lower reaches. Over the decades, empirical evidence from various river basins around the world has been piled up to unravel this issue, involving interdisciplinary efforts (Chau and Jiang 2003; Sigman 2005; Cai et al. 2016).

The other channel is the socio-economic interconnections between places that allow one place to utilize resources in another place and then exert environmental influences there. Particularly in the context of globalization, the increasing flows of resources, enlarging markets of goods, and deepening division of labours promote opportunities for development but simultaneously intensify the disturbance in the environment. As the socio-economic relationship shifts from local to global, the nature-society interactions become even more complicated than before. Just like the butterfly effect, a small change in one place can result in a large difference in remote places. For example, the US-China trade war changed the global trade pattern of soya-beans and the US soya-beans exports to China decreased significantly. On the other hand, Brazil is ready to increase its soya-bean production, thereby raising concerns about the deforestation in the Amazon rainforest (Fuchs et al. 2019).

There are already plenty of researches regarding environmental issues due to socio-economic interdependencies crossing national borders, such as the pollution embedded in trade (Peters and Hertwich 2006; Lin et al. 2016), carbon leakage (Babiker 2005; Jakob and Marschinski 2013), virtual water (Cai et al. 2019; Rosa et al. 2019), and tele-coupling/teleconnections of land use (Seto et al. 2012; Liu et al. 2013). These fields point to the fact that the spatial separation of production and consumption results in the global displacement of pollution, greenhouse gas emission, water consumption, land use, and resource utilisation.

Global displacement represents a burden-shifting effect that one place can transfer the resource-based, low-value-added, pollution-intensive activities to others so that this place can be free from environmental pressures. However, this is at the cost of sacrificing other places' environments. Environmental economics posits that this burden-shifting depends on the relative stringency of environmental regulation as well as the gap of economic development between two places. Taken together, the burden-shifting tends to occur from developed economies to developing ones (Cole 2004). However, this spatial trend is controversial in theory. Empirical studies also report various findings without consensus (Kahn 2003; Cave and Blomquist 2008; Zheng and Shi 2017). Advances in the literature highlight more complicated and interactive factors and conditions that may violate the pollution haven hypothesis (Dean et al. 2009; Cole et al. 2011; Jiborn et al. 2018).

Overall, the transboundary flows do not occur randomly but depend on both the differences and linkages between places. The incorporation of geographical wisdom may help to unravel the formation, configuration, and evolution of spatial interdependencies of environmental pollution. Understanding these spatial interdependencies depicts a big picture to trace the source of environmental problems, identify the stakeholders involved, and figure out the potential for joint actions.

Geography is a discipline centred on the nature-society interactions and insisting on exploring the surface of our planet with a holistic and systematic perspective. Hence, it is safe to say that geography is a discipline closely related to both environmental problems and human development. Geographers enjoy advantages of participating in the environmental-relevant studies, especially sustainability issues. In practice, there are geographers from various geography subdisciplines engaging in environmental-relevant studies and contributing to the theory and practice of sustainability. Physical geographers are active in recognizing, interpreting, and projecting environmental changes and their impacts (Day 2017). They follow the scientific paradigm and focus on the objective reality that can be observed, described, measured, understood, and managed (O'Brien 2010). By contrast, human geographers are more interested in the social construction of nature (Braun and Castree 2005). They highlight the ontological differences between different groups when facing and responding to environmental challenges (O'Brien 2010). A typical sub-discipline is the political ecology, which understands environmental change as a result of power relations.

In marked contrast to other sub-disciplines, the development of economic geography has surprisingly retreated from the environment issues during past decades. Although there are some conceptual and empirical works conducted by economic geographers in succession during past decades, the environment has seldom influenced the mainstream research of economic geography and also been absent from its widely used textbooks (Gibbs 2006; Soyez and Schulz 2008). As an old saying goes: Silence is true wisdom's best reply. However, is it really wise for economic geographers to keep silent in response to the increasing environmental challenges? There are some safe reasons why developing and practising environmental economic geography is of great necessity.

Firstly, the techno-economic paradigm is stepping into the next stage, which is manifested by "green", such as green innovation, green production, green consumption and green economy (Hayter 2008). It is imperative to incorporate environment into the already-existing theories of economic geography. Particularly, without a clear linkage with the environment, it is impossible to discuss the "green". Using the Environmental Kuznets Curve (EKC) hypothesis as an example, Stern (2004) proposes a "new toxics" scenario that while technological advances can decouple economic growth with pollution emissions, they are still likely to bring about new pollutants. "As the older pollutants are cleaned up, new ones emerge, so that overall environmental impact is not reduced (Stern 2004)". If this is the case, green innovation is less likely to be a panacea for all kinds of environmental challenges, and a better understanding of the environment is necessary.

Secondly, the environment is shifting from an external factor to an internal one for socio-economic activities. In the context of global environmental changes, socio-economic activities become increasingly vulnerable in response to various environmental challenges. Regarding the conventional interests of economic geographers in the locations, global production, and regional development, it becomes progressively more impossible to overlook the role of the environment or even simply treat the environment as a context. Environmental changes are becoming an internal (endogenous) determinant of socio-economic activities. For example, Gibson and Warren (2016) notice the role of resource scarcity and environmental regulation in the configuration of the global production network. They find that "scarcity of raw materials has spawned shifting economic geographies of new actors who influence the whole GPN (Gibson and Warren 2016)".

Thirdly, the complicated processes of global environmental change have both cultural and institutional dimensions, representing the reaction and responses of human society to global environmental change. That is, the shared value, behaviours, and rules of socio-economic activities would change accordingly. For instance, new consumption based on the changing environmental understanding will challenge and add to the literature of economic geography directly (Hobson 2003). Similarly, the rising demand for local environmental regulation and global environmental governance also exerts increasing influences in the classical theories of economic geography (Bush et al. 2015).

The three reasons above show that it is inevitable to incorporate the environment into economic geography. Besides, the development of EEG is much more than branching into a new sub-discipline. By focusing on the environmental challenges, which have widespread impacts on both physical and social sciences, EEG will allow economic geographers to participate in more science programs and expand their influence in supporting decision-making. In fact, various global scientific projects, such as the Land Use and Cover Change and the Global Land Project, have a bunch of topics closely related to the expertise of economic geographers. Scientists from other fields also start to highlight the value of economic geography. Munroe et al. (2014) propose that "economic geography can help land change science move beyond the neoclassical framing". Friis et al. (2016) also suggest that the theoretical perspective of economic geography allows land system science "to address the increasing importance of distal connections, flows and feedbacks".

Although economic geographers are relatively silent in this realm, the theoretical development of economic geography still exhibits its great potential for environmental studies. Economic geographers can distinctively contribute to addressing environmental problems and sustainable development.

First of all, economic geography highlights the spatiality of socio-economic activities. Why do the same environmental phenomena become troubles in some places but not in the others? Is one particular environmental issue inevitable in one place? Why are some places more vulnerable to environmental issues than others? Can a successful growth model of one place towards sustainability be replicated by other places? Answers to these questions are closely related to the spatiality of socio-economic activities. Advances in economic geography continue to answer the questions of why economic activities occur here but not there and how economic activities develop in one place, which points to the spatiality of economic activities directly.

Why does the spatiality of economic activities matter? Conventional environmental studies tend to provide various suggestions for selecting economic activities. The suggestions are fragmented, case-sensitive, and even conflicting. The spatiality of economic activities reminds environmental studies that socio-economic activities also have their inherently rules and rationales, which is subject to natural advantages as well as the techno-economic paradigm. As such, there is always some rationality of why environment-unfriendly activities come into being in one place. Also, it is impossible for one place to replace those undesirable activities with desirable ones all at once. In this regard, the participation of economic geography can make a difference.

Secondly, economic geography is keen on understanding the spatial configuration of economic activities at multiple scales, spanning from local clusters to global production networks. The spatial configuration is a crucial factor to moderate the environmental effects of economic activities. On the one hand, the spatial configuration represents the potential to improve environmental performance through resource reallocation and the joint action between economic agents, such as the local clusters and innovation systems. On the other hand, it also indicates the environmental performance between the global and the local. In some spatial configuration, a local benefit could be at the sacrifice of the others or would even result in a global loss. A better understanding of the spatial configuration is required to know how to coordinate the environmental performance between the local and the global. In this regard, the participation of economic geography in environmental studies can enrich the neoclassical framings of environmental studies by the theories on the spatial configuration. It will allow the "economy" in environmental studies to go beyond conventional thinking based on administrative confines.

Thirdly, economic geography tends to appreciate the capabilities of economic activities in reacting and responding to environmental challenges. Unlike the theoretical thinking of most environmental studies that seek for balance, economic geography cast doubt on the efforts to find an equilibrium between nature and society. Or put it another way, economic geography rejects that there is an optimal spatial configuration of the economic landscape. Instead, it sees more value in economic novelty. Thus, from the perspective of economic geography, responding to environmental challenges is not a matter of finding the limit of growth, such as the so-called environmental capacities. It is about to find sustained motivation for innovation, which is essential to alter environment-unfriendly ways. During past decades, advances in economic geography largely enrich the literature of innovation and transition studies. They seek to unravel how economic novelty can occur (or be prompted) in the spatially heterogeneous but interdependent world. Hence, the participation of economic geography in environmental studies will contribute to the studies on green innovation and sustainability transition directly.

1.2   Practicing Environmental Economic Geography in China

The development of China during the past decades provides a great opportunity to practice Environmental Economic Geography and reflect its development. The philosophy of economic development in China keeps changing during the last decades. It shifts from worshipping GDP in the previous period to increasingly highlighting sustainability and ecological civilization. And it also swings between guaranteeing spatial equality and prioritizing spatial efficiency. The interactions between the central government and local municipalities are looking for a dynamic equilibrium by fiscal and tax reform. The outward-oriented growth models also shift from scale expansion to quality improvement, which simultaneously promote the endogenous growth model. Grounded in such a context, the economic geography in contemporary China exhibit a wide range of dualities, such as the changing core-periphery pattern of the economic base, the central-local relationships of governments, the foreign-domestic interactions of resources and economic actors, and the local-nonlocal connections for regional integrations. Hence, we can use a couple of interactions to explain the main reasons why the case of China can contribute to the development of Environmental Economic Geography.

First of all, China is rich in the diversity of nature-society interactions. China is a huge country with an area of 9.6 million kilometres and a population of around 1.4 billion. It is also the second-largest economies around the world, whose gross domestic production reached 99.09 trillion Chinese yuan ($\approx$14.36 trillion US dollars) in 2019. Such a huge country has both eco-environmental and socio-economic diversity. Their interactions are even more complicated and diversified across places. For example, sustainable development of traditional resource-based regions is subject to the shifting socio-technique paradigm of industrialization, the depletion of local resource endowment, and the loss of attraction to human capital (Hao and Deng 2019). By contrast, sustainable development of outward-oriented regions is subject to the embeddedness of nonlocal inputs, the risks of external shocks, and the growth of endogenous forces. Overall, the rich diversity of nature-society interactions allows environmental economic geography to investigate the environmental performance of different development trajectories and explore the site-specific trajectory towards sustainability.

Secondly, China has interactions between the core and the periphery at multiple levels. Along with the rapid processes of industrialisation, globalization, and urbanization during the past decades, a coastal-inland division of development becomes clear. The economic take-offs in the coastal regions attract a wide range of economic factors to concentrate and increase the economic density there. A high economic density also generates great pressures on the local environment, which then results in crowding-out effects in recent years. However, the reverse trends of factor flow expose the inland regions to a risk of environmental degradation (Wu et al. 2017), since the inland regions tend to be more eco-fragile than the coastal. Hence, the coastal-inland division and their interactions provide a good case to reconsider the environmental effects of industrial transfer.

At the regional level, there are also clear core-periphery patterns within megaregions, city agglomerations, and provinces. The core-periphery pattern at the regional level represents the spatial division of labour, which is expected to improve environmental performance by sufficient integration. Thus, the core-periphery patterns at the regional level provide good cases to understand the environmental implication of regional integration (Mao et al. 2020). Besides, they also serve as a good spatial unit to examine the beggar-thy-neighbour behaviours of environment-unfriendly activities (Chen et al. 2018). At the local level, the core-periphery pattern tends to represent industrial agglomeration and urban–rural linkages (Xiao et al. 2020). The different industrialisation and urbanization models in China form various core-periphery patterns in Chinese urban spaces. These core-periphery patterns also provide great opportunities to exam the environmental effects of local agglomerations/clusters as well as urbanization.

Thirdly, the changing economic geography in China during the past decades depends on the interactions between domestic and foreign forces, especially foreign direct investment (FDI) and international trade. The integration with the global markets by the advantages of resources and labour forces has sparked China's economic miracle to some extent. However, it also exposes the internal regions to a higher level of environmental burden-shifting risks, which is embodied in the international socio-economic connections (Wang et al. 2017). There is no consensus on the net effects of China's involvement in globalization on its environment. On the one hand, China is used to serve as the "world factory" and get involved in the international division of labour with its comparative advantages of resources and labour. Empirical evidence supports that FDI and international trade are very likely to introduce polluting production into China and simultaneously accelerate the depletion of resources and environmental goods here (Lin 2017; Cheng et al. 2020). On the other hand, there is also empirical evidence that FDI and foreign trade contribute to the capacity building and economic upgrading of domestic industries (Huang et al. 2017). The overall improvement of economic performance will further benefit the environment. These seemingly conflicting findings from China require a better understanding of the conditions and mechanism of foreign-domestic interactions.

Fourthly, in addition to the opening-up, China's economic miracle also relies on the historic reform of marketization, which restructures the interactions between government and markets. The marketization reform covers a wide range of facets, such as reducing the state control over expenditures, prices, and investment, loosening the control over financial institutions, and reforming the state-owned enterprises (Wei 2001). With regard to the environmental issues, the interactions between government and markets may provide two perspectives for environmental economic geography. On the one hand, marketization reform also implies that the command-and-control mode of environmental regulation is not always efficient. In addition, the government starts to explore the market-oriented mode of regulation for environmental governance (Cheng et al. 2017). In such a setting, it is required to examine the effects of different modes of environmental regulation and their spatial variation.

On the other hand, the role of state-owned enterprises is quite special in response to environmental issues. On the economic side, empirical studies usually criticise that the state-owned enterprises are insensitive to costs and benefits so that they tend to be low efficient and less productive. As a result, state-owned enterprises tend to perform worse in the environment than others. However, on the environmental side, the state-owned enterprises may shoulder more social responsibilities and be more willing to pay for public goods under policy incentives (Zhu et al. 2016). Recent advances in economic geography are interested in the role of institutions, one of which is the institutional entrepreneurs. In fact, the reform of state-owned enterprises provides a good chance to investigate the role of institutional entrepreneurs in environmental improvement.

Last but not least, China has a rigid hierarchy consisting of multiple tiers. Governments at all tiers have defined duties, which would thereby lead to various interactions and gaming between the central and the local governments. Regarding economic side, the fiscal decentralization reform in China stimulate fierce inter-governmental competition for fiscal revenues, therefore benefiting local economic growth (Ma and Mao 2018). However, on the environmental side, the inter-governmental competition simultaneously impedes the enforcement of environmental regulation, which then results in an environmental "race-to-the-bottom" phenomenon. Recent efforts from the central government are twofold. Firstly, the central government seeks to strictly supervise the local environmental enforcement and monitor the key pollution sources. The central supervision has exhibited its effectiveness on pollution reduction (Zhang et al. 2018a, b). However, its effects on economic activities remain unclear. Secondly, the central government makes great efforts to propose a more integrated, flexible governance on the environment, breaking the hierarchical systems and the administrative confines. A typical case is the "river chief" systems for water resource protection since 2011, which enhances the unity of management within the same watershed. Overall, the interactions between the central and the local government directly intervene in the environmental effects of human activities. It responds to economic geographers' call for understanding how economic activities operate at the multi-level institutional contexts. Moreover, it also echoes the crucial role of environmental regulations in affecting nature-society interactions.

1.3 Overview of the Book

This book devotes to the practice of environmental economic geography and a better understanding of how China's economic geography interacts with its environmental issues. Instead of claiming new intellectual territory, this book focuses more on the advances in economic geography and rethinks their potential linkages with environmental issues. The main body of this book seeks to achieve two primary targets. Firstly, in response to the long-time absence of the environment in Economic Geography, it still requires in-depth discussions to figure out how to link the ongoing theoretical development of Economic Geography to emerging environmental challenges. This book seeks to exhibit how the incorporation of economic geography theory into environmental issues can contribute to theoretical debates without consensus currently or change our perceptions taken for granted usually. Secondly, the practice of environmental economic geography can have multiple facets as well as various approaches. Sometimes, there is a misconception that geographical studies require conducting a spatial analysis with the sample of different geographical units. Either for environmental scientists or environmental economist, the application of spatial analysis becomes more frequent. Then, what's left for environmental economic geography? This book seeks to demonstrate that practicing environmental economic geography is more than a matter of conducting spatial analysis for environmental issues. Instead, it proposes questions in different ways that concern how environmental issues correspond to local conditions and nonlocal interconnections. Then, it provides explanations as well as solutions for those no-one-fit-all problems.

Using China as a case, this book has to reflect two basic features of China's economic geography. The first one is simple but crucial. That is, China is a country with a huge territory, implying scale/scope economies or diseconomies. Its internal regions cover various types of growth models as well as eco-environmental conditions, which then form a wide array of nature-economy interactions. Understanding China's environmental economic geography requires to answer how the emergence and growth of economic activities interact with the local environment. Secondly, China has undergone a rapid process of economic transition and opening-up. Most regions in China benefit from the establishment of extra-regional linkages, such as the participation in foreign trade, the utilisation of foreign investment, and the acceleration of regional integration. How the exposure to (or integration with) global market affects the environment directly determines the development quality in socioeconomic terms as well as sustainability in eco-environmental terms. There are ongoing debates on this question, which requires more innovative thinking.

Also of note, since environmental issues take various forms and relate to almost all types of production and consumption, it is impossible to cover neither every environmental issue nor every type of economic activities. This book is also not intended to document a panorama of China's environmental economic geography. The authors of this book try to share our theoretical thinking of developing Environmental Economic Geography and make some attempts to practice it. The former is in response to the blind spot that economic geography leaves the environment out.

The latter seeks to change the status that current studies on Environmental Economic Geography tend to be abstract, where empirical practice is really scarce.

Correspondingly, the key environmental issue in this book is the geography of industrial environmental performance. Environmental performance is a relative term capturing both environmental outcomes and economic outputs. It serves as a proxy in this book to capture the levels of nature-economy interactions across spaces.

The data supporting our empirical practice of Environmental Economic Geography are from several nation-wide economic survey/census, such as China's Annual Survey of Industrial Firms (ASIF), the Statistics of China Customs, and the National Economic Census. These firm-specific/product-specific data provide a better chance to investigate the structural changes across industrial sectors as well as the changing spatial patterns of economic growth. On the other hand, these micro-level data provide another chance to crosscheck the quality of the macro-level statistics, which used to be criticised as being inaccurate or even manipulated (without convincing evidence actually, see Holz 2014). In practice, these survey/census data have been widely used in current empirical studies and supported various key findings.

The following chapters consist of one theoretical chapter, nine empirical chapters, and one summary chapter. The empirical chapters are further divided into two parts. Part I regards the internal forces that shape the geography of industrial environmental performance in China, regarding the role of local conditions (Chap. 3), core-periphery interactions (Chap. 4), government-market interactions (Chap. 5), and economic transition (Chaps. 6 and 7), respectively. Part II considers the geography of industrial environmental performance in a broader context, which is the global shift of environmental burdens. Based on the view of global–local interaction, it looks into how environmental burdens shift from other economies to China (Chap. 8), how the environmental burdens further shift between internal regions (Chap. 9), and how the local reacts to the shift of environmental burdens (Chaps. 10 and 11).

Chapter 2 devotes to the theoretical discussion. In this chapter, we scrutinise the loose grouping of studies on Environmental Economic Geography and reveal their great potential. Although current literature is fragmented and poly-vocal, the ongoing theoretical development of economic geography on regional development, globalization, innovation, institution, and transition can be well linked with environmental studies. The development of EEG can contribute to interdisciplinary environmental studies through various methods, such as identifying context-contingent causes and solutions of environmental issues, developing multi-scalar configurations of social-technique systems for environmental innovation and governance, and shaping co-evolutionary trajectories for coupling nature-society interactions and sustainable development. On this basis, we further review the continuous efforts from some economic geographers to develop a coherent EEG and posit four pillars for the conceptual foundation of EEG, namely, context-dependence, border-crossing, multi-scale, and co-evolution. These four pillars may help to ground EEG on the theoretical foundation of economic geography, representing the locational, relational, institutional, and evolutionary perspectives. Also, these four pillars further propose a research agenda for EEG, which consists of four geographical questions regarding the spatiality, horizontal interdependency, vertical interdependency, and evolution of the environment-economy interactions.

Chapter 3 looks into the local conditions that determine the geography of industrial environmental performance. At first, it provides an overview of the industrial geography of China and the spatial pattern of industrial pollution. Empirical results reveal a trend of industrial concentration and feature a coastal-inland division of the concentration level. However, traditional industries account for a larger proportion in inland regions. Such dependence on traditional industries is likely to aggravate the poor environmental performance in the inland regions, whose eco-environmental conditions are inherently fragile. According to the empirical evidence, in coastal regions, there is little reduction in the amount of industrial pollution but the pollution intensity has been remarkably improved. However, most inland regions exhibit neither obvious decline in quantity terms nor significant improvement in efficiency terms. This is particularly the case in the north-eastern and south-western China. Further investigation points to the fact that even the coastal regions represent better environmental performance, there is no strong evidence that could support the decoupling between economic growth and the environment. On the other hand, confronting the inland-coastal division of industrial development, without efficient environmental regulation, the inland regions would be exposed to a risk of becoming the "pollution haven".

From a perspective of core-periphery interactions, Chap. 4 revisits the "beggar-thy-neighbour" issue in the location choice of polluting firms. Conventional approaches highlight that polluting firms can make use of the transboundary pollution to escape environmental regulations. Consequently, they intend to locate in border regions to become a free-rider. In this chapter, we acknowledge that the border effect serves as an attractive determinant for the location choice of polluting industries. However, we argue that concerns for economic performance will trigger polluting firms to stay in regional centres so that they could benefit from various agglomeration externalities. The forces of agglomeration effects would be opposite to the ones of border effects only if the regional centres do not locate in the border regions. They work together to determine the location of polluting firms, which lead to uncertainties on the "beggar-thy-neighbour" phenomenon. We conduct Poisson and Tobit regression analysis to test the border effect and agglomeration effect using county-level data. Empirical results reveal that polluting firms avoid border and agglomeration effects dominating their location choice. Nevertheless, a lenient environmental regulation on borders will give rise to the production of dirty outputs. More evidence from China also supports that local governments lack strong incentives to take advantage of free-rider effects when dealing with environmental pollution, but value the agglomeration effects to boost economic growth.

Chapter 5 regards the government-market interactions and investigates the role of environmental regulations in pollution reduction. Based on the trios of scale-composition-technique decomposition, we explore how environmental regulation affects pollution reduction via these effects. On this basis, we compare the effects of different environmental regulations in China. Empirical findings suggest that environmental regulations are crucial for pollution reduction. Compared with other determinants, such as technological development, increasing scales, and inter-governmental competition, environmental regulation plays one of the most important roles. However, a strict regulation in one place is also likely to increase pollution in the surrounding areas. In the case of China, both the command-and-control regulation and the market-oriented regulation contribute to pollution reduction. They primarily restrict the scale effects and promote the technique effects. However, the market-oriented mode of regulation has more sustained effects than the command-and-control ones.

Chapter 6 investigates how economic transition in China affects environmental performance. In this chapter, we identify three key variables to capture the key facets of China's economic transition, namely, globalization, marketization, and decentralization. They affect the environment in different manners and the combination of their effects also varies from place to place. Using the city-level data on industrial sulphur dioxides and industrial dust emissions, we empirically examine the effects of globalization, marketization, and decentralization on the environment. Empirical results provide weak evidence for the positive role of economic liberalization in environmental improvement. But the dominance of SOEs has degraded the environment. Regarding globalization, no evidence supports the occurrence of pollution haven hypothesis. Instead, the domestic environment benefits largely from the integration with the global market. However, decentralization has induced the race-to-the-bottom phenomenon. Intergovernmental competition by lowering environmental standards or regulation stringency is emerging. Overall, the findings highlight that the institutional perspective provides an important angel to understand China's environmental issues and challenges.

Chapter 7 turns to an evolutionary perspective and regards how industrial restructuring affects the environment at the local level. In this chapter, we seek to ruin the illusion that industrial dynamics always represent replacing dirty industries with relatively clean ones. From the evolutionary perspective, we posit that industrial dynamics in China can be seen as a process of path-breaking amidst path dependence. Neither path dependence nor path-breaking indicates a unique effect on the environment. Empirical results show that path dependence does not always lead to poor environmental performance. On the contrary, it can also reduce pollution intensity through the accumulation of knowledge and the transformation of state-owned enterprises. On the other hand, path-breaking does not necessarily suggest an environmental improvement. The GDP-worship subsidies for new industries still increase the pollution intensity, while external linkages of pollution-intensive sectors would help to introduce new firms with better environmental performance.

Chapter 8 examines the changing position of China with the global shift of environmental burdens. This chapter revisits the pollution haven hypothesis from an innovative perspective. It argues that economic growth does not always reverse the pollution haven behaviour, even if one economy shifts from low income to high income. Instead, economic catching-up exposes one economy to a risk of converging pollution contents in trading products. Then, the trade of catching-up economies with both developed and developing economies generates significant pressures on the domestic environment. This chapter proposes a term "pollution trap" to define this phenomenon. Based on the case of China, this chapter applies the multi-regional input–output analysis to trace the distribution of pollution embodied in China's foreign trade. Then it further uses the fixed effect model to examine the causes of the "pollution trap" phenomenon. Empirical evidence from China supports the existence of the "pollution trap". Environmental regulation keeps China from becoming a pollution haven of developed economies rather than relocating polluting production to other developing economies. Furthermore, the current intra-industry trade of low value-added products worsens China's environmental performance. The diluting advantage of labour and the accumulating advantage of capital keep China specialising in polluting production. Overall, the "pollution trap" may alert catching-up economies to prepare for a sustainability transition.

Chapter 9 further explores how environmental burdens will be shifted among internal regions through looking into the regional division of labour. International trade reallocates resources among places, which thereby reshapes the spatial division of labour at multiple scales. Accordingly, there are rising concerns about the inequality issue associated with trade growth. One environmental-relevant concern is whether foreign trade promotes the growth of some places at the sacrifice of other places' resources. Or put it another way, whether trade-induced specialization in polluting/non-polluting industries leads to the polarization of environmental performance among regions. We answer this question based on the observation of China's outward-oriented development, which shows that the environmental performance among regions in China does not exhibit a trend towards polarization. To better understand the trade-induced spatial division of labour, we propose that it is essentially related to the process of path extension dynamically rather than the level of specialization statically. Path extension suggests the sustained growth of incumbent activities. Foreign trade is one of the driving forces for path extension. In such a setting, the occurrence of regional polarization is built on the premise that once created, the path for regional development will extend endlessly and does not change. From the evolutionary perspective, this is impossible. Both internal and external forces are restricting the path extension. Internal forces stem from the sources of scale diseconomies, such as congestion effects. External forces indicate the effects of exogenous factors, including institutions and nonlocal linkages. Empirical findings support the role of external and internal forces in restricting path extension of polluting industries, which explain why regions specialized in polluting industries are not locked.

Chapter 10 investigates how the local environmental performance will react to the global and interregional shift of environmental burdens. To achieve this goal, we propose a framework consisting of the global transfer, domestic transfer, and local absorption, representing the global–local interactions. According to the results, trade expansion into developed economies or surrounding markets will expose the local to a higher pressure of industrial pollution emission. Domestic transfer relies on the industrial mobility of low value-added sectors from the core to the periphery. Thus, the low-level economic efficiency in lagging regions tend to amplify the trade-environment effect. Industrial agglomeration can promote environmental performance and underlie the positive effect of local absorption. Overall, although exports provide lagging regions with more opportunities for growth, the growing global transfer, the emerging domestic transfer, and the weak local absorption also work together to amplify the trade-environment effect there. In contrast, the leading regions can offset global transfer through domestic transfer and local absorption.

Chapter 11 further explores how the local reacts to the shift of environmental burdens by export upgrading. Based on the global–local interactions, we propose that the environmental improvement by export upgrading depends on the combination of global linkages (the way to upgrade) and local linkages (agglomeration and institution). We apply the decomposition of export sophistication to quantify diverse upgrading ways. To indicate the local linkages, we categorize the sample into groups by their specialization in polluting sectors and their stringency of environmental regulation. Empirical findings suggest that environmental improvement by export upgrading in China manifests a displacement effect largely. Exporting sectors tend to change their product mix to avoid environmental costs. On the contrary, the role of efficiency promotion is still insignificant. Local linkages alter the environmental effects of export upgrading. Agglomeration externalities allow exporting sectors to be more capable of changing product mix, while environmental regulation imposes additional costs and thereby guarantees the desirable effects of upgrading on the environment.

Chapter 12 concludes by pulling together the main themes and key findings in the book. On this basis, this chapter further discusses how empirical findings raise more questions that remain to be addressed. Overall, this book seeks to reveal that environmental economic geography should be a fresh start for economic geography, but EEG does not necessarily need a fresh start.

References

Babiker, M. H. (2005). Climate change policy, market structure, and carbon leakage. *Journal of International Economics, 65*(2), 421–445.

Bush, S. R., Oosterveer, P., Bailey, M., et al. (2015). Sustainability governance of chains and networks: A review and future outlook. *Journal of Cleaner Production, 107,* 8–19.

Braun, B., & Castree, N. (2005). *Remaking reality: Nature at the millennium.* New York NY: Routledge.

Cai, B., Zhang, W., Hubacek, K., et al. (2019). Drivers of virtual water flows on regional water scarcity in China. *Journal of Cleaner Production, 207,* 1112–1122.

Cai, H., Chen, Y., & Gong, Q. (2016). Polluting thy neighbour: Unintended consequences of China's pollution reduction mandates. *Journal of Environmental Economics and Management, 76,* 86–104.

Cave, L. A., & Blomquist, G. C. (2008). Environmental policy in the European Union: Fostering the development of pollution havens? *Ecological Economics, 65*(2), 253–261.

Chau, K. W., & Jiang, Y. W. (2003). Simulation of transboundary pollutant transport action in the Pearl River Delta. *Chemosphere, 52*(9), 1615–1621.

Chen, Z., Kahn, M. E., Liu, Y., et al. (2018). The consequences of spatiality differentiated water pollution regulation in China. *Journal of Environmental Economics and Management, 88,* 468–485.

Cheng, Z., Li, L., & Liu, J. (2017). The emissions reduction effect and technical progress effect of environmental regulation policy tools. *Journal of Cleaner Production, 149,* 191–205.

Cheng, Z., Li, L., & Liu, J. (2020). The impact of foreign direct investment on urban $PM_{2.5}$ pollution in China. *Journal of Environmental Management, 265,* 110532.

Cole, M. A. (2004). Trade, the pollution haven hypothesis and the environmental Kuznets curve: Examining the linkages. *Ecological Economics, 48*(1), 71–81.

Cole, M. A., Elliott, R. J. R., & Zhang, J. (2011). Growth, foreign direct investment, and the environment: Evidence from Chinese cities. *Journal of Regional Science, 51*(1), 121–138.

Day, T. (2017). The contribution of physical geographers to sustainability research. *Sustainability, 9*(10), 1851.

Dean, J. M., Lovely, M. E., & Wang, H. (2009). Are foreign investors attracted to weak environmental regulations? Evaluating the evidence from China. *Journal of Development Economics, 90*(1), 1–13.

Friis, C., Nielsen, J., Otero, I., et al. (2016). From teleconnection to telecoupling: Taking stock of an emerging framework in land system science. *Journal of Land Use Science, 11*(2), 131–153.

Fuchs, R., Alexander, P., Brown, C., et al. (2019). Why the US-China trade war spells disaster for the Amazon. *Nature, 567,* 451–454.

Gibbs, D. (2006). Prospects for an environmental economic geography: Linking ecological modernization and regulationist approaches. *Economic Geography, 82*(2), 193–215.

Gibson, C., & Warren, A. (2016). Resource-sensitive global production networks: Reconfigured geographies of timber and acoustic guitar manufacturing. *Economic Geography, 92*(4), 430–454.

Hao, X., & Deng, F. (2019). The marginal and double threshold effects of regional innovation on energy consumption structure: Evidence from resource-based regions in China. *Energy Policy, 131,* 144–154.

Hayter, R. (2008). Environmental economic geography. *Geography Compass, 2*(3), 831–850.

Hobson, K. (2003). Consumption, environmental sustainability and human geography in Australia: A missing research agenda? *Australian Geographical Studies, 41*(2), 148–155.

Holz, C. A. (2014). The quality of China's GDP statistics. *China Economic Review, 30,* 309–338.

Huang, J., Chen, X., Huang, B., et al. (2017). Economic and environmental impatcs of foreign direct investment in China: A spatial spillover analysis. *China Economic Review, 45,* 289–309.

Jakob, M., & Marschinski, R. (2013). Interpreting trade-related CO_2 emission transfers. *Nature Climate Change, 3,* 19–23.

Jiborn, M., Kander, A., Kulionis, V., et al. (2018). Decoupling or delusion? Measuring emissions displacement in foreign trade. *Global Environmental Change, 49,* 27–34.

Kahn, M. E. (2003). The geography of US pollution intensive trade: Evidence from 1958 to 1994. *Regional Science and Urban Economics, 33*(4), 383–400.

Li, Y., Chang, M., Ding, S., et al. (2017). Monitoring and source apportionment of trace elements in $PM_{2.5}$: Implications for local air quality management. *Journal of Environmental Management, 196,* 16–25.

Lin, F. (2017). Trade openness and air pollution: City-level empirical evidence from China. *China Economic Review, 45,* 78–88.

Lin, J., Tong, D., Davis, S., et al. (2016). Global climate forcing of aerosols embodies in international trade. *Nature Geoscience, 9,* 790–794.

Liu, J., Hull, V., Batistella, M., et al. (2013). Framing sustainability in a telecoupled world. *Ecology and Society, 18*(2), 26.

Ma, G., & Mao, J. (2018). Fiscal decentralisation and local economic growth: Evidence from a fiscal reform in China. *Fiscal Studies, 39*(1), 159–187.

Mao, X., Huang, X., Song, Y., et al. (2020). Response to urban land scarcity in growing megacities: Urban containment or inter-city connection? *Cities, 96,* 102399.

Munroe, D. K., McSweeney, K., Olson, J. L., et al. (2014). Using economic geography to reinvigorate land-change science. *Geoforum, 52,* 12–21.

O'Brien, K. (2010). Responding to environmental change: A new age for human geography? *Progress in Human Geography, 35*(4), 542–549.

Peters, G. P., & Hertwich, E. G. (2006). Pollution embodied in trade: The Norwegian case. *Global Environmental Change, 16,* 379–387.

Rosa, L., Chiarelli, D. D., Tu, C., et al. (2019). Global unsustainable virtual water flows in agricultural trade. *Environmental Research Letters, 14*(11), 114001.

Seto, K. C., Reenberg, A., Boone, C. G., et al. (2012). Urban land teleconnections and sustainability. *Proceedings of the National Academy of Sciences, 109*(20), 7687–7692.

Shi, C., Nduka, I. C., Yang, Y., et al. (2020). Characteristics and meteorological mechanisms of transboundary air pollution in a persistent heavy $PM_{2.5}$ pollution episode in central-east China. *Atmospheric Environment, 223.* https://doi.org/10.1016/j.atmosenv.2019.117239.

Sigman, H. (2005). Transboundary spillovers and decentralization of environmental policies. *Journal of Environmental Economics and Management, 50*(1), 82–101.

Soyez, D., & Schulz, C. (2008). Facets of an emerging environmental economic geography (EEG). *Geoforum, 39,* 17–19.

Stern, D. I. (2004). The rise and fall of the Environmental Kuznets Curve. *World Development, 32*(8), 1419–1439.

Wang, H., Zhang, Y., Zhao, H., et al. (2017). Trade-driven relocation of air pollution and health impacts in China. *Nature Communications, 8,* 738.

Wei, Y. D. (2001). Decentralization, marketization, and globalization: The triple processes underlying regional development in China. *Asian Geographer, 20*(1–2), 7–23.

Wu, H., Guo, H., Zhang, B., et al. (2017). Westward movement of new polluting firms in China: Pollution reduction mandates and location choice. *Journal of Comparative Economics, 45*(1), 119–138.

Xiao, Q., Geng, G., Liang, F. et al. (2020). Changes in spatial patterns of $PM_{2.5}$ pollution in China 2000–2018: Impact of clean air policies. *Environment International, 141,* 105776.

Yang, H., Chen, J., & Wen, J. (2016). Composition and sources of $PM_{2.5}$ around the heating periods of 2013 and 2014 in Beijing: Implications for efficient mitigation measures. *Atmospheric Environment, 124,* 378–386.

Zhang, B., Chen, X., & Guo, H. (2018a). Does central supervision enhance local environmental enforcement? Quasi-experimental evidence from China. *Journal of Public Economics, 164,* 70–90.

Zhang, Y., Li, X., Nie, T., et al. (2018b). Source apportionment of $PM_{2.5}$ pollution in the central six districts of Beijing China. *Journal of Cleaner Production, 174,* 661–669.

Zheng, D., & Shi, M. (2017). Multiple environmental policies and pollution haven hypothesis: Evidence from China's polluting industries. *Journal of Cleaner Production, 141,* 295–304.

Zhu, Q., Liu, J., & Lai, K. (2016). Corporate social responsibility practices and performance improvement among Chinese national state-owned enterprises. *International Journal of Production Economics, 171*(3), 417–426.

Zimmerer, K. S. (2010). Retrospective on nature-society geography: Tracing trajectories (1911–2010) and reflecting on translations. *Annals of the Association of American Geographers, 100*(5), 1076–1094.

Chapter 2
Developing Environmental Economic Geography

Geography is a discipline of human-nature interactions regarding *"why of where and so what"*. The development of Environmental Economic Geography (EEG) is no exception. Environmental issues are consequences of economic processes, which in turn alter the foundation that underlies economic processes. EEG exactly delves into the reciprocal linkages between the economy and the environment, which are subject to the spatiality of economic processes and their interconnections between spaces. Hence, EEG closely relates to key themes and concepts of economic geography, such as location, agglomeration, globalization, innovation, institution, and evolution. In this regard, EEG represents economic geography's efforts to take the environment more seriously (Hayter 2008).

However, EEG consists of "a loose grouping of burgeoning but disparate studies" (Hayter 2008, p. 832). The analysis of EEG is still fragmented and polyvocal in epistemic terms so that Bridge called it a "topical contrivance" (Bridge 2008a, p. 76). Thus, this chapter sets out to review this loose grouping of studies by linking the various themes of economic geography to the environmental issues firstly. There are two rationales behind this. First of all, the interaction between the economy and the environment is an interdisciplinary topic. Related fields include environmental economics, ecological economics, environmental management, political economics, political ecology, and other neighbouring fields. Linking the environmental issues with themes and concepts of economic geography may help to identify the unique perspectives and the disciplinary boundary of EEG. Second, various studies have been involved in the interactions between environmental issues and economic geography. However, not all of them label themselves as EEG studies. Incorporating studies relating to themes of economic geography may enrich the research scope of EEG and then contribute to the development of a comprehensive research agenda.

Overall, EEG can articulate with key themes of economic geography in three distinctive ways. One popular theme is investigating the environmental outcomes/performances stemming from the spatiality and spatial interdependency of economic processes. There has been rich literature regarding the environmental

© Springer Nature Singapore Pte Ltd. 2020
C. He and X. Mao, *Environmental Economic Geography in China*,
Economic Geography, https://doi.org/10.1007/978-981-15-8991-1_2

effects of regional development, industrial agglomeration, and economic globalization. Unlike other related fields, EEG tends to apply the conceptual foundations of economic geography, such as places, flows, location, neighbourhood, inequality and interdependency, to understand the place-specific causes of environmental issues and trace the causes among places or even across multiple geographical scales.

Another common theme is to investigate the spatial pattern, organization, and dynamics of environment-relevant activities. Specifically, this literature can monitor the spatial dynamics of environmentally sensitive economic processes, such as the transform of resource-based development and the dynamic of polluting industries. This literature may also investigate the spatial organization of environmentally benign actions, such as green innovation and policy intervention. These processes or actions do not occur in a vacuum nor work on their own. They put forward new theoretical and empirical opportunities for economic geographers to revisit current theories and models (Aoyama et al. 2011). First, EEG studies look into the emergence and diffusion of green innovation. Second, considering the transboundary nature of environmental issues and the spatial interdependencies of economic activities, EEG studies are also interested in the neighbourhood effects of environment-relevant actions and the possibilities of transboundary cooperation. Third, concerning the public nature of the environment, recent interest of EEG also points to the role of multi-level institutions in environmentally benign activities. Fourth, from the evolutionary perspective, EEG studies may further reveal what drives regions towards a path with sustainability.

Last but not least, EEG articulates with the themes of economic geography by its focus on the spatially different capacities in response to environmental challenges. Places are not only carriers but also receptors of environmental issues. The place-specific physical, economic, social, cultural, and institutional characteristics equip places with different capacities for development. In this regard, EEG studies may be able to answer why and how regions act/react differently to environmental challenges. Recent advances in Economic Geography highlight the term of economic resilience, describing how places will react to or recover from external shocks (Martin 2012; Martin and Sunley 2015). Since environmental change is one of the major sources of external shocks, EEG studies accordingly can explore economic resilience to environmental challenges, spanning from the instant natural disasters to the long-term environmental changes.

The three ways EEG articulates with Economic Geography reveal potential conceptual foundations for EEG, including the critical concepts around regional development, globalization, innovation, institution, and evolution. Then, this chapter reviews the "loose grouping" of studies by these key concepts and discusses the potential facets of EEG. On this basis, this chapter further introduces the efforts of EEG in proposing its own coherent "epistemic project" and comprehensive research agendas. Moreover, this chapter puts forth four pillars of the conceptual foundations of EEG, based on which we propose a research agenda for the developing EEG in China.

2.1 Places, Regional Development, and the Environment

Environmental issues are different from place to place, representing various human–environment contradictions. EEG embeds these contradictions in the physical, cultural, institutional, and social context to understand the spatial disparity and diversity of environmental issues. Place and location are exactly the concepts that economic geographers used to describe the "context". Economic geographers have long been of interest in how the place-specific context affects economic growth and shapes the trajectories of regional development. This literature includes a wide array of research themes, such as location and concentration. On the basis of these efforts, EEG takes one step further to explore the place-specific effects of production and consumption on the environment. These efforts also allow EEG to revisit the spatial patterns and dynamics of environment-relevant economic processes.

2.1.1 Concentration of Economic Activities and the Environment

Economic activities tend to concentrate spatially, resulting in the uneven distribution of economic activities across the Earth surface. Spatial concentration can reduce the geographical distance between economic agents and increase the potential for interactions. Spatial concentration also extends the scale of economic activities by attracting more and various firms to co-locate, then generating the economies of scale and scope. A continuous concentration accumulates the advantages of economic growth, which are expected to produce a self-reinforcing process of economic concentration (Rosenthal and Strange 2004).

Conventionally, economic geographers focus on the nature of economic concentration, identifying the environment as one of the sources for economic concentration. The environment will equip one place with advantageous conditions for economic concentration, which is called natural advantages. Ellison and Glaeser (1999) posit that over half of the economic concentration is due to natural advantages based on the cases of the United States. Roos (2005) reports that at most 36% of the spatial variation of gross domestic production can be explained by direct and indirect effects of geography. On the other hand, new economic geography literature further demonstrates the fact that environmental degradation will produce congestion effects due to diseconomies of scale, then limiting the sustained development of concentration (Fu and Hong 2011; Saito and Wu 2016). However, economic geographers seldom ask how economic concentration will affect the environment.

In this regard, EEG may answer this question based on the overwhelming studies on economic concentration/industrial agglomeration. Answers to this question will help to understand why economic activities affect the environment in different ways among places. Nevertheless, there is no simple answer to this question.

A direct effect comes from scale expansion in the wake of economic concentration. Theoretical thinking posits the pros and cons of enlarging scales. On the negative side, enlarging scales suggest an increase in production capacity and energy consumption (Cheng 2016), either of which will significantly raise the volume of pollution emissions. However, on the positive side, scaling up embodies scale economies, which would increase production and reduce per-unit costs. Industrial agglomeration is especially expected to generate external scale economies, lowering the average costs by sharing facilities, efficient suppliers, skilled labour, and so on Duranton and Puga (2004). Consequently, emissions per unit may reduce. Scale expansion is likely to be negative in volume terms but positive in efficiency terms (Chen et al. 2018a).

However, the effects of industrial agglomeration on the environment are more than the interactions between quantity and efficiency. Many confounders can exist and then alter these effects. First, economic agents within the agglomeration are heterogeneous. Accordingly, their environmental demand and performance can be very different (Sun and Yuan 2015). In such a case, whether spill-over effects based on economic performance still work for environmental performance is suspicious. As for the scale diseconomies, economic agents also respond differently to the congestion effects. Non-manufacturing sectors would be more sensitive to the crowing-out effects (Hosoe and Naito 2006).

Moreover, firm heterogeneity also raises a question of selection/sorting effects (Baldwin and Okubo 2006). That is, the positive/negative effects of industrial agglomeration on firm performance are not because of "being there". Instead, it is because these firms are "qualified" to be there. For instance, environmental regulation does not necessarily encourage all firms to improve their environmental performance. It may also trigger some less competitive firms to relocate, resulting in a selection/sorting effect (Naughton 2014; Zhu and Ruth 2015). The stringency and enforcement of environmental regulation vary from one agglomeration to another (Wang et al. 2018). Industrial agglomerations with lax environmental regulation tend to become favourable destinations of polluting firms (Pflueger 2001; Neary 2006; Zeng and Zhao 2009). Taking advantage of this mechanism, developing regions within a nation would dismantle the regulatory standards to compete for the profitable but polluting firms (Woods 2006). In the Chinese case, this mechanism relocated water-polluting activities from downstream cities of rivers (also relatively developed cities) to upstream ones, exposing a larger proportion of the rivers to higher risks of pollution (Chen et al. 2018b).

Second, neither the processes of economic concentration and environmental change are necessarily confined to particular places. Economic concentration tends to affect the growth of surrounding areas, such as the spill-over effect, trickle-down effect, polarization effect, and the shadowing effect (Wagner and Timmins 2009; Cheng 2016). Technically, the environmental performance can be ascribed to not only regional economic concentration but also the spatial division of labour across places (Cheng et al. 2017a). Similarly, the environmental outcomes are also in a state of flux. Environmental issues can be generated in one place but become a challenge in another place, such as air pollution and water pollution (Cai et al. 2016). Taken together, examining the environmental effect of economic concentration requires the consideration of transboundary linkages (Naito 2010).

Third, the environmental effects of economic concentration keep changing as economic concentration is a dynamic process. At the industrial level, the growth of economic concentration tends to shift from its early specialization towards a new specialization or a diversification (Marrocu et al. 2013). Even if we suppose that the levels of concentration are the same, this shift can embody entirely different environmental outcomes. At the firm level, the change of economic concentration is associated with significant structural changes, including firm births, expansions, contractions, and closures (Dumais et al. 2002). In this regard, the environmental outcomes are not only affected by the level of concentration (da Schio et al. 2019), but also the structure of it (Sun and Yuan 2015).

These issues present new opportunities for both environmental economics and EEG to revisit the interactions between economic concentration and the environment. For environmental economics, studies seek to examine the causal effects of economic concentration on the environment. They tend to identify the above issues as the "endogeneity", which impacts the causal inference. Efforts are made to apply new proxies and analytical techniques to cope with the endogeneity. By contrast, EEG would be interested in why the environment would perform better in one concentration than the others. Accordingly, issues above require EEG to go beyond the level of economic concentration and explore the spatial organization and structural transformation of economic concentration. In this regard, EEG essentially links the environment to the "context" for economic concentration rather than the "level" of economic concentration.

2.1.2 Location of Economic Activities and the Environment

Economic geographers use the concept "location" to capture the properties of places that explain why economic activities occur here and predict the occurrence of particular activities. In this regard, the location of economic activities provides EEG with a set of new themes to investigate human–environment interactions.

The first theme regards the location of polluting industries. This kind of literature is interested in whether polluting firms tend to locate in border areas, also known as the "border effects". Pollutants can transport across jurisdictions, becoming a source of negative externalities. Thus, polluting firms can locate in the border areas to make use of this externality so that they can increase their environmental performance at the costs of their neighbours (Fernandez 2002; Cai et al. 2016). Empirical evidence supports such a preference of polluting firms (Chen et al. 2018b; Duvivier and Xiong 2013). Furthermore, this seemingly locational advantage of border areas may further result in lax environmental regulations, and then attracting more polluting firms to locate (Kahn 2004). Taken together, empirical findings also reveal that an increasing number of "borders" will exacerbate such pollution externalities (Lipscomb and Mobarak 2017).

Although empirical evidence has reported the concentration of polluting firms in border areas, whether the polluting firms prefer the border areas rather than the core ones is still suspicious. From the perspective of economic geography, an attractive location consists of various factors, including the endowments, accessibilities, facilities, and policies. The substantial differences between the border and non-border areas are not confined to pollution externalities, but also could include development bases, such as labour pools and facilities. In this regard, the location of polluting firms in the border areas should be a net effect of multiple forces rather than the pollution externalities solely. The preference for border areas is thereby sector-specific and place-specific. What kind of polluting firms and border areas tend to produce pollution externalities? The answers are left to EEG.

For instance, the borders are different from one to another, which are subject to their levels and neighbours. Empirical studies have found substantial differences in pollution externalities between national and sub-national jurisdictional boundaries (Konisky and Woods 2010). Moreover, studies on the United States found that pollution externalities perform differently between Mexican border and Canadian border (Kahn 2004). The trade policies and environmental policies also exhibit different environmental effects on border areas (Fernandez and Das 2011).

The second theme regards the relocation of polluting industries and identifies the determinants, especially the role of environmental regulation. At the global level, there are considerable studies on the so-called pollution haven phenomena that polluting industries tend to move from countries with strict environmental regulations to countries with lax ones. Empirical studies concern that pollution haven phenomena tend to occur between developed and developing economies, resulting in the overall loss of welfare (Cole 2004; He 2006). However, there is yet no consensus achieved towards the determinants of pollution haven phenomena. Despite the technique issues and data quality for empirical studies, current debates are primarily over the role of environmental regulation in industrial mobility (Millimet and List 2004; Kearsley and Riddel 2010). From a geographical perspective, this debate essentially loses a big picture of the reshaping economic geography across the world and a comprehensive framework to understand the location choice.

At the national level, empirical studies also concern the relocation of polluting industries from the core to the periphery. China is especially the case. Empirical studies in China have witnessed the westward move of polluting industries and carbon-intensive industries (Wang et al. 2019; Wu et al. 2017). The industrial upgrading and functional decentralization of core regions are also moving the polluting firms out (Li et al. 2019). This process is intertwined with the industrial restructuring process across the whole nation. This process is also in line with the changing philosophies of development policies from the central government, which are previously biased to the coastal regions. Also of note, the hinterland of China happens to the region that is more ecologically vulnerable and environmentally sensitive. As a result, the location choice of polluting firms can be a compromise between economic, environmental, social, technological, institutional, and even cultural forces, where the environmental regulation does not necessarily stand at the centre. Taken together, the theme of polluting firms' relocation leaves a key question to EEG. That is how to develop a more comprehensive framework to capture the location choice of polluting firms, going beyond the regulation-centred perspective.

Last but not least, industrial location provides an opportunity to apply the natural experiment to causal inference, looking into the effectiveness of environmental regulation. Through comparing the performance of industries across jurisdiction boundaries, along the river, and so on, studies are able to design a quasi-natural experiment to examine the efficiency of policies, *certeris paribus*. For instance, Kahn et al. (2015) use economic geography data of industrial water polluters and designed a natural experiment. They proved that changing local political promotion criteria from the central government can reduce, rather than increase, the pollution in border regions. Based on the difference-in-difference-in-difference method (DDD), Cai et al. (2016) evidence the downstream effects (move polluting activities to the most downstream places) of polluting activities in response to the enforcement of pollution fee collection. Zhang et al. (2018) apply the regression discontinuity (RD) approach and revealed that central supervision can enhance the local environmental enforcement significantly. Overall, EEG should notice the potential of designing natural experiments with geographical characteristics to explore the environmental outcomes driven by economic interferences.

2.1.3 Regional Development and the Environment

Unpacking the relationship between regional development and the environment is a classical and interdisciplinary topic, requiring the efforts of the environmental scientist, economists, geographers, and so on. Why do geographers concern this issue? The reason is that the development-environment relationship is one of the crucial facets of human–environment/nature-society interactions. Economic geographers have long been interested in the various models of regional development, including their determinants and trajectories. Surprisingly, they prefer to investigate how regional development is subject to its external environment, but seldom concern how regional development will affect the environment in turn (Hayter 2008). One of the primary tasks of EEG is to shift the theoretical perspective from how the environment underlies regional development to how the environment interacts with regional development.

Regarding this issue, there are already interdisciplinary efforts unravelling the environmental impacts of regional development. Environmental scientists seek to trace anthropogenic impacts on the environment and compare their differences. One of the widely used analytical tools is the IPAT equation, based on the debate between Commoner and Ehrlich and Holdren in the 1970s (Commoner et al. 1971; Enrlich and Holdren 1971). This debate is over the question of whether the technological transformation of economic activities leads to environmental degradation. Commoner et al. (1971) insist that technological transformation took responsibility for environmental challenges. Ehrlich and Holdren (1971) highlight the role of increasing population and the accumulating affluence, rejecting the ascription to technological transformation. As such, the IPAT equation considers the environmental impacts as the product of population (P), affluence (A), and technology (T), identifying the crucial components of development that result in the environmental crisis.

In analogy with the Cobb–Douglas production function, the IPAT equation has been widely used in empirical studies, deriving rich variants (Chertow 2000). This equation provides opportunities to incorporate more factors as divisors as well as dividends and then conducts decomposition analysis. This approach underlies the literature on energy consumption and greenhouse gas emissions (Song et al. 2011; Yue et al. 2013). Another important variant is the STIRPAT (Stochastic Impacts by Regression on Population, Affluence, and Technology) proposed by Dietz and Rosa (1997). They argue that conventional IPAT equation overlooks the non-monotonic, disproportional changes between I and PAT. The STIRPAT allows for the non-monotonicity and disproportionality by reformulating the equation as $I = aP^bA^cT^de$. Such a formulation makes it possible to follow the hypothesis test approach by regression analysis, and connects the quantitative approaches of physical sciences with the social sciences' (Chertow 2000).

On the other hand, environmental economists propose an inverted-U-shaped curve for the relationship between economic growth and the environment, which is the famous Environmental Kuznets Curve hypothesis (Grossman and Krueger 1993; de Bruyn 1997; Dasgupta et al. 2002). EKC posits that economic growth will increase pollution emissions and deteriorate the environment in the early stages. However, when going beyond certain level of economic growth, economic growth will benefit the environment. How could such a reversal be formed? Empirical studies provide various explanations. For example, as one country becomes rich, it affords activities relating to environmental improvements, such as investments in green innovation and employment of green technology (Lindmark 2002; Carson 2010). Besides, residents in rich countries require environment with higher quality, which may also force environmental improvement (Dinda 2004).

Regarding this issue, Grossman and Krueger (1993) also provide a decomposition framework consisting of scale, composition, and technique effects. This is a precise way to understand the dynamics of the development-environment relationship. It figures out that productivity and the changes in economic size, output and input mix, work together to determine the pollution (Stern 2004). Empirically, scale effects tend to worsen the environment due to the enlarging scale of economic activities. On the contrary, technique effects are expected to benefit the environment. Composition effects indicate the redistribution of factors across industries, representing ambiguous effects on the environment (Cui et al. 2016). This decomposition approach has been widely used in studies on pollution emissions.

However, empirical studies on the EKC and the decomposition have reported highly inconsistent results. It is inevitable in part because of the analytical technique issues (Stern 2004; Galeotti et al. 2006). But more than that, the EKC is also theoretically ambiguous in some ways. Critiques span from its temporal dimension to the spatial one. Temporally, the EKC is just one of the likely development-environment relationships in practice (He and Richard 2010). Just like Stern (2004) argues in his critical review on the EKC, that "an effort to reduce some environmental impacts may just aggravate other problems". Most of the studies on the EKC consider neither the substitution among different pollutions nor the feedbacks of environment on growth, which are likely to make the growth-environment interaction a spiral rather than an inverted-U shaped curve. Critiques from the spatial perspective argue that fostering economic growth is not the only channel for development-environment effect. The increasing mobility of commodities, capitals, and labours allows pollution transfer between countries (Cole 2004). The EKC relationship can be an outcome of the trade-induced redistribution of "dirty" production (Arrow et al. 1995).

No matter the IPAT equation or the EKC and scale-composition-technique decomposition, empirical studies highlight the case-sensitive results due to spatial heterogeneity and spatial diversity. Incorporating the spatial perspectives into the development-environment linkages is of great significance. Thus, considering the rich development models proposed by economic geographers, EEG may further link the models of regional development with the environment, and compare the differences of their environment.

2.2 Flows, Globalization, and the Environment

Both economic activities and environmental issues are interconnected across spaces. As a result, the human–environment interactions are subject to local-nonlocal inter-actions. In other words, the sources of environmental outcomes can be traced to not only local determinants but also nonlocal ones. Since the accelerative twin processes of globalization and localization in the 1980s, Economic Geography has witnessed a "relational turn" that focuses on how economic agents interact in geographical spaces (Boggs and Rantisi 2003), especially the spatial organization of multi-national enterprises.

The rise of multi-national enterprises reallocates resources, capitals, goods, migrants, and information flows across the world, promoting the spatial separation of production and consumption through offshore outsourcing (Copeland and Taylor 1999). Such cross-border activities will affect the environment of both host and home places (Wyckoff and Roop 1994; Unteroberdoerster 2001; Lin et al. 2014). The spatial pattern of various flows and the spatial organization of transnational agents have the potential to enrich the scope of EEG. From the local perspective, this literature concerns the environmental effects of various flows during globalization, such as foreign investment and trade. As for the global aspect, the development of the global production network (GPN) provides a framework to explore how firms could transform environmental constrains to competitive advantages and then upgrade along GPN in environmental terms.

2.2.1 Factor Flows and the Environment

Factor flows reallocate the resources among their origins and destinations, resulting in the spatial mismatch between consumption and production. Also, the inflows of resources and factors will further interact with local resources and factors, changing the current ways of resource utilization. Hence, factor flows will significantly affect the environment. From the view of geography-related, previous studies primarily focus on four facets of the interactions between factor flows and environment.

First, the connections between the host and home places play a key role in altering the environmental effects of factor flows. On the one hand, transnational factor flows depend on the attractiveness of host economies. On the other hand, the flows are also subject to home economies' supportiveness and unfavourableness (Gaur et al. 2018). Thus, the origins of factor flows imply differentiated motivations, such as exploring new markets and utilizing cheap factors (Pan 2003), then determining their diversified behaviour and capacity in host places (Xu and Yeh 2013).

For instance, empirical studies find that firms tend to perform better in foreign markets that are more similar to their home markets (Head and Ries 2008). Support from home government and advanced institutions of host country can reduce factor flow's dependence on prior entry experience (Lu et al. 2014). Specifically, firms from developed economies can cope more easily with the costs of such dissimilarities (Cezar and Escobar 2015).

As for environmental performance, empirical evidence also supports that factor flows from developed and developing economies have different impacts on the environment. The factor flows from developed economies tend to exhibit the pollution haven phenomenon (Cole et al. 2011), where the gap of environmental regulation matters. By contrast, the factor flows from developing economies tend to follow the factor endowment hypothesis, which is affected significantly by the trade pattern (Mao and He 2018).

Economic geography has developed a framework of multiple proximities, including geographical, cognitive, organizational, institutional, and cultural ones (Boschma 2005; Hansen 2015). This framework also applies to the connections of transnational factor flows between the host and home countries. Economic geographers use this framework to explore the quantity and quality of transnational factor flows, especially in terms of knowledge (Capone and Lazzeretti 2018; Makkonen and Williams 2018). Based on this approach, EEG may further explore the environmental effects of transnational flows, looking into the interactions between different proximities.

Second, studies concern how the interactions between factor inflows and incumbent factors will modify the environmental effects of inflows. Such interactions can be either between local and nonlocal firms or the recombination of local and nonlocal resources (Tödtling and Trippl 2018). According to previous findings, the introduction of nonlocal flows can produce spill-over effects that encourage indigenous innovation (Wang and Wu 2016). It is believed that nonlocal flows can introduce complementary resources which would then recombine with incumbent ones and lead to incremental innovation (Boschma et al. 2017). On the other hand, nonlocal flows may also introduce utterly different technology and facilitate drastic innovation (Grillitsch et al. 2018). Either way, the technological gap between local and nonlocal firms matters. A narrow gap allows local firms to absorb new technology transfer more effectively and efficiently, and would further promote environmental performance (Bu et al. 2019). The interaction between local and nonlocal firms can also help to upgrade the production towards an environmentally benign mode (Pazienza 2019). Moreover, empirical studies also indicate that nonlocal inflows may further stimulate the demand for higher standards of environmental regulation, then benefit the indigenous environmental performance (Zeng and Eastin 2007).

Another facet of the factor flows and environment interactions is the effect of environmental regulation on factor flows. Particularly, there are concerns about whether strict environmental regulation will impact the inflow of foreign investment (Naughton 2014). It is believed that environmental regulations can force institutions to promote their economic activities, so that they can sustain their economic competence in response to the increasing production costs imposed by environmental regulation (Lanoie et al. 2008; Zhu and Ruth 2015). However, the stringency of environmental regulation is spatially uneven. This fact allows economic activities to escape through relocation, which is known as the pollution haven phenomenon. To upgrade or to escape is a hot topic in the literature.

On this basis, there is another well-known hypothesis regarding this phenomenon, known as the "race to the bottom" phenomenon. This phenomenon concerns that, due to the GDP worship, economies are willing to lower their environmental regulation standards to gain comparative advantages and compete for polluting industries. This phenomenon can occur between either nations or sub-national regions (Sheldon 2006; Konisky 2007). New advances in the literature further delve into the spatial structure of factor inflows, accounting for the "third-country" effects (Baltagi et al. 2007). Their empirical findings report that the complex factor flows from developed to developing economies, highlighting the existence of competition for factor inflows by lax environmental regulation (Rasli et al. 2018).

However, empirical studies also point out that there is still a chance to witness the "race to the top" phenomenon rather than the "race to the bottom" (Millimet 2003; Prakash and Potoski 2006). The relationship between environmental regulation and firm relocation may be offset by the intercountry institutional distance and firms' environmental capabilities (Madsen 2009). This relationship may also be contingent on the market sizes of the host and home places. If the market is small, foreign investment is likely to increase the standards of regulation in the host country so that the race to the top would occur. Otherwise, if the market is large enough, the continuous increase of regulation standards may impact the inflow of foreign investment. Then, the race to the top phenomenon can be replaced by the "regulatory chill" (Dong et al. 2012). Also of note, the "race to the bottom" phenomenon can be affected by the types of pollutants. Assuming that the pollutants are border-crossing rather than local, economic activities would be less sensitive to local environmental regulation. Hence, environmental regulation is no longer a straightforward way to gain comparative advantages.

Last, studies are also interested in the neighbourhood effect of factor inflows. Economic geographers have noticed that transnational factor inflows, such as foreign investment and foreign trade, may lead to spill-over effects on surrounding regions (Fredriksson and Millimet 2002). In this regard, current studies tend to apply the spatial dependence model for empirical studies to cope with the likely endogeneity issues, which may have more reliable causal inference between factor inflows and the environment (Huang et al. 2017; Liu et al. 2018). However, for EEG, it should be more important to understand the rationale behind such spatial dependence. Do transnational factor inflows incur environmental gain or loss for surrounding regions? What if the pollutants are also transboundary? Answering these questions requires more geographical wisdom, where EEG can contribute to.

2.2.2 Global Production Network and the Environment

Globalized production is a particularly visible manifestation of globalization, separating production from consumption spatially. In the wake of rising transnational corporations and offshore outsourcing, global production has also been organizationally fragmented and spatially dispersed (Yeung and Coe 2015). The fragmented and dispersed production leads to challenges for tackling environmental issues, whose sources can be traced to transboundary agents and activities. As a result, whether the identification of the responsibility for environmental issues or implementation of environmental regulation becomes an issue with interregional dependency. No place can manage alone or stand aloof. In this regard, studies develop new concepts, like virtual water flows (Hoekstra and Hung 2005), emissions embodied in trade (Wyckoff and Roop 1994; Peters and Hertwich 2006), carbon leakage (Kuik and Gerlagh 2003), and tele-coupling of land use (Seto et al. 2012), to capture the environmental burden-shifting effect caused by the spatial separation between consumption and production. New progress in this literature further notices that the spatial environmental burden shift is closely related to the pattern of global production (Jakob and Marschinski 2013).

Since the 1990s, the literature successively develops network approaches to depict the pattern of global production, such as the global supply chain, global commodity chain, global value chain, and global production network. Although not designed for environmental issues, these approaches can still contribute to a better understanding of how global production is related to sustainability. Global supply chain portrays the inter-/intra-firm relations of design, manufacture, assembling, distribution, and retailing, exploring their functional and geographical integration (Meixell and Gargeya 2005). The modern management of the global supply chain especially focuses on the effects of the supply chain on the environment, social welfare, and labour right. Particularly, there is increasing awareness of sustainable supply chain management (Gopalakrishnan et al. 2012; Frostenson and Prenkert 2015). Unlike global supply chain, which is based on business administration studies, global commodity chain and global value chain stem from the political economy, and focus on the evolution of global division of labour and the unequal distribution of values across different processes and places (Gereffi et al. 2005; Gibbon et al. 2008). GCC and GVC specially highlight the role of leading firms in industrial organization. Their relation to the environment is largely manifested by their attention to natural resources and resource-based sectors (Gibbon 2001; Stringer 2006).

As for the global production network, it is applied by economic geographers to capture the "organizational arrangement", which comprises interconnected economic and non-economic actors coordinated by a global lead firm and produce goods and services across multiple geographic locations for worldwide markets (Yeung and Coe 2015, 32)". Compared with the GCC/GVC, the GPN further embeds the economic relations into their spatial context, emphasizing the role of non-firm actors and institutions in shaping global production networks (Coe 2012). However, like the studies on GCC/GVC, "GPN researchers haven't extended their approach to sustainability and the environment other than the relations of natural resource exploitation (Bush et al. 2015, 10)".

Overall, in the context of the global environmental burden shift, these approaches provide opportunities for understanding environmental issues and their solutions by separating production-consumption. For EEG researches, efforts are primarily made to understand the industrial organization of resource-based industries. Research questions include the nexus of material resources and socio-economic development (Bridge 2008a; Gibson and Warren 2016), and the network configurations of resource-based industries (Bridge 2017; Bridge and Bradshaw 2017). These questions are in response to the fact that "the influence of materiality that exerts on the industrial organization has long been underplayed (Bridge 2008a, 415)". For instance, Ciccantell and Smith (2009) highlight that material resources and the location of resource-based sectors played very different roles in the production network. Gibson and Warren (2016) incorporate new actors into the GPN and emphasized the capacity of raw materials to shape GPNs and attend geopolitical relations. Using the global gas market as a case, Bridge and Bradshaw (2017) recognize the role of markets in determining the spatial configuration of GPNs, which is not only responsive to markets. Afewerki et al. (2019) find that the configuration of the oil and gas production network relies on the interaction between firms' practice on network development and host states' strategies on market development.

Another potential facet for EEG is the sustainability governance *in*, *of*, and *through* GPN (Bush et al. 2015), which allows this literature to go beyond resource-based industries. The governance in and of GPN refers to an internal perspective, which looks for the sustainability practices internal to GPN. Studies may explore how firms can promote a green strategy to get upgraded in GPN and improve their environmental performance. For example, De Marchi et al. (2013) put forth an integrated framework for environmental upgrading strategies, including functional upgrading, process upgrading, product upgrading, and intersectoral upgrading. They explore how firms can transform environmental constraints into new drivers of competitive advantage. By contrast, the governance through GPN represents an external perspective, emphasizing the role of non-firm actors. Studies may incorporate a wider array of networked actors and activities, particularly the institutional agents. For example,

MacKinnon et al. (2019) compare the development of the offshore wind sector in Germany, UK and Norway. They find national states play a crucial role in coordinating regional and national assets to support the development of the offshore wind sector.

As one of the most important advances in Economic Geography, GPN provides an important framework to depict the dispersed distribution and networked relations of global production. This feature corresponds to the transboundary nature of global environmental challenges. As such, the development of EEG would benefit from its full use of GPN to trace the source of environmental issues, to understand the configuration of resource-based sectors, and to explore green strategies and relevant stakeholders in order to transform GPN towards sustainability.

2.3 Innovation, Institution, Transition, and the Environment

Why can economic activities in one place outperform those in other places? Answering this question is one of the basic tasks of economic geographers. Innovation, institution, and transition are three crucial components of the conceptual foundation adopted by economic geographers to explore the answers. Equipped with innovation, one place has the capacities to grow, and could benefit from the growth and get self-reinforced in turn. Institution rules the economic processes in one place, which affects the efficiencies for economic growth. Transition triggers a set of structural changes, which alters the trajectories of economic growth and underlies the sustained growth of one place. Besides, these three components are interwoven with each other. Transition essentially represents the outcome of economic novelty, which involves innovation processes. Although innovation may lay the foundation for transition, both of them are internal (or endogenous). Thus they are both subject to the institutional arrangement. The institution never stands still but coevolves with the processes of innovation and transition.

These three components correspond to the adaptive and public nature of environmental issues. Environmental issues interact with economic processes. Conventionally, studies focus more on the environmental outcomes of economic processes, which treat the relationship between these two as a zero-sum game (Hayter 2008). However, environmental issues are not external to economic processes. They are transformative and adaptive in association with the changes in economic processes. In this regard, innovation and transition capture the changes that may lead to a reciprocal relationship between economic activities and the environment. As for public nature, environmental issues have significant negative externalities, which means that economic benefits are gained by the private while social costs are bear by the public. Institution thereby plays a key role in curbing the negative externalities of environmental issues. From this perspective, conventional studies focus on the implementation and efficiencies of environmental regulation. Nevertheless, economic processes operate at a multi-scalar institutional system that consists of multiple agents and various arrangements (Gertler 2010; Dawley 2014). Hence, institutional studies in economic geography provide a broader view of how institutional agents at different scales interact with environmental issues, which may go beyond the sole focus on environmental regulation.

Overall, these three interrelated terms, namely, innovation, institution, and evolution, are important conceptual foundations used by economic geographers to answer why some regions can outperform others. They also provide the unique EEG theoretical base to investigate regional sustainable development in response to global environmental challenges. From the innovation view, EEG may look into the spatiality and the spatial dimensions of networks of green innovation. Incorporating institutional variables, EEG may further explore how the multi-scalar institutional context affect the green innovation, the environmental upgrading of incumbent industries, and the development of green industries. On this basis, through applying theories and approaches to the evolutionary economic geography and transition studies, EEG may answer how one region creates the path towards sustainability.

2.3.1 Geography of Innovation and the Environment

Innovation has long been identified as a crucial moderator to alleviate the impacts of socio-economic activities on the environment. Innovation is expected to provide efficient alternatives to meet the same demand for socio-economic activities while at the same time reduce the utilization of material resources and the undesirable outcomes on the environment. For economic agents at the micro-level, in response to severe environmental conditions and strict environmental regulations, promoting innovation is also a feasible way to reduce costs, gain comparative advantages, and then increase both economic and environmental performance (Lanoie et al. 2011). For regions at the meso-/macro-level, promoting innovation serves as an engine that drives regional development towards a sustainable trajectory.

Economic geographers are also actively involved in innovation studies. Innovation essentially represents the production of knowledge or the creation of new knowledge, which requires specific factors and conditions (Cooke 2001; Asheim and Isaksen 2002; Gertler 2003). Therefore, the appearance of innovation is particularly concentrated in specific places, even if globalization has broadened the geographical options for economic activities (Asheim and Gertler 2005). In other words, economic geographers explore the place-specific context for innovation. Besides, the changing nature of scientific research shifts the innovation towards decentralized and collaborative activities. Innovation occurs not only in laboratories or institutes but also in the interactions among a wide array of actors (Asheim et al. 2011). As a result, despite the spatial context, innovation also displays a network-based feature consisting of both private and public agents (Lau and Lo 2015). Taken together, the concept of innovation system has developed to emphasize the regional capacities for promoting both systemic learning and interactive innovation, involving a set of networks between public and private agents (Cooke et al. 1997). Regional innovation systems show great potential to understand the evolution of renewable energy consumption and production and the regional exemplar for such a "green turn" (Cooke 2010).

Regarding the network-based feature of innovation, besides the innovation systems approach, economic geographers also apply the concept of proximity to explore the interactions for innovation, which determines the probability of knowledge diffusion and absorption (Balland et al. 2015; Hansen 2015). Studies on proximities are thereby closely related to the topic of industrial clusters. Theoretical thinking focuses on the distinct but interrelated role of geographical, cognitive, organizational, institutional, and social proximities in promoting innovation (Boschma 2005). Such proximities imply the probability that learning can occur between actors.

Organizational, institutional and social proximity fundamentally capture the similarity of relations between agents in different places. More similarity allows actors to absorb new knowledge and information more easily. For instance, Antonioli and Mazzanti (2017) find that trade union involvement of firms is relevant for their adoption of more complex and radical environmental innovation. Cognitive proximity indicates the relevance between new knowledge and the current base. Such relevance allows new knowledge to be recombined with the current base to support innovation. As such, cognitive proximity should be neither too close nor too far. Too much closeness can not introduce sufficient increments for innovation, while too little will hamper the recombination due to irrelevance (Nooteboom 2000). Finally, geographical proximity highlights the distance costs for interactions. Interactions for innovation would be easier to occur between co-located agents (Sonn and Storper 2008). Moreover, it is also believed that face-to-face interactions tend to benefit the innovation process. Compared with other proximities, the effects of geographical proximity can be sensitive to spatial and temporal scales. For example, geographical proximity can be achieved by temporal activities. It does not necessarily depend on stable co-location (Torre 2008). Likewise, how large is the geographical scale that can promote knowledge spill-overs by distance reduction? Answers to this question are still controversial.

Overall, innovation studies in economic geography allow EEG to explore the linkage between innovation and environment further. Three research facets have emerged in this kind of literature. First of all, studies may explore the specific spatiality for green-innovation (Horbach 2014). Green innovation, also known as environmental innovation, eco-innovation, or sustainable innovation, represents the novel economic activities that can reduce the impacts on natural resources, ecosystems, and the environment, compared to their alternatives (Kemp 2010). In contrast to other types of innovation, green innovation is favoured by both public and private sectors since it may help to break the zero-sum game between economic and environmental performance.

Consequently, a set of questions follows this unique feature: what kinds of determinants does green innovation share with other innovation? Are there any specific determinants for green innovation? Are the specific determinants for green innovation place-contingent? How is green innovation related to the national/regional innovation systems? For instance, Horbach et al. (2013) compare French and German industries and find that green innovation tends to be driven by the motivation of cost-saving and be more subject to regulations. Compared with other innovations, green innovation requires more nonlocal knowledge and information. Moreover, they find the roles of universities in green innovation are quite different between German and French industries, suggesting the place-contingent nature of green innovation. Antonioli et al. (2018) highlight the effects of the joint adoption of two radical innovations, namely, information communication technology and green technology. They find that green innovation strategies are subject to regional innovation systems. More polluting sectors tend to adopt joint innovation. Although green innovation may help polluting sectors to increase performance, it may simultaneously increase divergences within regions.

The second facet may study how spatial concentration of industries promotes green innovation and whether green innovation can trigger spatial concentration in turn, like other innovations. A new term "cleantech cluster" has emerged and attracted some researchers' attention. (Casper 2007; Caprotti 2012). Cleantech cluster is based on the expectation that environmental industries can also benefit from spatial concentration as those traditional ones. Moreover, compared with traditional clusters, cleantech clusters may further represent the basic idea of ecological modernization and generate regional development towards a green economy (Marra et al. 2017). Tvedt (2019) looks into the formation and structure of cleantech clusters with the case in San Diego, Graz, and Dublin. Empirical findings report that the formation of cleantech clusters is not a chance event but relies on the local inherited experience and knowledge. Accordingly, the motivation for developing cleantech clusters is place-specific with retrospective nature.

The third facet may explore the interactions between local and nonlocal resources and linkages for green innovation. Regions and firms are not isolated organizations but embedded in multiple relational networks at different scales. From an evolutionary perspective, innovation utilizing only local resources will suffer from insufficient increments of knowledge and result in lock-in effects (Bathelt et al. 2004; Grillitsch et al. 2018). In contrast, the introduction of nonlocal resources is likely to make a difference. In other words, one specific place is less likely, also unnecessarily, to cover all the resources needed for innovation. Instead, it has to interact with agents in different places at different levels so that to integrate sufficient resources for innovation. For example, Mothe et al. (2018) further reveal the role of open innovation systems. They find that the type of green innovation relies on how the knowledge is acquired. External acquisition by R&D tends to promote more complex product innovations, while internal acquisition results in process innovations.

2.3.2 Multi-level Institutions and the Environment

Solutions for environmental issues require efforts from not only private sectors but also public ones, considering the public nature of environmental issues. In the context of global environmental change and globalization, the transboundary or border-crossing features of environmental issues appear to be significantly increasing. As a result, the governance of global environmental issues requires the formation of international institutions (Paterson 2001), which has developed into a complex poly-centric structure (Bulkeley 2005; Bulkeley et al. 2014). Such institutional systems involve the agents at the supra-national, national, and sub-national levels, formu-lating spatially diversified practices in response to environmental challenges (Gupta et al. 2007). For instance, in the literature on climate change, the multi-level gover-nance approaches are proposed to enhance collective action across scales and foster the diffusion of practices (Gregorio et al. 2019).

Such theoretical thinking exactly corresponds to the institutional approaches in economic geography, which seeks to understand how the distinctive institutional configurations shape different economic behaviours and development paths across places, particularly the spatially varied competitive advantages and innovative poten-tials (Gertler 2010). Early studies in this literature primarily compare the different institutional architectures among nations and explore how they produce economic strengths and shape the trajectory for growth, which is also interwoven with the studies on national innovation systems. Similar to the rising of studies on industrial agglomeration and regional innovation systems, there is an increasing awareness that trajectories of economic development at the sub-national level are divergent. Thus, we should not overlook the sub-national constellations of institutions. Economic geog-raphers set out to explore the formation of regional institutions and their interactions with the national ones (Pike et al. 2015; Schröder and Voelzkow 2015).

Overall, institutions represent rules that will encourage particular types of economic behaviours while simultaneously discourage those undesirable ones. When it comes to environmental terms, conventional thinking tends to focus more on envi-ronmental regulation. Both theoretical and empirical studies make great efforts to examine whether environmental regulation follows the rules of the zero-sum game or creates a win–win situation for economy and the environment. The latter one is also known as the famous "Porter Hypothesis", which posits that well-designed envi-ronmental regulation can stimulate innovations and then improve both the economic and environmental performance of firms (Porter and van der Linde 1995). Jaffe and Palmer (1997) further put forth three different versions of the Porter Hypothesis, namely, weak, narrow, and strong version. The weak version explores the linkage between regulation and innovation, and posits that properly designed environmental regulation will promote innovation. The narrow version focuses on the performance-based regulation and postulates that environmental regulation with more flexibility is more likely to trigger firms to innovate. The strong version emphasizes the impact of regulation-induced innovation on firm competitiveness. There are many empirical studies on the Porter Hypothesis based on various techniques, cases, and datasets, while reporting conflicting findings (Ambec et al. 2013). Studies also investigate the effects of different modes of environmental regulation (Cheng et al. 2017b), responding to the narrow version of the Porter Hypothesis.

Another strand of the literature relating to the institution focuses on the debate over the institutional architecture of environmental governance, which is known as environmental federalism. The question is whether centralization or decentralization of environmental regulation would improve the environmental performance more efficiently. The centralization of environmental regulation will internalize the likely externalities between agents and regions, avoiding the race-to-the-bottom phenomenon and the tragedy of the commons (Zhao and Percival 2017). However, such a top-down command-and-control system also exhibits significant deficiencies in transboundary environmental issues, exposing regions to a risk of increasing pollution from their neighbours (Eaton and Kostka 2018). In contrast, the decentralization of environmental regulation responds to local conditions and allows more flexibility to the context-contingent environmental issues (Banzhaf and Chupp 2012). However, if lacking efficient interregional cooperation, this bottom-up system may also expose regions to a risk of destructive competition. There is still no consensus on this debate. Both theoretical and empirical studies doubt the thinking that decentralization of environmental regulation will result in a destructive competition (Ferrara et al. 2014; Sjöberg and Xu 2018). Researchers start to revisit these issues from a broader perspective, which embeds environmental regulation in the institutional systems and explores the linkages between environmental regulation and other institutions. For instance, Fredriksson and Wollscheid (2014) investigate the connection between environmental decentralization and political centralization and find that the effectiveness of environmental decentralization is conditioned on the level of political centralization.

Overall, findings for the Porter Hypothesis suggest what is well-designed environmental regulation, particularly focusing on its modes and instruments. Studies on the environmental federalism point to the institutional interaction between the nation and the local. The literature in global environmental governance, particularly the climate change, has further expanded our understanding of institutional system to the global level (Gregorio et al. 2019). These studies share a similar research question of how to design/form an efficient and effective system of environmental regulation (Mol 2009). In contrast, EEG may contribute to this realm from a different perspective.

As mentioned above, economic geographers are curious about how institutions shape economic behaviours and development paths. In this regard, the research question for EEG in the institution-relevant realm should go beyond the environmental regulation itself. Its specific focus may shift from the formation of environmental regulation to the identification of institutions that are relevant for regional development towards sustainability. Thus, EEG can either explore how the environment-relevant economic activities operate simultaneously in the institutional context at the local, national, and even international level simultaneously, or examine how the local development path is linked to the environment-relevant institutions at multiple levels. For instance, Dawley (2014) looks into the development of the offshore wind industry in the periphery regions. The findings reveal that national regulatory institutions and policy architectures provide broader technological and market conditions for the industry's growth. They also change the capacity and form of local development institutions. Moreover, the findings also display that local policy interventions provide an enabling environment for industrial development at an early stage. Mackinnon et al. (2019) compare the development of the same industry (i.e. the offshore wind industry) between Germany, the UK, and Norway. Their findings also emphasize the role of multi-scalar institutional environments and policy initiatives.

2.3.3   Transition Towards Sustainability

Economic restructuring, transition, upgrading and other words related to economic novelty are most frequently used in the solutions and targets for tackling environmental issues. However, those approaches generally treat economic novelty as external factors to development, while overlooking the evolutionary nature of economic processes (Boschma and Martin 2010).

The evolutionary nature of economic processes can be derived from the fact that, as insisted by Schumpeter, economic novelty is endogenous rather than exogenous. Thus, the emergence of new economic activities, the adoption of new technology and the development of new products are not pre-given for economic growth (Boschma and Martin 2010). Instead, they continue to accumulate and get adapted to the economic process. On this basis, evolutionary economic geographers apply the generalized Darwinism theory, the complexity theory, and the path dependence theory to capture the evolutionary nature of economic processes (Boschma and Martin 2007).

According to the Generalized Darwinism theory, regions provide selection environment for evolutionary processes. One population of economic agents will affect the dynamics of others within the environment. Also, the population in one place will interact with populations in other places. Regional development towards different trajectories is thereby an outcome of the co-evolution between agents and places (Essletzbichler and Rigby 2007). This perspective is similar to the theoretical thinking of environmental studies on socio-environmental relations. That is, the economic and environmental processes are intertwined and coevolved, indicating a relationship of reciprocal causation.

The complexity theory helps to capture the openness of both economic and geographical spaces. There are constant interactions between economic agents and their environment. Thus, the development of regions is constantly in flux rather than reaching an equilibrium (Martin and Sunley 2007). The complexity theory has been widely used in environmental studies, such as land-use change, ecosystem management, and landscape planning, especially in incorporating anthropogenic factors into the evolution of physical processes (Veldkamp and Verburg 2004). In this regard, the complexity theory provides a likely common basis to link the evolution of economic and environmental processes.

The path dependence framework is supposed to be the most widely used theory in regional development from an evolutionary perspective. Evolutionary economic geographers apply the path dependence thinking to argue that path creation for regional development is not a random event or historical accident. In contrast, path creation is an outcome of economic novelty, which is subject to the already-existing accumulation to some extent (Martin and Sunley 2006). They also posit four basic stages of development, namely, pre-formation, path creation, path lock-in, and path dissolution (Martin and Simmie 2008). Thus, regarding path creation and regional development, a set of questions are raised: how can one region end the current path and move to a new one? Can various paths co-exist? How will particular industries evolve spatially?

This research avenue is closely related to the sustainability transition. Sustainability transition considers industrial sectors as a social-technical system, which is composed of tightly interrelated actors and institutions (Geels 2004). This system keeps dynamic so long as the technological, material, organizational, institutional, political, economic, or socio-cultural relationships changes and alter the structure of this system. Sustainability transition exactly represents the shift of this system towards a more sustainable model and the way to promote it (Markard et al. 2012). Based on the theoretical foundation of evolutionary economics, transition studies, and innovation studies, there are two basic conceptual frameworks for sustainability transition, which are the technological innovation system (TIS) and the multi-level perspective (MLP). TIS is similar to the approach of regional innovation systems and national innovation systems, highlighting the role of new technology in facilitating sustainability. Empirical studies tend to apply this approach to identify the barriers to promote the application of clean technology (Coenen and Truffer 2012).

In contrast, the MLP approach consists of three levels, namely, niche, regime, and landscape. Niche represents the basic units for technological innovation, such as laboratories, and serves as incubation spaces with uncertainty and disorder. On the contrary, the regime is the complex of scientific knowledge, engineering practices, production technology, product characteristics, user needs, institutions, and infrastructures (Rip and Kemp 1998). The regime captures the forces that sustain the existing configuration of this complex. The landscape is the external environment to the regime, which puts pressure on it. MLP believes that the transition of socio-technical systems is subject to the co-evolution of these three levels.

However, Connen et al. (2012) argue that the sustainability transition somehow overlooks the geographical unevenness of transition processes. They clarify the need to incorporate spatial perspective into sustainability transition, which is to provide spatial contexts to investigate diversity in transition processes, and multi-scalar, trans-scalar nature of transition processes. Hansen and Connen (2015) further propose the geography of sustainability transitions, which should focus on how place-specificity and scale influence transition processes. Particularly, they also emphasize that theoretical frameworks and modes for sustainable transition studies should also be context-contingent, noticing the differences between developed and developing economies. In this regard, the combination of evolutionary economic geography and transition studies is likely to provide a potential research avenue for EEG.

We should also notice that the evolutionary perspective raises a research question regarding the linkage between regional resilience and the environment. Usually, resilience can be defined from three different respects. The first is from engineering systems, which considers resilience as the capability to recover from an external shock and get back to its original equilibrium status (Meerow et al. 2016). The second is from ecology view, considering that one system may have multiple equilibrium statuses along with its development. The external shocks will alter the structure and function of the system so that resilience concerns the ability of this system to shift from one equilibrium to another one (Côté and Darling 2010; Mumby et al. 2014). The third is the evolutionary one. As mentioned above, from an evolutionary perspective, regions tend to be an open system with complexity. Their development is less likely to reach a stable equilibrium state. Therefore, evolutionary economic geographers define regional resilience as the capacities of regions to reconfigure their socio-economic structure to sustain long-term development in response to the external crisis (Boschma 2015).

That is, from the evolutionary perspective, regional resilience is an ongoing process of development rather than a recovery back to either old or new equilibrium (Simmie and Martin 2010). Interestingly, regarding environmental shocks like natural hazards and environmental degradation, studies tend to apply the first two perspectives (the recovery perspective) to define regional resilience, assess the socio-economic risks, and enhance capacity to recover as soon as possible. However, such regional resilience tends to capture socio-economic processes in the short term, since their base on equilibrium perspective. For the long-term development of sustainability, the evolutionary perspective would be of great value. As such, the combination of evolutionary economic geography and environmental studies on resilience may provide another potential research avenue for EEG.

Generally, we suppose that the further development of EEG is likely to benefit from its combination with another EEG, which is evolutionary economic geography. In fact, Patchell and Hayter (2013) have encouraged one combination of environmental and evolutionary economic geography, labelled as EEG-square. Such a combination allows both EEGs to explore the environmental issues with a multi-scalar co-evolutionary approach.

2.4 Moving Forward: From Polyvocal EEGs to a Coherent EEG?

The seemingly fragmented and polyvocal EEG studies have already exhibited great potential. If we look inward, EEG will help to recall the importance of the environment in economic geography, extend the franchise of economic geography, and unsettle some theoretical thinking that is taken for granted previously (Bridge 2004). But more importantly, if looking outward instead, EEG would be an important platform for economic geographers to participate in and communicate with the interdisciplinary studies on sustainability, which has already attracted extensive participation, especially within the discipline of geography. However, regarding sustainable development and global environmental challenges, economic geographers who are supposed to be experts on regional development and globalization are somewhat silent. For example, Gibbs (2006) points to the fact that economic geography exhibits limited capacity in response to the rapidly evolving studies on environmental issues. In this regard, EEG can make a difference and contribute to interdisciplinary environmental studies. EEG identifies context-contingent causes and solutions of environmental issues, develops multi-scalar configurations of social-technique systems for environmental innovation and governance, and shapes co-evolutionary trajectories for coupling nature-society interactions and sustainable development.

Although the loose grouping of EEG studies suggests bright prospects, it reflects more difficulties in achieving its potential. Various types of economy-nature dualities have made economic geographer retreat from environment-relevant issues for a long time. In most conceptual foundations of economic geography, the environment is external to both economic processes and spatial context. Consequently, the burgeoning EEG studies embrace diverse conceptual frameworks and approaches from related fields, such as institutional economics, regulation theory, ecological modernization, and industrial ecology (Gibbs 2006; Bridge 2008a; Hayter 2008). Thus, it raises a question of how EEG may bridge the conceptual and methodological gap between economic geography and environmental studies, which is supposed to be a core of EEG. Answers to this question point to the demand of a coherent EEG with unique research questions and clear conceptual foundations. Put it another way, the burgeoning and fledgling EEG is still suffering from the absence of a consistent research agenda.

2.4.1   *Continuous Efforts to Developing a Research Agenda for EEG*

Developing a consistent research agenda for EEG encounters a set of primary questions at first. The fundamental one is what kind of research questions EEG should commit to, distinguished from its neighbouring fields. Following this, EEG scholars should clarify what kind of environmental issues should be incorporated in. Also, to bridge the gap between economic geography and environmental studies, should EEG apply notions and approaches of economic geography to revisit environmental issues or incorporate new environmental study knowledge to redevelop notions and approaches?

Back to the very beginning of this chapter, we introduce that geography is a discipline regarding "why of where and so what". EEG should be without exception. For economic geography, it is relatively easy to figure out the meanings of "where" and "why". However, when it comes to EEG, there are several more options. Broadly defined, EEG concerns environment-economy interactions, which is not the only discipline regarding this issue. "Where" can represent location, spatial distribution, and spatial pattern of environmental issues. Correspondingly, "why" refers to the sources and causes of their occurrence and spatial patterns. On the other hand, "where" can also indicate environmental-relevant economic activities. If so, "why" explains the unique spatiality of environmental-relevant economic activities. At the same time, from an institutional perspective, "where" can further represent the spatial configurations of those activities in response to environmental issues, such as regulation and governance. "Why" thereby indicates the formation of these spatial configurations.

In this regard, Soyez and Schulz (2008) suppose that EEG should answer the crucial question of "who governs by which means in which context to which geographic effect". They hold the opinion that environmental governance should be the core component of EEG. Accordingly, "where" represents the spatiality and spatial configuration of environmental governance. EEG should explain how the relations between actors, multi-level institutions, and place-specific environment work together to achieve the environmental targets. Building on this premise, they emphasized the value of revisiting the environmental issues with traditional notions of economic geography. Besides, they also call attention to the construction of nature and the environment, which is discursive and includes all manner of phenomena. Similarly, Bridge (2008a) also criticizes that current definition of the environment is overly broad. Complex topical categorization is one of the main reasons that EEG is fragmented. In this respect, EEG should pay more attention to "the connectivity and spatial interdependence of physical environmental processes, their spatial and temporal variation, and/or their capacity for autonomous action make a difference to the functioning of economic processes." (Bridge 2008a, 79).

On this basis, Bridge (2008a) further put forth three lines of inquiry for EEG. The first one regards the effects of environmental changes on economic behaviours. Environmental changes are spatially disparate but simultaneously interdependent. Therefore, economic actors may have various strategies towards these changes, including geographical relocation and integration. This provides plenty of geographical scopes to investigate how changing environmental processes will reshape the economic landscape. The second one considers the opposite, which is the effects of economic processes on environmental conditions. EEG should focus more on the environmental conditions that underlie people's livelihood, economic growth, and social welfare. In this way, EEG would contribute to the causes more than sources of environmental issues. Specific questions regard why economic activities perform differently in environmental terms, and what environmental outcomes would be in the wake of changing spatial configurations of economic activities. The third line of inquiry relates to the forms of environmental governance. Based on the three lines above, EEG should also investigate the reciprocal relationship between the environment and the economy that could support the governance of environmental issues.

Unlike Bridge who tends to propose an agenda based on a more precision term of "environment" that is fit for economic geography, Hayter (2008) adopts an alternative way that applies theoretical and conceptual foundations to environmental studies. Hayter (2008) suggests three proposals for the EEG agenda, namely, regions, remapping, value chains. First, EEG should focus on the regions embedded in a wide range of global–local interactions, especially in institutional terms. EEG should document the regional impacts of environmental changes on economic activities and the corresponding policy and behavioural responses, involving the efforts from global, national, and regional agents. Second, EEG should remap the core and the periphery of resource geography and its neighbouring sub-disciplines like trade geography. Conventional thinking focuses more on resource availability. Nevertheless, there is an increasing awareness of the relational, political, cultural construct of resources. More conflicts and paradoxes around resource production and consumption are developing (Bridge 2009). Hence, understanding the changing resource geography beyond the availability perspective and the associated changes of economic geography should be a meaningful component for EEG, which would also allow it to apply theories and approaches of economic geography. The third proposal advises the application of global value chain and global production network through extending their environmental elaboration. This is also closely related to the second proposal since GPN provides an integrated framework to depict the spatial configuration and organization of global production, including the resource-based sectors (Bridge 2008b; Gibson and Warren 2016; Breul and Diez 2018). Advances in this literature have further combined the studies on agglomeration to understand the transformation and upgrading of economic activities. Accordingly, the third proposal also suggests that EEG should figure out the proper distribution, organization, and location of economic activities to resolve environmental issues.

In contrast, Gibbs (2006) seeks to link economic geography with insights from ecological modernization theory and regulation approaches. The tasks of EEG are, therefore, to figure out how to shift the economy towards a more ecologically rational one. From this perspective, EEG will focus more on the role of institutions in regulating the interactions between economic activities and environmental processes. Gibbs (2006) further proposes four foci for this perspective, including institutional forms, neoliberalism, globalization, and consumption.

Patchell and Hayter (2013) propose to combine Environmental Economic Geography with Evolutionary Economic Geography. They posit that these two EEGs share the same roots in institutional and path-dependent approaches, and hold strong connections with innovation processes. Grounded in such theoretical thinking, they outline a co-evolutionary and multi-scalar framework for EEG[2] to capture the role of transition, innovation, multi-national corporations, and value chains. Based on this framework, three research themes emerge. They are "extending the place-based analysis of localized clusters, broadening the scope of global value chain analysis, and re-engaging the analysis of core-periphery relations (Patchell and Hayter 2013, 111)".

The efforts to develop a consistent research agenda for EEG continue. These early efforts provide useful proposals to clarify the tasks and define the boundary of EEG. However, successive empirical studies are still relatively scarce and scattered. By far, through scrutinizing current thinking and their successive empirical efforts, two significant features have developed. First, a common way of EEG studies is to apply theories, concepts and approaches of economic geography to analyse environmental issues, such as the agglomeration, innovation systems, and global production networks (Afewerki et al. 2019; Mackinnon et al. 2019; Calignano et al. 2019). They interact with the institutional economic geography, relational economic geography, and evolutionary economic geography significantly. Second, regarding the meaning of "environment", studies are more interested in three facets, which are environmental governance, green innovation, and sustainability transition. They do not mean to monitor and trace the sources of specific environmental problems. Instead, they focus on how to shift the economic activities towards a more sustainable way, exploring the spatially various efforts of public and private agents at multiple levels.

2.4.2 Building Four Pillars for EEG

Based on the review above, we acknowledge that the primary task of EEG is to bridge the gap between economic geography and environmental studies. Particularly, the "bridge" here tends to convey the meaning of providing sustainable issues with the expertise in economic geography, which has focused on development-relevant issues for a long time. To achieve this goal, we suppose that EEG should recombine the relevant knowledge on environment-economy interactions and socio-technique systems at first, preparing a conceptual foundation for EEG. Then this foundation can underlie our revisit to the spatiality and spatial interdependencies of environmental problems. Besides, it will also support our examination and design of the spatial configuration and organization of policies and behaviours in response to environmental problems.

Grounded on the theoretical context of economic geography, we postulate four pillars for the conceptual foundation of EEG, namely, context-dependence, border-crossing, multi-scale, and co-evolution.

Context-dependence captures the spatiality of both economic and environmental processes, which is related to the theoretical foundation of regional geography and regional science. As for EEG, using context-dependence as one pillar of EEG's conceptual foundation means a likely topical categorisation by contexts. Thus, regarding various environmental problems, it requires EEG to identify and remap the core-periphery regions, capturing their various features and urgent demands. For example, in response to climate changes, not all regions are expected to suffer from risks and losses. The coastal regions will become more vulnerable due to the increasing sea level, while the arid and semi-arid regions may benefit from the changing spatial–temporal patterns of precipitation. Such differences result in utterly different attitudes and feedbacks to the so-called global challenges. Thus, context-dependence allows EEG to develop its first basic question: why do particular kinds of environmental changes trouble the economy here rather than there?

Border-crossing captures the spatial interdependencies of both economic and environmental processes, considering the relation between actors and spaces at the same level. As for EEG, border-crossing is manifested by the term of "flows", including both physical flows and socio-economic flows. Physical flows engender the externalities of environmental problems and associated risks of free-rider issues, resulting in the transboundary pollution problems and requiring interregional governance cooperation. Economic flows shift the environmental burden across spaces by separating production and consumption spatially. Therefore, border-crossing provides EEG with the second basic question: how do environmental and economic flows redistribute the environmental problems across spaces?

Multi-scale points to the fact that environmental conditions, institutional context, and economic actors perform differently across multiple levels. Global environmental changes and economic globalization have further highlighted the need to combine related processes from the global to the local (O'Brien 2010). Regarding the environmental processes, local changes may impact not only the local environment and economy but also the whole human–environment/social-technique systems, involving various nations and regions. On the contrary, global changes may also impact the nations and local places in different ways and result in divergence of their development. Therefore, understanding the interdependencies of environmental and economic processes needs not only the horizontal perspective of transboundary flows (*border-crossing*) but also the vertical one of trans-scale interactions (*multi-scale*). Advances in most research avenues of economic geography have also developed multi-scalar approaches, including the institutional economic geography, relational economic geography, innovation geography, and evolutionary economic geography. Specific frameworks contain the global production network, national/regional innovation systems, multi-level governance, global–local linkages, local-nonlocal interactions, and so on. Thus, EEG can apply these approaches to answer the third basic question: how do environmental governance and sustainable development operations in the multi-level systems consisting of actors and institutions?

Co-evolution suggests two types of causation relationships in the development of the environment and the economy. Considering in isolation, the environmental/economic processes are subject to cumulative causation. That is to say, their development does not occur at random but is influenced by the already-existing bases. This essentially suggests the co-evolution between actors and contexts. Regarding the environmental processes, its development manifests as the shift from one equilibrium state to another. By contrast, the development of economic processes, if following the Schumpeterian approach, represents the emergence of economic novelty. That is, economic processes focus on changes, while environmental processes are to take equilibrium (Dosi 1990). Thus, co-evolution also points to the reciprocal causation between the environmental and economic processes. The economic novelty will shock the previous equilibrium of the environmental process, while the changing equilibrium state of the environment will impact the capabilities of economic novelty. New development in the Evolutionary Economic Geography is likely to support this research avenue. In this regard, co-evolution posits the fourth basic question of EEG: how can actors in one region coordinate the interactions between economic novelty and environmental equilibrium, especially in the long-run?

Overall, we suppose that these four pillars may help to ground EEG on the theoretical foundation of economic geography as well as to make it in line with the development trends of economic geography. Also, these four pillars may further propose a research agenda for EEG consisting of four geographical questions regarding the spatiality, horizontal interdependency, vertical interdependency, and evolution of the environment-economy interactions.

2.4.3 Future Foci of EEG on the Transiting and Opening China

Great efforts have been made to propose a conceptual foundation and research foci for EEG. However, empirical studies on EEG is still scarce and scattered. Empirical studies will provide various research agendas of EEG experiences and lessons, so that to enrich their scopes and improve their theoretical foundation. In this regard, we posit that China should be an inevitable case for conducting EEG empirical studies, corresponding to the four pillars above.

First, China is a nation with a huge territory. Its environmental, economic, social, institutional, and cultural conditions exhibit high levels of heterogeneities, respectively. Their combination results in rich types of development contexts at the local level. For instance, previous studies get used to comparing the east, central, west areas of China by their distinct status on resource endowments, comparative advantages, and environmental vulnerability. In economic terms, there are also hot-spot contexts, such as the Beijing-Tianjin-Hebei regions, the Pearl River Delta, and the Yangtze River Delta. Recently, there is an increasing awareness of the green development of the "Belt-and-Road" initiative regions, the sustainable development of the Yangtze River Economic Belt regions, the revitalisation of northeast provinces and the transition of resource-based areas. In environmental terms, the development of river basin, mountain areas, semi-arid areas, coastal areas, karst areas, and resource-based areas have also drawn wide attention.

Also, there are various types of "borders" within the territory, either physical or socio-economical, continuing to influence the interdependencies between regions. These borders provide plenty of scopes to investigate the spatial interdependencies across spaces. Previously, scholars have noticed the connections between upper reaches and lower reaches along a river, between two administrative units sharing the same boundary, between urban and rural areas, between places inside and outside a policy area, and so on.

Third, there are multiple hierarchical systems of institutions within the nations, such as the "village-town/county-prefecture-province-central" five-tiers system for administration. At the same time, the continuous opening-up processes allow China to couple with the global production network and participate in global cooperation. As such, the global level in various terms is especially meaningful in the multi-level systems of China. We should also notice that the development policy in China prefers local pilot projects, also known as the local experiment. Thus, it allows to set these experimental zones as the local regarding the role of multi-level institutions. In economic terms, such multi-level systems would be more flexible, based on the "local–regional-national-global" four-tiers systems. The regional level can refer to physical/economic compartments, urban agglomerations, mega-city regions, special development zones, and so on.

Lastly, regional development in China displays diversified trajectories, varying from one place to another. Their driving forces for cumulation causation and economic novelty can be very different. Accordingly, they also have different urgent demands and difficulties for economic restructuring and sustainable development. For example, sustainable development of traditional resource-based regions is subject to the shifting socio-technique paradigm of industrialization, the depletion of local resource endowment, and the loss of attraction to human capital. On the contrary, sustainable development of outward-oriented regions is affected by the embeddedness of nonlocal inputs, the risks of external shocks, and the growth of endogenous forces.

Taken together, we posit that empirical studies of EEG in China are able to address issues from four primary perspectives as below.

The locational perspective

- Remapping the core-periphery patterns of development-relevant environmental problems at different scales (Chaps. 3, 4 and 5).

The relational perspective

- Exploring the role of transboundary physical and economic flows in shifting environmental burdens across spaces and scales (Chaps. 8, 9 and 11)

The institutional perspective

- Identifying the configuration of multi-level governance of the environment, including the changing central-local interactions of governments and the incorporating of global forces (Chap. 6).

The evolutionary perspective

- Investigating the capabilities and modes that regions steer their development trajectories toward sustainability (Chaps. 7 and 10).

The perspective of political economy

- Investigating the green governance of both production network and territorial spaces. Although this book does not conduct empirical studies from this perspective, it is undoubtly meaningful for environmental issues and has already attracted some attention.

2.5 Conclusion and Discussion

It is essential to get insights on human–environment interactions to respond to the environmental challenges, which has drawn interdisciplinary efforts. Regarding this issue, economic geography seems to be silent for a long time, compared with other sub-disciplines in Geography and their neighbouring disciplines. Developing the Environmental Economic Geography thereby shoulders the responsibilities to bridge the knowledge gap between economic geography and environmental studies in spite of their ontological differences between "objective science" and "social construction of nature" (O'Brien 2010).

In fact, advances in the theoretical and conceptual foundations of economic geography have already exhibited great potential in understanding interactions between the environment and the economy. Particularly, economic geographers are adept at spatial configuration and transformation of economic landscape across spaces and scales, which can provide an in-depth understanding of some crucial questions, including but not limited to followings.

- Why do environmental processes trouble the economy here rather than there?
- Why is the environmental performance of economic processes better in some spaces than others?
- How do the environmental/socio-economic changes of one place tele-connect with those of others?
- Why are cross-border cooperations for environmental challenges always difficult?
- Who should take more environmental responsibilities along with economic development?
- How are the environmental responsibilities for economic development differentiated across agents at various levels?
- Is it possible to replicate practices for sustainable development from one space to another?
- How can one space steer its current development trajectories towards a sustainable one?
- How can one space build the capacity for continuous economic novelty?

Despite its great potential, the road ahead EEG is still long. EEG is still suffering from its fragmented bound and polyvocal topics, striving for a consistent research agenda. Great efforts had been made to clarify the meaning of "environment" to EEG and to propose research foci. Scholars emphasize the rejection of an "all-in" term. Instead, turning to the environmental processes that are internal (endo) to economic processes would be meaningful. They are also in favour of the research avenue that combines theoretical and conceptual foundations in economic geography to revisit the environmental problems, rather than the opposite. Correspondingly, the approaches of innovation systems, global production networks, economic restructuring, and path creation prevail in the successive empirical studies.

Through scrutinizing early efforts on research agenda and advances in the empirical studies, we appeal for building four pillars of the conceptual foundation of EEG, representing four inherent tasks of EEG as well as four common interests between economic geography and environmental studies. These four pillars are context-dependence, border-crossing, multi-scale, and co-evolution, representing the spatiality, horizontal spatial interdependency, vertical spatial interdependency, and evolution, respectively. Through combining the theories and concepts within these four pillars, we are likely to revisit environmental problems from the EEG perspective. Nevertheless, considering the empirical works on EEG is still scarce, further development of EEG requires not only proposals but also the experiences and lessons from empirical studies.

References

Afewerki, S., Karlsen, A., & MacKinnon, D. (2019). Configuring floating production networks: A case study of a new offshore wind technology across two oil and gas economies. *Norsk Geografisk Tidsskrift–Norwegian Journal of Geography, 73*(1), 4–15.

Ambec, S., Cohen, M. A., Elgie, S., et al. (2013). The Porter Hypothesis at 20: Can environmental regulation enhance innovation and competitiveness? *Review of Environmental Economics and Poilicy, 7*(1), 2–22.

Antonioli, D., Cecere, G., & Mazzanti, M. (2018). Information communication technologies and environmental innovations in firms: Joint adoptions and productivity effects. *Journal of Environmental Planning and Management, 61*(11), 1905–1933.

Antonioli, D., & Mazzanti, M. (2017). Towards a green economy through innovations: The role of trade union involvement. *Ecological Economics, 131,* 286–299.

Aoyama, Y., Berndt, C., Glückler, J., et al. (2011). Emerging themes in Economic Geography: Outcomes of the Economic Geography 2010 workshop. *Economic Geography, 87*(2), 111–126.

Arrow, K., Bolin, B., Costanza, R., et al. (1995). Economic growth, carrying capacity, and the environment. *Ecological Economics, 15*(2), 91–95.

Asheim, B. T., & Gertler, M. (2005). The geography of innovation: Regional innovation systems. In J. Fagerberg, D. Mowery, & R. Nelson (Eds.), *The Oxford Handbook of Innovation* (pp. 291–317). Oxford: Oxford University Press.

Asheim, B. T., & Isaksen, A. (2002). Regional innovation systems: The integration of local 'sticky' and global 'ubiquitous' knowledge. *Journal of Technology Transfer, 27*(1), 77–86.

Asheim, B. T., Smith, H. L., & Oughton, C. (2011). Regional Innovation Systems: Theory, empirics and policy. *Regional Studies, 45*(7), 875–891.

Baldwin, R. E., & Okubo, T. (2006). Heterogeneous firms, agglomeration, and economic geography: Spatial selection and sorting. *Journal of Economic Geography, 6*(3), 323–346.

Balland, P.-A., Boschma, R., & Frenken, K. (2015). Proximity and innovation: From statics to dynamics. *49*(6), 907–920.

Baltagi, B. H., Egger, P., & Pfaffermayr, M. (2007). Estimating models of complex FDI: Are there third-country effects? *Journal of Econometrics, 140*(1), 260–281.

Banzhaf, H. S., & Chupp, B. A. (2012). Fiscal federalism and interjurisdictional externalities: New results and an application to US Air pollution. *Journal of Public Economics, 96*(5–6), 449–464.

Bathelt, H., Malmberg, A., & Maskell, P. (2004). Clusters and knowledge: Local buzz, global pipelines and the process of knowledge creation. *Progress in Human Geography, 28*(1), 31–56.

Boggs, J. S., & Rantisi, N. M. (2003). The 'relational turn' in economic geography. *Journal of Economic Geography, 3,* 109–116.

Boschma, R. (2005). Proximity and innovation: A critical assessment. *Regional Studies, 39*(1), 61–74.

Boschma, R. (2015). Towards an evolutionary perspective on regional resilience. *Regional Studies, 49*(5), 733–751.

Boschma, R., Coenen, L., Frenken, K., et al. (2017). Towards a theory of regional diversification: Combining insights from evolutionary economic geography and transition studies. *Regional Studies, 51*(1), 31–45.

Boschma, R., & Martin, R. (2007). Editorial: Constructing an evolutionary economic geography. *Journal of Economic Geography, 7*(5), 537–548.

Boschma, R., & Martin, R. (2010). *The Handbook of Evolutionary Economic Geography.* Cheltenham, UK: Edward Elgar.

Breul, M., & Diez, J. R. (2018). An intermediate step to resource peripheries: The strategic coupling of gateway cities in the upstream oil and gas GPN. *Geoforum, 92,* 9–17.

Bridge, G. (2004). Contested terrain: Mining and the environment. *Annual Review of Environment and Resouces, 29,* 205–259.

Bridge, G. (2008a). Environmental economic geography: A sympathetic critique. *Geoforum, 39,* 76–81.

Bridge, G. (2008b). Global production networks and the extractive sector: Governing resource-based development. *Journal of Economic Geography, 8,* 389–419.

Bridge, G. (2009). Material worlds: Natural resources, resource geography and the material economy. *Geography Compass, 3,* 1217–1244.

Bridge, G. (2017). The map is not the territory: A sympathetic critique of energy research's spatial turn. *Energy Research & Social Science, 36,* 11–20.

Bridge, G., & Bradshaw, M. (2017). Making a global gas market: Territoriality and production networks in liquefied natural gas. *Economic Geography, 93*(3), 215–240.

Bu, M. L., Li, S., & Jiang, L. (2019). Foreign direct investment and energy intensity in China: Firm level evidence. *Energy Economics, 80,* 366–376.

Bulkeley, H. (2005). Reconfiguring environmental governance: Towards a politics of scales and networks. *Political Geography, 24,* 875–902.

Bulkeley, H., Andonova, L. B., Betsill, M. M., et al. (2014). *Transnational climate change governance.* Cambridge: Cambridge University Press.

Bush, S. R., Oosterveer, P., Bailey, M., et al. (2015). Sustainability governance of chains and networks: A review and future outlook. *Journal of Cleaner Production, 107,* 8–19.

Cai, H., Chen, Y., & Gong, Q. (2016). Polluting thy neighbour: Unintended consequences of China's pollution reduction mandates. *Journal of Environmental Economics and Management, 76,* 86–104.

Calignano, G., Fitjar, R. D., & Hjertvikrem, N. (2019). Innovation networks and green restructuring: Which path development can EU Framework Programmes stimulate in Norway. *Norsk Geografisk Tidsskrift–Norwegian Journal of Geography, 73*(1), 65–78.

Capone, F., & Lazzeretti, L. (2018). The different roles of proximity in multiple informal network relationships: Evidence from the cluster of high technology applied to cultural goods in Tuscany. *Industry and Innovation, 25*(9), 897–917.

Caprotti, F. (2012). The cultural economy of cleantech: Environmental discourse and the emergence of a new technology sector. *Transactions, 37*(3), 370–385.

Carson, R. T. (2010). The environmental Kuznets curve: Seeking empirical regularity and theoretical structure. *Review of Environmental Economics and Policy, 4*(1), 3–23.

Casper, S. (2007). How do technology clusters emerge and become sustainable? Social network formation and inter-firm mobility within the San Diego biotechnology cluster. *Research Policy, 36,* 438–455.

Cezar, R., & Escobar, O. R. (2015). Institutional distance and foreign direct investment. *Review of World Economics, 151*(4), 713–733.

Chen, D., Chen, S., & Jin, H. (2018a). Industrial agglomeration and CO_2 emissions: Evidence from 187 Chinese prefecture-level cities over 2005–2013. *Journal of Cleaner Production, 170*(20), 993–1003.

Chen, Z., Kahn, M. E., Liu, Y., et al. (2018b). The consequences of spatially differentiated water pollution regulation in China. *Journal of Environmental Economics and Management, 88,* 468–485.

Cheng, Z. (2016). The spatial correlation and interaction between manufacturing agglomeration and environmental pollution. *Ecological Indicators, 61*(2), 1024–1032.

Cheng, Z., Li, L., & Liu, J. (2017a). Identifying the spatial effects and driving factors of urban PM2.5 pollution in China. *Ecological Indicators, 82,* 61–75.

Cheng, Z., Li, L., & Liu, J. (2017b). The emissions reduction effect and technical progress effect of environmental regulation policy tools. *Journal of Cleaner Production, 149,* 191–205.

Chertow, M. R. (2000). The IPAT equation and its variants. *Journal of Industrial Ecology, 4*(4), 13–29.

Ciccantell, P., & Smith, D. A. (2009). Rethinking global commodity chains: Integrating extraction, transport, and manufacturing. *International Journal of Comparative Sociology, 50*(3–4), 361–384.

Coe, N. M. (2012). Geographies of production II: A global production network A-Z. *Progress in Human Geography, 36*(3), 389–402.

Coenen, L., & Truffer, B. (2012). Places and spaces of sustainability transitions: Geographical contributions to an emerging research and policy field. *European Planning Studies, 20*(3), 367–374.

Cole, M. A. (2004). Trade, the pollution haven hypothesis and the environmental Kuznets curve: Examining the linkages. *Ecological Economics, 48*(1), 71–81.

Cole, M. A., Elliott, R. J. R., & Zhang, J. (2011). Growth, foreign direct investment, and the environment: Evidence from Chinese cities. *Journal of Regional Science, 51*(1), 121–138.

Commoner, B., Corr, M., & Stamler, P. J. (1971). The causes of pollution. *Environment, 13*(3), 2–19.

Connen, L., Benneworth, P., & Truffer, B. (2012). Toward a spatial perspective on sustainability transitions. *Research Policy, 41*(6), 968–979.

Cooke, P. (2001). Regional innovation systems, clusters, and the knowledge economy. *Industrial and Corporate Change, 10*(4), 945–974.

Cooke, P. (2010). Regional innovation systems: Development opportunities from the 'green turn.' *Technology Analysis and Strategic Management, 22*(7), 831–844.

Cooke, P., Uranga, M. G., & Etxebarria, G. (1997). Regional innovation systems: Institutional and organisational dimensions. *Research Policy, 26*(4–5), 475–491.

Copeland, B., & Taylor, S. (1999). Trade, spatial separation, and the environment. *Journal of International Economics, 47,* 137–168.

Côté, I. M., & Darling, E. S. (2010). Rethinking ecosystem resilience in the face of climate change. *Plos Biology, 8*(7), e1000438.

Cui, J., Lapan, H., & Moschini, G. (2016). Productivity, export, and environmental performance: Air pollutants in the United States. *American Journal of Agricultural Economics, 98,* 447–467.

Da Schio, N., Boussauw, K., & Sansen, J. (2019). Accessibility versus air pollution: A geography of externalities in the Brussels agglomeration. *Cities, 84,* 178–189.

Dasgupta, S., Laplante, B., Wang, H., et al. (2002). Confronting the environmental Kuznets curve. *Journal of Economic Perspectives, 16*(1), 147–168.

Dawley, S. (2014). Creating new paths? Offshore wind, policy activism, and peripheral regions development. *Economic Geography, 90*(1), 91–112.

de Bruyn, S. M. (1997). Explaining the environmental Kuznets curve: Structural change and international agreements in reducing sulphur emissions. *Environment and Development Economics, 2*(4), 485–503.

De Marchi, V., Di Maria, E., & Ponte, S. (2013). The greening of global value chains: Insights from the furniture industry. *Competition and Change, 17*(4), 299–318.

Dietz, T., & Rosa, E. A. (1997). Effects of population and affluence on CO2 emissions. *Proceedings of the National Academy of Sciences, 94,* 175–179.

Dinda, S. (2004). Environmental Kuznets curve hypothesis: A survey. *Ecological Economics, 49*(4), 431–455.

Dong, B., Gong, J., & Zhao, X. (2012). FDI and environmental regulation: Pollution haven or a race to the top? *Journal of Regulatory Economics, 41*(2), 216–237.

Dosi, G. (1990). Economic change and its interpretation, or is there a Schumpeterian Approach? In A. Heertje & M. Perlman (Eds.), *Evolving technology and market structure.* Ann Arbor: University of Michigan Press.

Dumais, G., Ellison, G., & Glaeser, E. L. (2002). Geographic concentration as a dynamic process. *Review of Economics and Statistics, 84*(2), 193–204.

Duranton, G., & Puga, D. (2004). Micro-foundations of urban agglomeration economies. *Handbook of Regional and Urban Economics, 4,* 2063–2117.

Duvivier, C., & Xiong, H. (2013). Transboundary pollution in China: A study of polluting firms location choices in Hebei province. *Environment and Development Economics, 18*(4), 459–483.

Eaton, S., & Kostka, G. (2018). What makes for good and bad neighbours? An emerging research agenda in the study of Chinese environmental politics. *Environmental Politics, 27*(5), 782–803.

Ellison, G., & Glaeser, E. L. (1999). The geographic concentration of industry: Does natural advantage explain agglomeration? *the American Economic Review, 89*(2), 311–316.

Enrlich, P., & Holdren, J. (1971). Impact of population growth. *Science, 171,* 1212–1217.

Essletzbichler, J., & Rigby, D. (2007). Exploring evolutionary economic geographies. *Journal of Economic Geography, 7*(5), 549–571.

Fernandez, L. (2002). Trade's dynamic solutions to transboundary pollution. *Journal of Environmental Economics and Management, 43*(3), 386–411.

Fernandez, L., & Das, M. (2011). Trade transport and environment linkages at the US-Mexico border: Which policies matter? *Journal of Environmental Management, 92*(3), 508–521.

Ferrara, I., Missios, P., & Yildiz, H. M. (2014). Inter-regional competition, comparative advantage and environmental federalism. *Canadian Journal of Economics, 47*(3), 905–952.

Fredriksson, P. G., & Millimet, D. L. (2002). Strategic interaction and the determination of environmental policy across U.S. states. *Journal of Urban Economics, 51*(1), 101–122.

Fredriksson, P. G., & Wollscheid, J. R. (2014). Environmental decentralization and political centralization. *Ecological Economics, 107,* 402–410.

Frostenson, M., & Prenkert, F. (2015). Sustainable supply chain management when focal firms are complex: A network perspective. *Journal of Cleaner Production, 107,* 85–94.

Fu, S., & Hong, J. (2011). Testing urbanization economies in manufacturing industries: Urban diversity or urban size? *Journal of Regional Science, 51*(3), 585–603.

Galeotti, M., Lanza, A., & Pauli, F. (2006). Reassessing the environmental Kuznets curve for CO2 emissions: A robustness exercise. *Ecological Economics, 57*(1), 152–163.

Gaur, A. S., Ma, X., & Ding, Z. (2018). Home country supportiveness/unfavourableness and outward foreign direct investment from China. *Journal of International Business Studies, 49*(3), 324–345.

Geels, F. W. (2004). From sectoral systems of innovation to socio-technical systems: Insights about dynamics and change from sociology and institutional theory. *Research Policy, 33*(6–7), 897–920.

Gereffi, G., Humphrey, J., & Strugeon, T. (2005). The governance of global value chains. *Review of International Political Economy, 12*(1), 78–104.

Gertler, M. S. (2003). Tacit knowledge and the economic geography of context, or the undefinable tacitness of being (there). *Journal of Economic Geography, 3*(1), 75–99.

Gertler, M. S. (2010). Rule of the game: The place of institutions in regional economic change. *Regional Studies, 44*(1), 1–15.

Gibbon, P. (2001). Upgrading primary production: A global commodity chain approach. *World Development, 29*(2), 345–363.

Gibbon, P., Bair, J., & Ponte, S. (2008). Governing global value chains: An introduction. *Economy and Society, 37*(3), 315–338.

Gibbs, D. (2006). Prospects for an environmental economic geography: Linking ecological modernization and regulationist approaches. *Economic Geography, 82*(2), 193–215.

Gibson, C., & Warren, A. (2016). Resource-sensitive global production networks: Reconfigured geographies of timber and acoustic guitar manufacturing. *Economic Geography, 92*(4), 430–454.

Gopalakrishnan, K., Yusuf, Y. Y., Musa, A., et al. (2012). Sustainable supply chain management: A case study of British Aerospace (BAe) Systems. *International Journal of Production Economics, 140*(1), 193–203.

Gregorio, M. D., Fatorelli, L., Paavola, J., et al. (2019). Multi-level governance and power in climate change policy networks. *Global Environmental Change, 54,* 64–77.

Grillitsch, M., Asheim, B., & Trippl, M. (2018). Unrelated knowledge combinations: The unexplored potential for regional industrial path development. *Cambridge Journal of Regions, Economy, and Society, 11*(2), 257–274.

Grossman, G. M., & Krueger, A. B. (1993) Environmental impacts of a North American Free Trade Agreement. In P. M. Garber (Ed.), *The U.S.-Mexico free trade agreement* (pp. 13–56). Cambridge MA: MIT Press.

Gupta, J., van der Leeuw, K., & de Moel, H. (2007). Climate change: A 'glocal' problem requiring 'glocal' action. *Environmental Sciences, 4*(3), 139–148.

Hansen, T. (2015). Substitution or overlap? The relations between geographical and non-spatial proximity dimensions in collaborative innovation projects. *Regional Studies, 49*(10), 1672–1684.

Hansen, T., & Connen, L. (2015). The geography of sustainability transitions: Review, synthesis and reflections on an emergent research field. *Environmental Innovation and Societal Transitions, 17,* 92–109.

Hayter, R. (2008). Environmental Economic Geography. Geography. *Compass, 2,* 831–850.

He, J. (2006). Pollution haven hypothesis and environmental impacts of foreign direct investment: The case of industrial emission of sulfur dioxide (SO_2) in Chinese provinces. *Ecological Economics, 60*(1), 228–245.

He, J., & Richard, P. (2010). Environmental Kuznets curve for CO_2 in Canada. *Ecological Economics, 69*(5), 1083–1093.

Head, K., & Ries, J. (2008). FDI as an outcome of the market for corporate control: Theory and evidence. *Journal of International Economics, 74*(1), 2–20.

Hoekstra, A. Y., & Hung, P. Q. (2005). Globalisation of water resources: International virtual water flows in relation to crop trade. *Global Environmental Change, 15*(1), 45–56.

Horbach, J. (2014). Do eco-innovations need specific regional characteristics? An econometric analysis for Germany. *Review of Regional Research, 34*(1), 23–38.

Horbach, J., Oltra, V., & Belin, J. (2013). Determinants and specificities of eco-innovations compared to other innovations–an econometric analysis for the French and German industry based on the community innovation survey. *Industry and Innovation, 20*(6), 523–543.

Hosoe, M., & Naito, T. (2006). Trans-boundary pollution transmission and regional agglomeration effects. *Papers in Regional Science, 85*(1), 99–120.

Huang, J., Chen, X., Huang, B., et al. (2017). Economic and environmental impacts of foreign direct investment in China: A spatial spillover analysis. *China Economic Review, 45,* 289–309.

Jaffe, A. B., & Palmer, K. (1997). Environmental regulation and innovation: A panel data study. *Review of Economics and Statistics, 79*(4), 610–619.

Jakob, M., & Marschinski, R. (2013). Interpreting trade-related CO2 emission transfers. *Nature Climate Change, 3,* 19–23.

Kahn, M. E. (2004). Domestic pollution havens: Evidence from cancer deaths in border countries. *Journal of Urban Economics, 56*(1), 51–69.

Kahn, M., Li, P., & Zhao, D. (2015). Water pollution progress at borders: The role of changes in China's political promotion incentives. *American Economic Journal: Economic Policy, 7*(4), 223–242.

Kearsley, A., & Riddel, M. (2010). A further inquiry into the pollution haven hypothesis and the Environmental Kuznets Curve. *Ecological Economics, 69*(4), 905–919.

Kemp, R. (2010). Eco-innovation: Definition, measurement and open research issues. *Economia Politica, 27,* 397–420.

Konisky, D. M. (2007). Regulatory competition and environmental enforcement: Is there a race to the bottom? *American Journal of Political Science, 51*(4), 853–872.

Konisky, D. M., & Woods, N. D. (2010). Exporting air pollution? Regulatory enforcement and environmental free riding in the United States. *Political Research Quarterly, 63*(4), 771–782.

Kuik, O., & Gerlagh, R. (2003). Trade liberalization and carbon leakage. *the Energy Journal, 24*(3), 97–120.

Lanoie, P., Party, M., & Lajeunesse, R. (2008). Environmental regulation and productivity: Testing the porter hypothesis. *Journal of Productivity Analysis, 30*(2), 121–128.

Lanoie, P., Laurent-Lucchetti, J., Johnstone, N., et al. (2011). Environmental policy, innovation and performance: New insights on the Porter Hypothesis. *Journal of Economics and Management Strategy, 20*(3), 803–842.

Lau, A. K. W., & Lo, W. (2015). Regional innovation system, absorptive capacity and innovation performance: An empirical study. *Technological Forecasting & Social Change, 92,* 99–114.

Li, T. K., Liu, Y., Wang, C. R., et al. (2019). Decentralization of the non-capital functions of Beijing: Industrial relocation and its environmental effects. *Journal of Cleaner Production, 224,* 545–556.

Lin, J., Pan, D., Davis, S. J., et al. (2014). China's international trade and air pollution in the United States. *Proceedings of the National Academy of Sciences of the United States of America, 111*(5), 1736–1741.

Lindmark, M. (2002). An EKC-pattern in historical perspective: Carbon dioxide emissions, technology, fuel prices and growth in Sweden 1870–1997. *Ecological Economics, 42*(1–2), 333–347.

Lipscomb, M., & Mobarak, A. M. (2017). Decentralization and pollution spillovers: Evidence from the re-drawing of county borders in Brazil. *Review of Economic Studies., 84*(1), 464–502.

Liu, Q., Wang, S., Zhang, W., et al. (2018). Does foreign direct investment affect environmental pollution in China's cities? A spatial econometric perspective. *Science of the Total Environment, 613–614*, 521–529.

Lu, J. Y., Liu, X. H., Wright, M., et al. (2014). International experience and FDI location choices of Chinese firms: The moderating effects of home country government support and host country institutions. *Journal of International Business Studies, 45*(4), 428–449.

MacKinnon, D., Dawley, S., Steen, M., et al. (2019). Path creation, global production networks and regional development: A comparative international analysis of the offshore wind sector. *Progress in Planning, 130*, 1–32.

Madsen, P. M. (2009). Does corporate investment drive a "race to the bottom" in environmental protection? A re-examination of the effect of environmental regulation on investment. *Academy of Management Journal, 52*(6), 1297–1318.

Makkonen, T., & Williams, A. M. (2018). Developing survey metrics for analysing cross-border proximity. *Geografisk Tidsskrift–Danish Journal of Geography, 118*(1), 114–121.

Mao, X., & He, C. (2018). A trade-related pollution trap for economies in transition? Evidence from China. *Journal of Cleaner Production, 200*, 781–790.

Markard, J., Raven, R., & Truffer, B. (2012). Sustainability transitions: An emerging field of research and its prospects. *Research Policy, 41*(6), 955–967.

Marra, A., Antonelli, P., & Pozzi, C. (2017). Emerging clean-tech specializations and clusters–a network analysis on technological innovation at the metropolitan level. *Renewable and Sustainable Energy Reviews, 67*, 1037–1046.

Marrocu, E., Paci, R., & Usai, S. (2013). Productivity growth in the old and new Europe: The role of agglomeration externalities. *Journal of Regional Science, 53*(3), 418–442.

Martin, R. (2012). Regional economic resilience, hysteresis and recessionary shocks. *Journal of Economic Geography, 12*(1), 1–32.

Martin, R., & Simmie, J. (2008). Path dependence and local innovation systems in city-regions. *Innovation, 10*(2–3), 183–196.

Martin, R., & Sunley, P. (2006). Path dependence and regional economic evolution. *Journal of Economic Geography, 6*(4), 395–437.

Martin, R., & Sunley, P. (2007). Complexity thinking and evolutionary economic geography. *Journal of Economic Geography., 7*(5), 573–601.

Martin, R., & Sunley, P. (2015). On the notion of regional economic resilience: Conceptualization and explanation. *Journal of Economic Geography, 15*(1), 1–42.

Meerow, S., Newell, J. P., & Stults, M. (2016). Defining urban resilience: A review. *Landscape and Urban Planning, 147*, 38–49.

Meixell, M. J., & Gargeya, V. B. (2005). Global supply chain design: A literature review and critique. *Transportation Research Part E: Logistics and Transportation Review, 41*(6), 531–550.

Millimet, D. L. (2003). Assessing the empirical impact of environmental pollution. *Journal of Regional Science., 43*(4), 711–733.

Millimet, D. L., & List, J. A. (2004). The cast of the missing pollution haven hypothesis. *Journal of Regulatory Economics, 26*(3), 239–262.

Mol, A. P. J. (2009). Environmental Governance through information: China and Vietnam. *Singapore Journal of Tropical Geography, 30*(1), 114–129.

Mothe, C., Nguyen-Thi, U. T., & Triguero, A. (2018). Innovative products and services with environmental benefits: Design of search strategies for external knowledge and absorptive capacity. *Journal of Environmental Planning and Management, 61*(11), 1934–1954.

Mumby, P. J., Chollett, I., Bozec, Y.-M., et al. (2014). Ecological resilience, robustness and vulnerability: How do these concepts benefit ecosystem management? *Current Opinion in Environmental Sustainability, 7*, 22–27.

Naito, T. (2010). Regional agglomeration and transfer of pollution reduction technology under the presence of transboundary pollution. *2*(2), 157–175.

Naughton, H. T. (2014). To shut down or to shift: Multinationals and environmental regulation. *Ecological Economics, 102,* 113–117.

Neary, J. P. (2006). International trade and the environment-theoretical and policy linkages. *Environment and Resource Economics, 33*(1), 95–118.

Nooteboom, B. (2000). Learning by interaction: Absorptive capacity, cognitive distance and governance. *Journal of Management and Governance, 4,* 69–92.

O'Brien, K. (2010). Responding to environmental change: A new age for human geography? *Progress in Human Geography, 35*(4), 542–549.

Pan, Y. Y. (2003). The inflow of foreign direct investment to China: The impact of country-specific factors. *Journal of Business Research, 56*(10), 829–833.

Patchell, J., & Hayter, R. (2013). Environmental and evolutionary economic geography: Time for EEG2? *Geografiska Annaler: Series B, Human Geography, 95*(2), 111–130.

Paterson, M. (2001). *Understanding global environmental politics: Domination, accumulation, resistance.* Basingstoke: Palgrave.

Pazienza, P. (2019). The impact of FDI in the OECD manufacturing sector on CO2 emission: Evidence and policy issues. *Environmental Impact Assessment Review, 77,* 60–68.

Peters, G. P., & Hertwich, E. G. (2006). Pollution embodied in trade: The Norwegian case. *Global Environmental Change, 16,* 379–387.

Pflueger, M. (2001). Ecological dumping under monopolistic competition. *the Scandinavian Journal of Economics, 103*(4), 689–706.

Pike, A., Marlow, D., McCarthy, A., et al. (2015). Local institutions and local economic development: The local enterprise partnerships in England, 2010-. *Cambridge Journal of Regions, Economy, and Society, 8*(2), 185–204.

Porter, M., & van der Linde, C. (1995). Toward a new conception of the environment-competitiveness relationship. *Journal of Economic Perspective, 9*(4), 97–118.

Prakash, A., & Potoski, M. (2006). Racing to the bottom? Trade, environmental governance, and ISO14001. *American Journal of Political Science, 50*(2), 350–364.

Rasli, A. M., Qureshi, M. I., Isah-Chikaji, A., et al. (2018). New toxics, race to the bottom and revised environmental Kuznets curve: The case of local and global pollutants. *Renewable and Sustainable Energy Reviews, 81*(2), 3120–3130.

Rip, A., & Kemp., R. . (1998). Technological change. *Human Choice and Climate Change, 2*(2), 327–399.

Roos, M. W. M. (2005). How important is geography for agglomeration? *Journal of Economic Geography, 5*(5), 605–620.

Rosenthal, S. S., & Strange, W. C. (2004) Evidence on the nature and sources of agglomeration economies. In J. V. Henderson & J.-F. Thisse (Eds.), *Handbook of regional and urban economics* (Vol. 4, pp. 2119–2171).

Saito, H, Wu, J. (2016) Agglomeration, congestion, and U.S. regional disparaities in employment growth. *Journal of Regional Science, 56*(1), 53–71.

Schröder, M., & Voelzkow, H. (2015). Varieties of regulation: How to combine sectoral, regional and national levels. *Regional Studies, 50*(1), 7–19.

Seto, K., Reenberg, A., Boone, C. G., et al. (2012). Urban land teleconnections and sustainability. *Proceedings of the National Academy of Sciences of the United States of America, 109*(20), 7687–7692.

Sheldon, I. (2006). Trade and environmental policy: A race to the bottom? *Journal of Agricultural Economics, 57*(3), 365–392.

Simmie, J., & Martin, R. (2010). The economic resilience of regions: Towards an evolutionary approach. *Cambridge Journal of Regions, Economy and Society, 3*(1), 27–43.

Sjöberg, E., & Xu, J. (2018). An empirical study of US environmental federalism: RCRA enforcement from 1998–2011. *Ecological Economics, 147,* 253–263.

Song, M., Wang, S., Yu, H., et al. (2011). To reduce energy consumption and to maintain rapid economic growth: Analysis of the condition in China based on expended IPAT model. *Renewable and Sustainable Energy Reviews, 15*(9), 5129–5134.

Sonn, J. W., & Storper, M. (2008). The increasing importance of geographical proximity in knowledge production: An analysis of US patent citations, 1975–1997. *Environment and Planning a, 40*(5), 1020–1039.

Soyez, D., & Schulz, C. (2008). Facets of an emerging Environmental Economic Geography (EEG). *Geoforum, 39,* 17–19.

Stern, D. I. (2004). The rise and fall of the Environmental Kuznets Curve. *World Development, 32*(8), 1419–1439.

Stringer, C. (2006). Forest certification and changing global commodity chains. *Journal of Economic Geography, 6*(5), 701–722.

Sun, P., & Yuan, Y. (2015). Industrial agglomeration and environmental degradation: Empirical evidence in Chinse cities. *Pacific Economic Review, 20*(4), 544–568.

Torre, A. (2008). On the role played by temporary geographical proximity in knowledge transmission. *Regional Studies, 42*(6), 869–889.

Tödtling, F., & Trippl, M. (2018). Regional innovation policies for new path development: Beyond neo-liberal and traditional systemic views. *European Planning Studies, 26*(9), 1779–1795.

Tvedt, H. L. (2019). The formation and structure of cleantech clusters: Insights from San Diego, Dublin, and Graz. *Norsk Geografisk Tidsskrift–Norwegian Journal of Geography, 73*(1), 53–64.

Unteroberdoerster, O. (2001). Trade and transboundary pollution: Spatial separation reconsidered. *Journal of Environmental Economics and Management, 41*(3), 269–285.

Veldkamp, A., & Verburg, P. H. (2004). Modelling land use change and environmental impact. *Journal of Environmental Management, 72*(1–2), 1–3.

Wang, Z., Chen, S. T., Cui, C., et al. (2019). Industry relocation or emission relocation? Visualizing and decomposing the dislocation between China's economy and carbon emissions. *Journal of Cleaner Production, 208,* 1109–1119.

Wang, C. C., & Wu, A. (2016). Geographical FDI knowledge spillover and innovation of indigenous firms in China. *International Business Review, 25*(4), 895–906.

Wang, Y., Yan, W., Ma, D., et al. (2018). Carbon emissions and optimal scale of China's manufacturing agglomeration under heterogeneous environmental regulation. *Journal of Cleaner Production, 176*(1), 140–150.

Wagner, U. J., & Timmins, C. D. (2009). Agglomeration effects in foreign direct investment and the pollution haven hypothesis. *Environmental and Resource Economics, 43*(2), 231–256.

Woods, N. D. (2006). Interstate competition and environmental regulation: A test of the race-to-the-bottom thesis. *Social Science Quarterly, 87*(1), 174–189.

Wu, H., Guo, H., Zhang, B., et al. (2017). Westward movement of new polluting firms in China: Pollution reduction mandates and location choice. *Journal of Comparative Economics, 45,* 119–138.

Wyckoff, A. W., & Roop, J. M. (1994). The embodiment of carbon in imports of manufactured products: Implications for international agreements on greenhouse gas emissions. *Energy Policy, 22,* 187–194.

Xu, Z., & Yeh, A. (2013). Origin effects, spatial dynamics and redistribution of FDI in Guangdong, China. *Tijdschrift Voor Economische En Sociale Geografie, 104*(4), 439–455.

Yeung, H. W. C., & Coe, N. M. (2015). Towards a dynamic theory of global production networks. *Economic Geography, 91*(1), 29–58.

Yue, T., Long, R., Chen, H., et al. (2013). The optimal CO_2 emissions reduction path in Jiangsu province: An expanded IPAT approach. *Applied Energy, 112,* 1510–1517.

Zeng, D., & Zhao, L. (2009). Pollution havens and industrial agglomeration. *Journal of Environmental Economics and Management, 58*(2), 141–153.

Zeng, K., & Eastin, J. (2007). International economic integration and environmental protection: The case of China. *International Studies Quarterly, 51*(4), 972–995.

Zhang, B., Chen, X., & Guo, H. (2018). Does central supervision enhance local environmental enforcement? Quasi-experimental evidence from China. *Journal of Public Economics, 164,* 70–90.

Zhao, H. Y., & Percival, R. (2017). Comparative environmental federalism: Subsidiarity and central regulation in the United States and China. *Transnational Environmental Law, 6*(3), 531–549.

Zhu, J., & Ruth, M. (2015). Relocation or reallocation: Impacts of differentiated energy saving regulation on manufacturing industries in China. *Ecological Economics, 110,* 119–133.

Part I
Internal Development and the Geography of Industrial Environmental Performance in China

Chapter 3
How Is Geography of Industries Related to Industrial Pollution?

China has achieved remarkable industrial growth since the implementation of the reform and opening-up policy, which is prominently reflected in the enormous growth of gross industrial output and the top global rankings of some manufacturing products. In particular, since China became a member of WTO in 2001, the industrialization has significantly speeded up. China has gradually built up its industrial system and has jumped to the world's largest manufacturer. In 2017, China's industrial added value is approximately 28 trillion yuan, which is an increase of 53 times than 1978, with an annual growth rate of 10.8%. However, the prosperity of China's industrial development in the past few decades has happened at the expense of deteriorating environment. Increasing pollution incidents put threats not only on human health but also the economic development to some extent, such as the Songhua River water pollution and Cadmium pollution in Beijiang River, Guangdong province in 2005, Lake Taihu blue algae crisis in 2007, and the emerging haze and acid rain.

Fortunately, China has recognized the importance of environmental protection since the 1980s when conflicts between economic growth and ecological conservation draw more attention. For example, the improvement of the ecological environment is taken as one of the primary developmental goals in the National 13th Five-year Plan during 2016–2020. Besides, the central government has emphasized that China should accelerate structural reform of ecological civilization and build a beautiful country. Therefore, environmental protection, green development, and ecological civilization have become the critical development objectives for China.

Following our understanding of Environmental Economic Geography (EEG) in Chap. 2, a basic perspective that distinguishes EEG from other environment-relevant fields is its highlights in locational conditions. In the case of industrial pollution emissions, why industrial production troubles the environment is a matter of what it is as well as where it locates. The word "where" here includes not only the physical conditions for environmental capacities and vulnerabilities (Tian and Sun 2018), but also the social-economic conditions for polluting intensities and abatement capacities (Pei et al. 2020). Thus, from a perspective of EEG, there must be an inconsistency between the geography of industrial types and the geography of their environmental

© Springer Nature Singapore Pte Ltd. 2020
C. He and X. Mao, *Environmental Economic Geography in China*,
Economic Geography, https://doi.org/10.1007/978-981-15-8991-1_3

performance. Such an inconsistency manifests the nature-economy interactions on the one hand. On the other, it further points to the fact that "green" development of industries is a relative term. Despite the improvement of production processes (e.g. the cleaner production), finding the right place for production also makes a difference. Then the question is what determines a place the right place. Answers to this question require to investigate the determinants of environmental performance spatially.

On this basis, this chapter first investigates the geographies and their dynamics of Chinese industries as well as their pollution emission intensity from 2004 to 2013. Then, we further explore the spatial relationship between the two geographies.

3.1 Industrial Geography in China

Since the reform and opening-up, industrial geography in China has undergone a significant transformation. From the aspect of industrial agglomeration, most Chinese manufacturing industries are highly concentrated in the coastal areas, especially in the Pearl River Delta, the Yangtze River Delta and the Bohai Rim region (He et al. 2010). Among them, the agglomeration effect of export enterprises is much more obvious than that of non-export enterprises (He et al. 2016). Market-oriented and globalized enterprises, and high-technology firms are mainly located on the coast region, while low technology and resource-based enterprises tend to be distributed in inland areas (He and Wang 2012). Recently, Chinese industries have shown a spatial shift from the coastal region to the central, northeast, and western regions (Chen et al. 2015). Hu and Zhang (2015) analyse the spatial evolution pattern of China's industrial development from 1996 to 2012 and conclude that although the gravity centre of industry gradually shifts to the west during this period, there is still a phenomenon of excessive industry concentration in the coastal region. There is also significant spatial auto-correlation in the industrial development of Chinese provinces and regions, with high-high agglomeration and low-low agglomeration occupying a dominant position (Wei and Ye 2013). The existing studies show that a high–high type of agglomeration is mainly distributed in the coastal region, whose trend is further enhanced, while low-low agglomeration usually appears in the central and western regions (Liu and Zeng 2016).

Specifically, we focus on Pollution-Intensive Industries (PII), which refer to industries that directly or indirectly produce a large number of pollutants if they are not treated properly in the production process (Xia 1999). There are three main characteristics for identifying pollution-intensive industries, including the cost of pollution abatement, the intensity of pollution emissions, and the scale of pollution emissions (Li et al. 2016). Since the twenty-first century, the spatial layout of China's pollution-intensive industries has seen significant changes. Overall, the coastal region is still the main agglomeration area of PIIs, although some of them show a trend of shifting to the central and western regions (Liu et al. 2012). Zhang et al. (2015) examine the spatial transfer pattern of China's PIIs from 2005 to 2011, and report that the northeast, southeast, and southern coasts are the areas with the most pollution-intensive industries while the southwest and north coasts are with the most PIIs transfers.

Factors affecting Chinese industrial geography are multifaceted, including the industrial characteristics, regional conditions, as well as the regional revitalization strategy of China over the past few decades (He et al. 2010; Hong et al. 2014). The economic development, financial environment, infrastructure, technological innovation, and market institutions are important factors determining the landscape of industrial enterprises in China (Song and Liu 2013). Shi et al. (2013) argue that industrial enterprises have spread from developed coastal areas to coastal underdeveloped cities and the central and western cities with better market accessibility and stronger industrial supporting capacity. China's national development strategy, policies, and institutions are also important factors shaping the geography of industries (He et al. 2008). In order to promote the development of regional industries, local governments' efforts in infrastructure supply, production subsidies, and minimum wage restrictions have determined the speed at which the differences between the eastern and western regions have narrowed (Lemoine et al. 2015). The transfer and redistribution of such inter-regional industries not only stimulate their industrial upgrading but also trigger regional development, so that China can maintain the international competitiveness of labour-intensive industries (Ito 2014).

To further support the findings above, we conduct the spatial analysis in this section to show the industrial geography in China during 2004 to 2013.

3.1.1 Data and Methods

This chapter is mainly based on the data of two national economic censuses in 2004 and 2013, from which the number of industrial employees is extracted as the indicator of industry size. At the sectoral level, the data cover mining, manufacturing, production, and supply of electric power, gas, and water, and are subdivided into 39 two-digit categories, which are adjusted according to the national industry classification in 2002 (GB/T4754-2002). At the regional scale, this chapter focuses on prefecture-level cities and regions, including 340 prefecture-level cities, municipalities and regions.

We examine the spatial distribution and migration of pollution-intensive industries. The identification of pollution-intensive industries is mainly based on the cost of pollution control, the intensity of pollution emission, and the scale of pollutant emission. The classification of pollution-intensive industries is different, according to different standards. This chapter extracts 10 pollution intensive industries following Zhou et al. (2015), including Papermaking and Paper Products (22), Farm and Sideline Products Processing (13), Chemical Raw Materials and Chemical Products Manufacturing (26), Textile Manufacturing (17), Ferrous Metal Smelting and Rolling Processing (32), Food Manufacturing (14), Leather, Fur, Feather (Down) and Its Products Industry (19), Petroleum Processing, Coking and Nuclear Fuel Processing

(25), Non-metal Mineral Products (31) and non-Ferrous Metal Smelting and Rolling Processing (33). We then examine the geography and its spatial transformation of 10 pollution-intensive industries.

We aggregate the employment distribution of individual industries as well as the whole industry at each prefecture-level region and display the spatial distribution of industries in 2004 and 2013. We further calculate the Gini coefficient and Global Moran's I for different industries, through which we are able to demonstrate their concentration and agglomeration. Meanwhile, in order to estimate the overall industrial agglomeration, Local Indicators of Spatial Association (LISA) of each spatial unit are calculated and LISA distribution maps are drawn with Geoda. The Gini coefficient is computed as follows:

$$Gini = \frac{1}{2N^2\mu} \sum_i \sum_j |x_i - x_j| \tag{3.1}$$

where N represents the number of spatial units, x_i and x_j, respectively, represent the proportion of the employment of city i and city j in the total employment of the country, and μ represents the average proportion of the employment of all spatial units. The value Gini coefficient ranges between 0 and 1. If the value is closer to 1, the distribution of industrial employment would be more concentrated, while if the value is closer to 0, the distribution would be more even.

The Global Moran's I is calculated as follows:

$$I = \frac{n \sum_{i=1}^{n} \sum_{j=1}^{n} w_{ij} (x_i - \bar{x})(x_j - \bar{x})}{\left(\sum_{i=1}^{n} \sum_{j=1}^{n} w_{ij} \right) \sum_{i=1}^{n} (x_i - \bar{x})^2} \tag{3.2}$$

where x_i and x_j denote the employment figures of spatial unit i and j, n refers to the total number of spatial units, $\bar{x}$ represents the average employment of every spatial unit and w_{ij} means the spatial weight matrix. In this chapter, Geoda is employed to calculate the Global Moran's I, and the spatial weight matrix is constructed by the principle of adjacency relationship. If two spatial units are adjacent, the weight would be 1, and if not, the weight would be 0. The value range of Global Moran's I is $-1 \leq I \leq 1$, within which I less than 0 declares a negatively spatial relationship of different units, indicating that units with different values tend to gather in space. On the contrary, if I is greater than 0, it is a positive spatial auto-correlation, that is, units with similar attribute values tend to agglomerate.

Local Moran's I is calculated as follows:

$$I_i = \sum_j w_{ij} z_i z_j \tag{3.3}$$

where I_i denotes Local Moran's I of spatial unit i, n represents the total number of spatial units, w_{ij} refers to the spatial weight matrix, z_i and z_j are attribute values normalized to spatial unit i and j, respectively. By observing the distribution of Local

Moran's *I*, we can find the clusters of high values and low values, and then get an overall view o f industrial agglomeration.

3.1.2 Spatial Distribution of China's Industries

China's industrial distribution decreases as it goes from east to west, presenting a spatial pattern of "ladder-like" (Fig. 3.1). This result is roughly the same as previous studies, in which the most concentrated areas of industries are in the eastern coast, such as the Yangtze River Delta, the Pearl River Delta, the Shandong Peninsula, and the Beijing-Tianjin-Hebei region. In addition, Chongqing also enjoys a high level of industrial agglomeration. Due to the disadvantages of the natural environment, transportation infrastructure, and human capital, the course of industrialization of the northwest and southwest regions is significantly behind that of the central and eastern regions. As the old industrial base in China, the northeast region has no advantages compared with the coastal provinces. Industries in the Northeast are mainly located in some key cities like Harbin, Changchun, Shenyang, and Dalian.

Examining the overall industrial agglomeration and concentration (Table 3.1), we find that the Gini coefficient of China's industries reaches 0.6 in both 2004 and 2013, which indicates that there is a highly concentrated distribution at the prefecture-level region, and industrial development disequilibrium is considerable. In addition, we calculate the Global Moran's I, which is about 0.5, suggesting that the

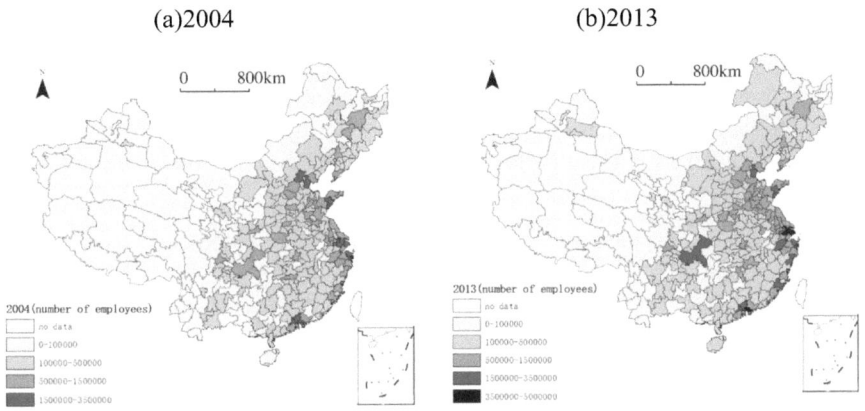

Fig. 3.1 Spatial distribution of Chinese industries in 2004 and 2013

Table 3.1 Indexes of industrial distribution

Year	Gini Coefficient	Moran's I
2004	0.606	0.507
2013	0.607	0.501

industrial distribution has a significant positive spatial auto-correlation. The LISA maps indicate the industrial distribution is dominated by high–high and low–low agglomerations (Fig. 3.2). The high–high concentration areas are mainly located in Jiangsu, Zhejiang, the Pearl River Delta in Guangdong Province and the Shandong Peninsula. While the low–low concentration areas primarily gather in the vast western region, as well as some cities in Hainan, Guangxi, and Hunan Provinces. By 2013, some low–low clusters emerge in the north-eastern regions. High-low agglomeration means that the industrial concentration of the region is significantly higher than the surrounding areas, such as Chongqing in 2004 and Harbin in 2013. These cities enjoy a high level of industrial development, but triggering down effects for neighbouring regions are weak.

Comparing industrial distribution of 2004 and 2013, we find that although in absolute terms, most of the country, especially the western region, have experienced a certain degree of growth, but the geographical profile of "ladder-like" and "high in the eastern and low in the west" has not changed significantly. In the past decade, the overall industrial scale in most cities has increased, and the areas with negative growth are mainly distributed in the Northeast region and Gansu, Hebei, Hunan as well as Beijing (Fig. 3.3). Among them, the declining appearance of manufacturing in Beijing is mainly due to the industrial structure upgrading, with the secondary industry replaced by the tertiary industry. Cities with higher growth rates are located in some relatively backward provinces like Tibet, Xinjiang, Qinghai, Yunnan, and Inner Mongolia, except which they are also concentrated in some central regions such as Jiangxi and Anhui Province. From the perspective of changes in the proportion of industrial-scale (Fig. 3.4), the industry as a whole has further agglomerated in the eastern coast, where Jiangsu, Zhejiang, Fujian, and Guangdong have witnessed a tremendous increase in their proportions. Besides, Chongqing is an important city where industries are more concentrated. Moreover, the proportion of industrial scale in the north-western and southwestern regions has also risen to a certain extent, which reflects the fact that China's strategy of developing the west since the twenty-first century has played a significant role in their industrial advancement. The areas where

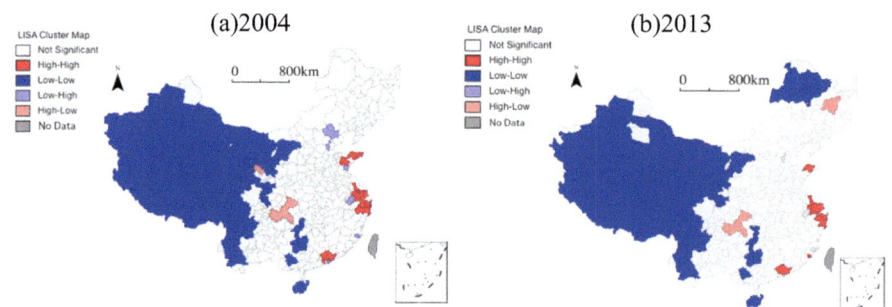

Fig. 3.2 LISA cluster map of industry

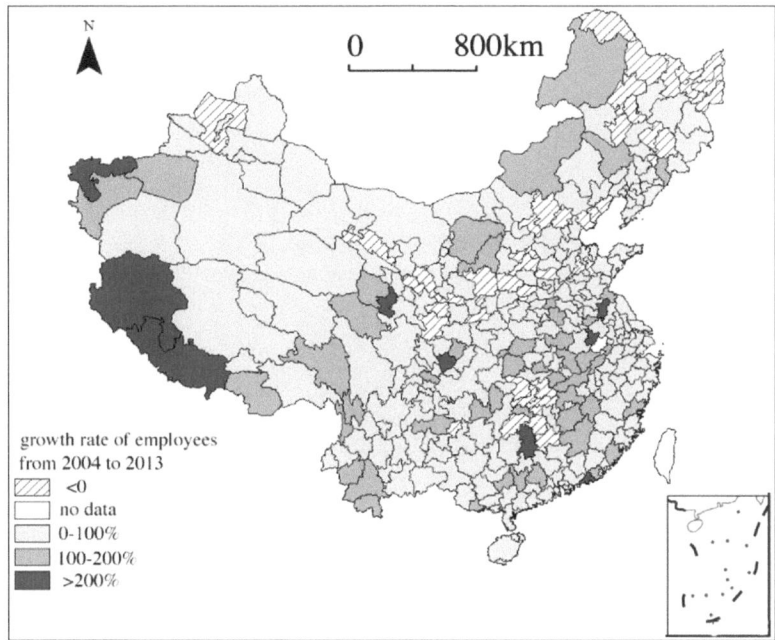

Fig. 3.3 Spatial distribution of industrial growth rate during 2004–2013

the proportion of industry has declined are mainly in provinces like Gansu, Hebei, Shaanxi, Shanxi, and the most part of the Northeast. These areas are gradually falling behind in industrial development throughout the country.

3.1.3 Structural Characteristics of Industrial Distribution

Based on the Gini coefficient of individual industries, we sort industries in 2004 and 2013 by descending order (Table 3.2). We can find that the most and least concentrated industries between 2004 and 2013 are highly similar. The high concentration of these industries stems from the fact that some advantages that those industries require are primarily distributed in a few cities. For example, the petroleum and gas extraction (07) is a resource-intensive industry, mainly distributed in provinces like Xinjiang, Shaanxi, Gansu and the Northeast part, which enjoy outstanding resource advantages. Communication equipment, computers and other electronic equipment manufacturing (40), instrumentation and culture, office machinery manufacturing (41) and chemical fibre manufacturing (28) are technology-intensive industries, thus often concentrate in the eastern coast. The tobacco products industry (16) has the nature of monopoly and enjoys local protection. It is mainly distributed in Yunnan, Fujian, and Shanghai, where there are many local industry oligarchs. In addition,

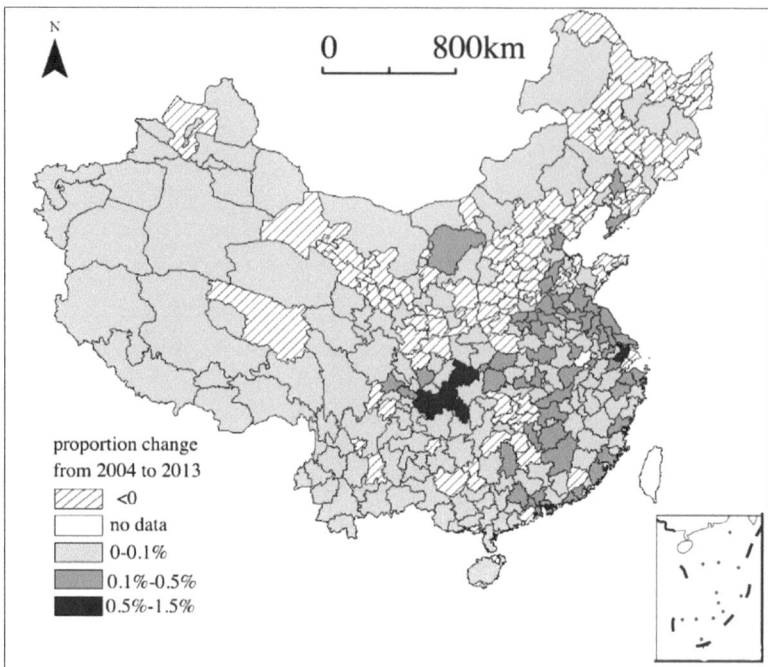

Fig. 3.4 Spatial distribution of industrial proportion change during 2004–2013

comparing the Gini coefficients for the two years, we can see that the overall Gini coefficient in 2013 is lower than that in 2004, which indicates that the distribution of most industries tends to be scattered. This is because each city constantly improves its competitiveness to develop industries. The period from 2004 is considered to be a time of China's industrial migration (Fan and Li 2011), during which the coastal industries gradually spread to the central and western regions. However, the reduction of the Gini coefficient is not that significant, and there is no substantial change in the concentration of some industries.

To further study the spatial pattern of individual industries, we follow He et al. (2007) to link the Gini coefficient with the Global Moran's I, respectively, as the horizontal and vertical coordinates, and take the average value as the intersection to draw the coordinate maps. In the figures, industries can be divided into four categories: concentrated-agglomerated, concentrated-discrete, disperse-agglomerated, disperse-discrete (Figs. 3.5 and 3.6, Table 3.3).

Sectors that belong to concentrated-agglomerated types, such as communication equipment, computer, and other electronic equipment manufacturing (40), electrical machinery and equipment manufacturing (39), textiles and clothing, shoes, cap manufacturing (18), and chemical fibre manufacturing (28) are mostly export-oriented or high-tech industries and they are concentrated in Zhejiang, Jiangsu, Shandong, Guangdong, and some other coastal provinces. These industries take transportation

Table 3.2 Gini coefficients of 39 industrial sectors

2004	Gini Coef.	2013	Gini Coef.
Petroleum and gas extraction	0.964	Petroleum and gas extraction	0.950
Communication equipment, computers and other electronic equipment manufacturing	0.913	Tobacco products industry	0.901
Cultural educational and sports goods	0.905	Communication equipment, computers and other electronic equipment manufacturing	0.900
Mining of other minerals	0.896	Instrumentation and culture, office machinery manufacturing	0.900
Leather, fur, feather, and their products	0.893	Chemical fibre manufacturing	0.888
Instrumentation and culture, office machinery manufacturing	0.865	Leather, fur, feather, and their products	0.864
Chemical fibre manufacturing	0.864	Cultural educational and sports goods	0.863
Textiles and clothing, shoes, cap manufacturing	0.844	Electrical machinery and equipment manufacturing	0.807
Tobacco products industry	0.835	Rubber products industry	0.800
Electrical machinery and equipment manufacturing	0.829	Coal mining and washing industry	0.799
Ferrous metals mining and dressing	0.814	Furniture manufacturing	0.797
Non-ferrous metals mining and dressing	0.809	Transportation equipment manufacturing	0.795
Handicrafts and other manufacturing	0.802	Non-ferrous metals mining and dressing	0.795
Furniture manufacturing	0.802	Ferrous metals mining and dressing	0.794
Rubber products industry	0.802	Textiles and clothing, shoes, cap manufacturing	0.782
Petroleum processing, coking and nuclear fuel processing industries	0.801	Textile manufacturing	0.773
Waste resources and waste recycling	0.790	Petroleum processing, coking and nuclear fuel processing industries	0.771
Metal products	0.790	Ordinary machinery manufacturing	0.765
Coal mining and washing industry	0.787	Metal products	0.763
Plastic products industry	0.787	Plastic products industry	0.762
Transportation equipment manufacturing	0.774	Mining of other mineral	0.761
Production and supply of gas	0.765	Handicrafts and other manufacturing	0.756
Textile manufacturing	0.750	Special equipment manufacturing	0.735
Ordinary machinery manufacturing	0.748	Printing and recording media copying	0.732

(continued)

Table 3.2 (continued)

2004	Gini Coef.	2013	Gini Coef.
Special equipment manufacturing	0.734	Non-ferrous metal smelting and rolling processing industry,	0.708
Ferrous metal smelting and rolling processing industry	0.733	Papermaking and paper products	0.706
Printing and recording media copying	0.731	Ferrous metal smelting and rolling processing industry,	0.690
Non-ferrous metal smelting and rolling processing industry	0.725	Wood processing and wood, bamboo, rattan, palm, grass products	0.685
Wood processing and wood, bamboo, rattan, palm, grass products	0.697	Waste resources and waste materials recycling	0.684
Papermaking and paper products	0.676	Medical and pharmaceutical products	0.678
Medical and pharmaceutical products	0.669	Food manufacturing	0.630
Food manufacturing	0.656	Production and supply of gas	0.609
Chemical raw materials and chemical products manufacturing	0.604	Chemical raw materials and chemical products manufacturing	0.602
Agricultural and sideline food processing	0.592	Beverage manufacturing	0.573
Non-metallic mining and dressing	0.592	Agricultural and sideline food processing	0.567
Beverage manufacturing	0.547	Non-metallic mining and dressing	0.547
Non-metallic mineral products	0.538	Non-metallic mineral products	0.515
Electricity and heat production	0.520	Electricity and heat production	0.476
Production and supply of water	0.458	Production and supply of water	0.459

Fig. 3.5 2004 Spatial distribution characteristics of 39 industrial sectors

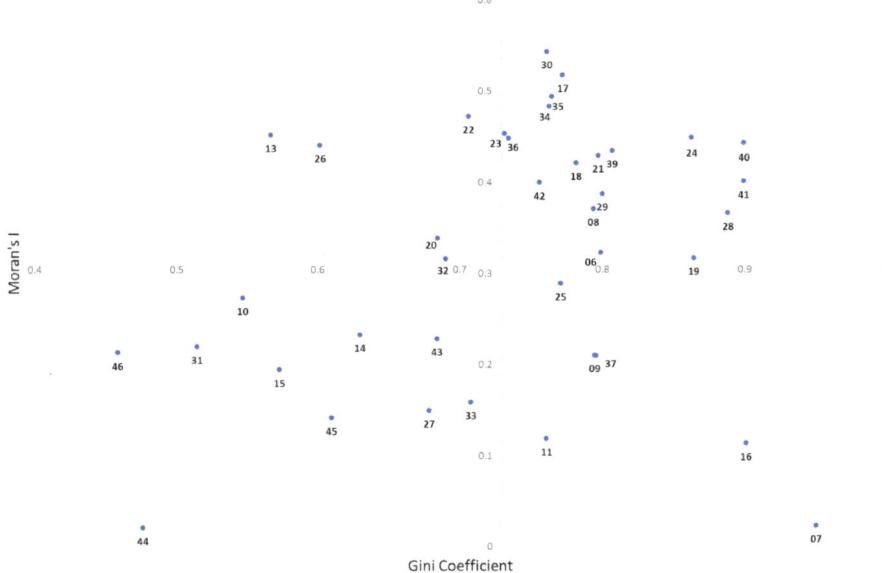

Fig. 3.6 2013 Spatial Distribution characteristics of 39 industrial sectors

infrastructure and substantial intellectual resources as advantages, and thus form highly agglomerated industrial clusters along coastal regions. Concentrated-discrete industries are mainly resource-oriented enterprises, such as petroleum and gas extraction (07), ferrous metal mining and dressing (08), and coal mining and washing industry (06), which mostly concentrate in the cities with rich resources, such as Shanxi, which is rich in coal resources, Xinjiang and Shaanxi, which is abundant in oil and natural gas. For industries of the dispersed-agglomerated type, like agricultural and sideline food processing (13), papermaking and paper products (22), wood processing and wood, bamboo, rattan, palm, grass products (20), their geographical distribution is more related to the access of raw materials and transportation infrastructure. Therefore, most of them are located in Shandong, Jiangsu, Zhejiang and Guangdong Provinces on the eastern coast. Dispersed-discrete industries are most evident in the production and supply of water (46), electricity and heat production and supply (44), which are developed in a variety of regions due to their rigid demand.

3.1.4 Spatial Distribution of Polluting Industries

Pollution-intensive industries are mainly concentrated in North China, Shandong Peninsula, Jiangsu-Zhejiang, Fujian, Pearl River Delta and Chongqing, whose scope has been further expanded in 2013 (Fig. 3.7). Those industries are less distributed in Northeast China, Southwest China, and the vast western region. Most of the polluting

Table 3.3 Code for 39 industrial sectors

Code	Industrial sector
06	Coal mining and washing industry
07	Petroleum and gas extraction
08	Ferrous metals mining and dressing
09	Non-ferrous metals mining and dressing
10	Non-metallic mining and dressing
11	Mining of other mineral
13	Agricultural and sideline food processing
14	Food manufacturing
15	Beverage manufacturing
16	Tobacco products industry
17	Textile manufacturing
18	Textiles and clothing, shoes, cap manufacturing
19	Leather, fur, feather, and their products
20	Wood processing and wood, bamboo, rattan, palm, grass products
21	Furniture manufacturing
22	Papermaking and paper products
23	Printing and recording media copying
24	Cultural educational and sports goods
25	Petroleum processing, coking and nuclear fuel processing industries
26	Chemical raw materials and chemical products manufacturing
27	Medical and pharmaceutical products
28	Chemical fibre manufacturing
29	Rubber products industry
30	Plastic products industry
31	Non-metallic mineral products industry
32	Ferrous metal smelting and rolling processing industry,
33	Non-ferrous metal smelting and rolling processing industry,
34	Metal products
35	Ordinary machinery manufacturing
36	Special equipment manufacturing
37	Transportation equipment manufacturing
39	Electrical machinery and equipment manufacturing
40	Communication equipment, computers and other electronic equipment manufacturing
41	Instrumentation and culture, office machinery manufacturing
42	Handicrafts and other manufacturing
43	Waste resources and waste materials recycling

(continued)

Table 3.3 (continued)

Code	Industrial sector
44	Electricity and heat production and supply
45	Production and supply of gas
46	Production and supply of water

(a)2004   (b)2013

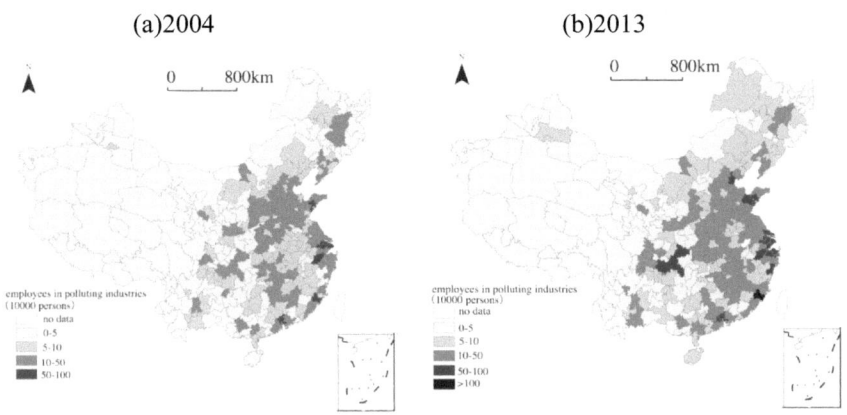

Fig. 3.7 Spatial of distribution of pollution-intensive industries in 2004 and 2013

industries in Northeast China only concentrate in several cities including Harbin, Jilin, Changchun, Dalian, and Shenyang. Besides, due to the locational disadvantage in the western region, the overall industrial development is relatively low. For this reason, it cannot attract large-scale pollution-intensive industries although its environmental regulation is lower than that of the more developed provinces in the east.

On the other hand, from the view of the proportion of pollution-intensive industry in the overall industry, the areas with relatively high pollution-intensive industries are more likely to be distributed in the western region and a part of central regions, while the eastern region has relatively lower proportion (Fig. 3.8). The reason for developing this pattern is that the eastern region has more diversified industrial structure and clean industries account for a larger share. On the contrary, in the western region, the industrial structure is relatively simple, where pollution-intensive industries can often dominate their economy. For example, the petroleum industry in Xinjiang accounts for a large proportion. Moreover, to attract industrial enterprises, the western region would appropriately lax environmental regulations, reduce the pressure for pollutant emission, and offer preferential treatment for enterprises. During 2004 and 2013, it is found that the proportion of polluting industries in the western region has generally increased. The stricter environmental regulation in the eastern region and the increasing land and labour costs have encouraged some pollution-intensive

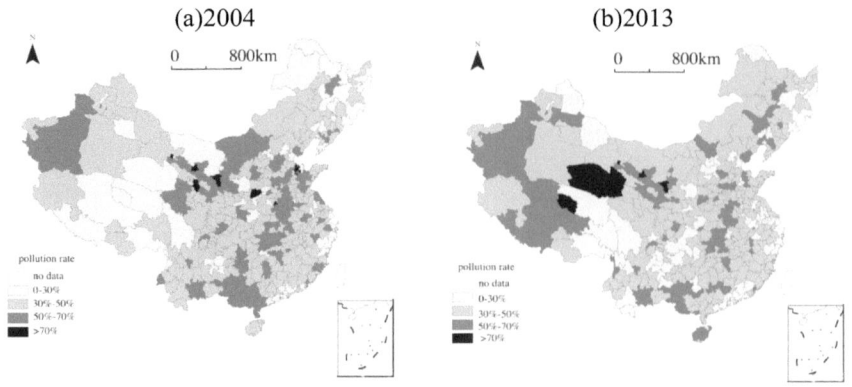

Fig. 3.8 Spatial distribution of proportion of pollution-intensive industries in 2004 and 2013

industries to move to the west, forming a "Pollution Heaven" phenomenon to some extent.

3.1.5 Changes of Spatial Distribution of Polluting Industries

Between 2004 and 2013, China's polluting industries have experienced a certain degree of growth in most areas, while only witnessed declines in a small number of regions (Fig. 3.9). Among them, the eastern region and Chongqing have the highest increase in polluting industries. The overall industrial scale in these areas is relatively large and the development advantages are more significant, consequently leading to faster growth. The Gini coefficient of the polluting industries is 0.57 and 0.55 in 2004 and 2013, respectively, and the concentration is lower than that of the overall industry (Table 3.4). The Global Moran's I of polluting industries is 0.49 and 0.47 in 2004 and 2013, respectively, indicating that they are significantly spatially autocorrelated.

Based on the change in the proportion of polluting industries in total industrial scale, it is found that the proportion in most central and eastern regions is declining. Areas with a large increase in the proportion are mainly concentrated in the northwest, north-eastern region and a few cities in the central region (Fig. 3.10), while places with a rather significant decrease in the proportion are more fragmented. The western region is in the early stage of industrial development, where it is often at the expense of the environment to attract some polluting enterprises from the eastern and central regions in order to promote economic development. In the past decade, the transfer of pollution-intensive industries to the western region largely results from the rise of environmental and production costs of the more developed regions. On the other hand, the Chinese central government has promoted regional revitalization strategies, improving development conditions for the backward areas, enhancing the attractiveness of the western region.

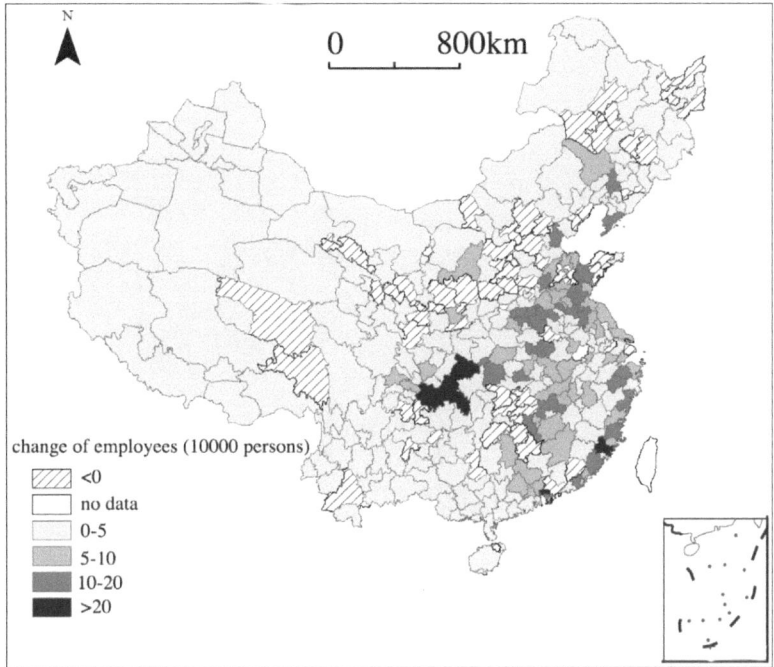

Fig. 3.9 Change in pollution-intensive industries during 2004–2013

Table 3.4 Indexes of pollution-intensive industries	Year	Gini Coefficient	Moran's I
	2004	0.572	0.498
	2013	0.552	0.477

3.2 Geography of Industrial Pollution Emission in China

China's industrial development has produced considerable industrial pollution. From the perspective of total pollution emissions, which largely complies with industries, the pollutants concentrate in the eastern region, especially in Shandong, Jiangsu and Hebei Provinces (Liu et al. 2010). It is also reported that industrial pollution is more agglomerated in the core cities, as well as provincial capitals in the provinces (He et al. 2014). Chen and Li (2018) explore the pollution situation of the Yangtze River Economic Belt during 2000 and 2015 and find that the reduction of industrial pollution is not satisfactory, and industrial pollution in eastern coastal cities is the most serious. Some studies claim that the distribution of industrial pollution has significant spatial agglomeration and spatial spill-over effects among regions (Hu et al. 2016; Zhou 2016). Overall, the spatial distribution of total industrial pollution

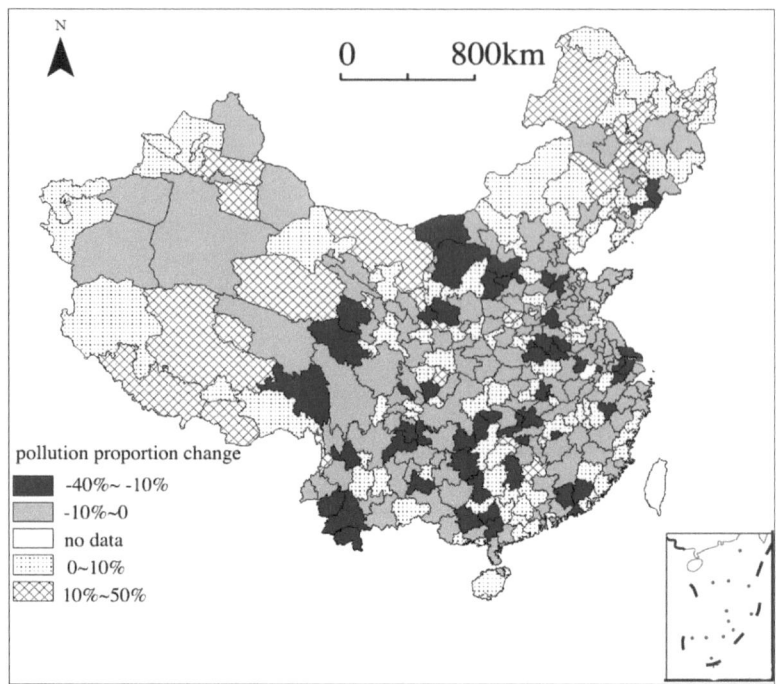

Fig. 3.10   Change in proportion of pollution-intensive industries during 2004 and 2013

emissions is consistent with the layout of industrial enterprises, especially pollution-intensive industries (Ren and Ma 2018; Li 2016). Zhu et al. (2018) find that air pollution emission has shifted to the east coast along with polluting industries during 2003 and 2013. However, the increase in industrial scale and productivity has reduced pollution emission intensity. Hu et al. (2018) report that industrial agglomeration is negatively correlated with industrial pollution intensity. Previous empirical studies have also shown that in the eastern region with high industrial density, the emission intensity of industrial wastewater and sulphur dioxide is relatively low.

In this section, we collect the data from China City Statistical Yearbooks 2004 and 2013 to describe the situation of industrial pollution emissions, extracting the discharges of industrial wastewater, industrial sulphur dioxide, and corresponding industrial output value. The data include pollutant discharge for more than 280 cities in 30 provincial-level administrative regions of China. Since the total industrial output value in 2004 and 2013 is calculated at their current price, in order to unify the price index, we adjust the total industrial output value in 2013 to the 2004 level according to the historical consumer price index.

3.2.1 Total Pollution Emission

We compare the spatial distribution of the two selected industrial pollutants (Fig. 3.11). The geography of different pollutant discharges shows significant differences. In general, the discharge of industrial pollutants in the northern region is higher than that in the southern region. Comparing the year of 2004 and 2013, we can find that the spatial structure of pollutant emissions has been adjusted to some extent. The pattern of high-pollution and low-pollution discharge areas has not changed greatly during 2004–2013. The distribution of industrial wastewater discharge is relatively scattered. However, we can still find that the industrial water discharge is high in certain areas like Jiangsu and Zhejiang Provinces, and conditions are also serious in Chongqing, Shandong, Guangxi, and Hebei Provinces. In North China (except Hebei) and the north-western region, due to the relatively dry climate, there are

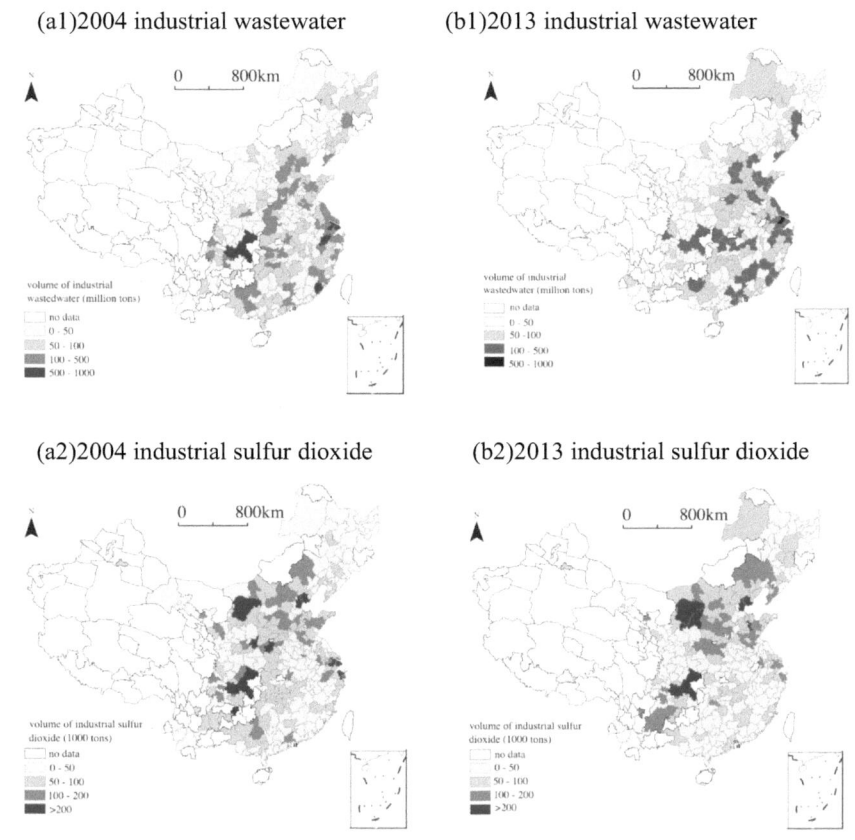

Fig. 3.11 Spatial patterns of industrial pollution in China between 2004 and 2013

fewer high-water-demanding industries. Thus, the discharge of industrial wastewater is also lower. The high emission area of industrial sulphur dioxide concentrates in the north region of the Qinling Mountains-Huaihe River line. As we can see, the industrial sulphur dioxide emissions in the south are significantly lower than those in the north, but its emissions in the southern region have further declined in 2013. Specifically, Hebei, Shanxi, Inner Mongolia and other provinces in North China are the main regions of high industrial sulphur dioxide emission. In addition, the more developed industrial areas such as Shandong Peninsula, Yangtze River Delta and Chongqing are also areas with high levels of industrial sulphur dioxide emissions.

Comparing the total emissions of the two industrial pollutants in 2004 and 2013, we can see that the discharge of industrial wastewater and industrial sulphur dioxide has only slightly decreased during this decade (Fig. 3.12). More specifically, industrial wastewater discharge has decreased from 21.2 billion tons to 20.5 billion tons and industrial sulphur dioxide emissions are reduced from 17.42 million tons to 16.53 million tons. In the past ten years, China's industry has developed at a relatively high growth rate, during which the industrial added value has grown from 6577.68 billion yuan in 2004 to 22237.6 billion yuan in 2013, with an increase of about 238%. In contrast, the discharge of industrial wastewater and industrial sulphur dioxide does not increase, indicating a great improvement of the overall industry pollution emission efficiency, and various pollution control methods are proved effective. Of course, from the total discharge of pollutants, industrial pollution has not been significantly mitigated during the decade. China still needs to upgrade its environmental regulation, promote technological progress and build an environment-friendly industrial structure to gradually break away from the developing path of high pollution.

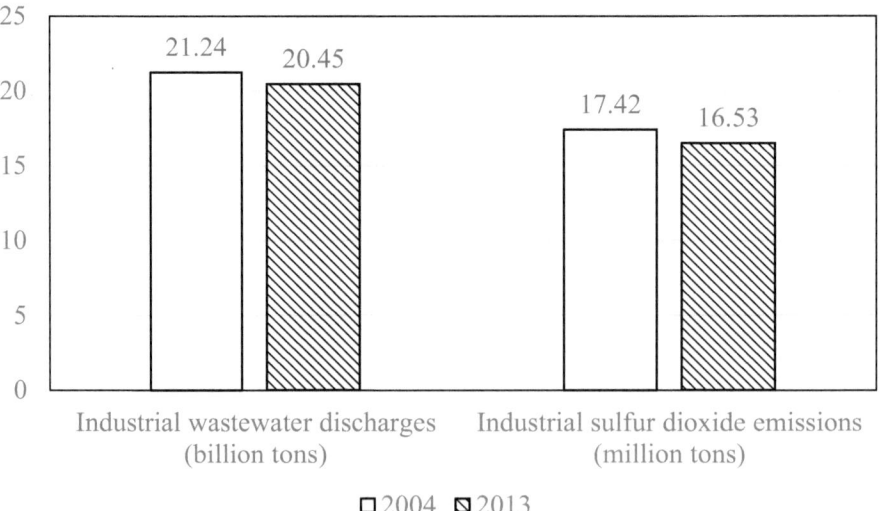

Fig. 3.12 Industrial pollution in China between 2004 and 2013

3.2.2 *Pollution Emission Intensity*

The intensity of pollution discharge reflects the environmental efficiency of industrial production in different regions to a certain degree. The higher the intensity is, the more pollution emissions per unit of output in the region and the lower is the environmental efficiency. It can be seen that the spatial pattern of pollutant emission intensity is rather different from that of pollutant discharge (Fig. 3.13). Industrially developed coastal areas enjoy higher pollution emission efficiency while lower environmental efficiency often occurs in the central and north-eastern regions. Cities with high industrial wastewater discharge intensity are mainly located in north-eastern and southwestern regions like Heilongjiang, Yunnan, Guangxi, and Sichuan. The regions with a high emission intensity of industrial sulphur dioxide concentrate in the north and southwestern provinces, including Inner Mongolia, Shaanxi, Shanxi, Guangxi, Sichuan and Chongqing, and the north-western region such as provinces Gansu and Ningxia has lower efficiency as well.

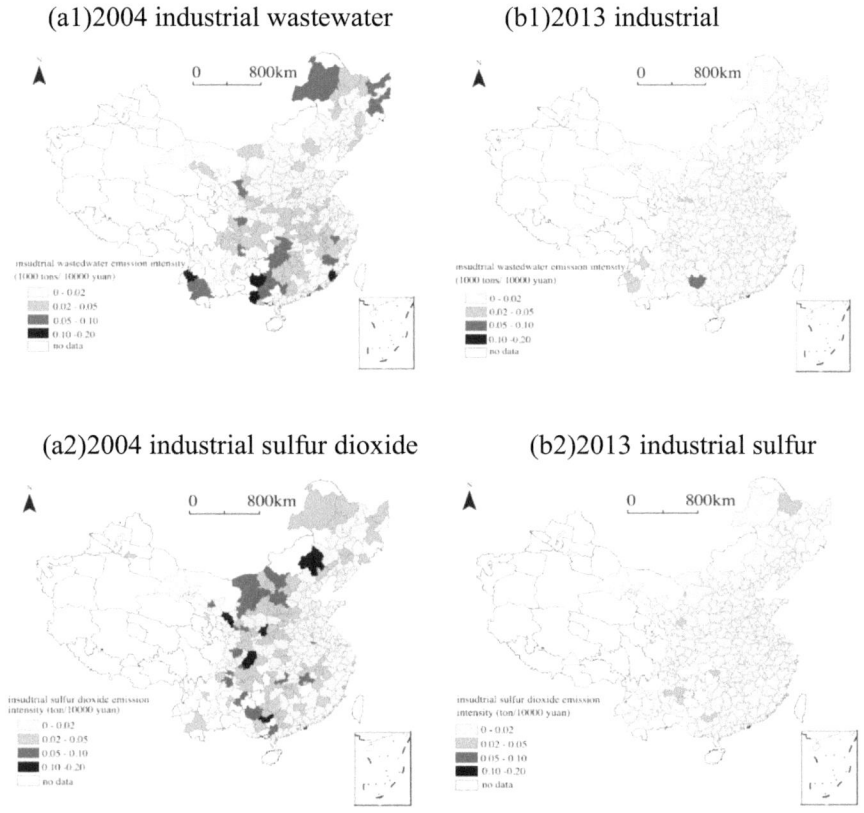

Fig. 3.13 Spatial patterns of industrial pollution intensity in China between 2004 and 2013

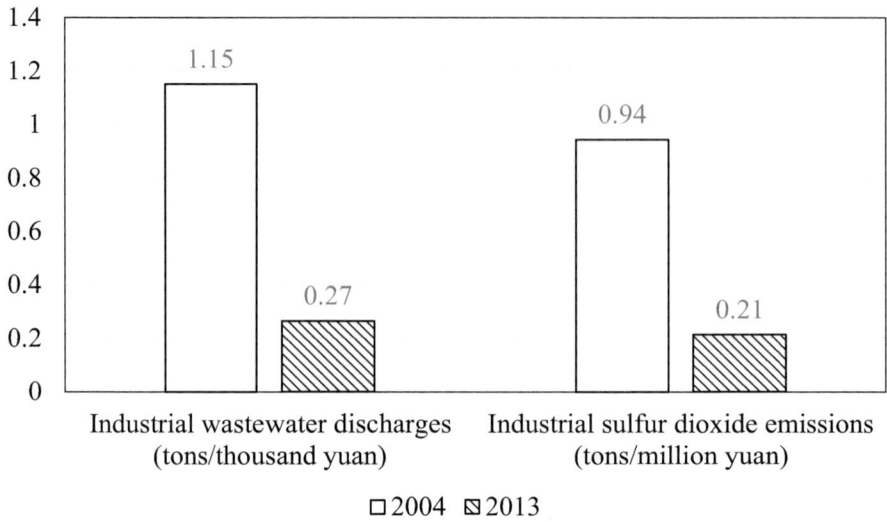

Fig. 3.14 Industrial pollution intensities in China between 2004 and 2013

Judging from the large decrease of pollution discharge intensity (Fig. 3.14), the overall pollution discharge efficiency has been greatly improved during 2004 and 2013. In 2004, the emission intensity of industrial wastewater and industrial sulphur dioxide is 115,100 tons per 100 million yuan and 94.34 tons per billion yuan, respectively. By 2013, the emission intensity of two pollutants decreases to 26,500 tons per 100 million and 21.38 tons per 100 million yuan, respectively, 23% and 22% of that in 2004. In addition to the technologies and local environmental regulations, the intensity of pollutant emissions is also closely related to industrial structure. For example, textile industry and paper industry have a greater impact on industrial wastewater discharge, in which wastewater discharge is rather intensive in Jiangsu and Zhejiang where the industries agglomerate. Industrial sulphur dioxide emissions are mainly from the petrochemical industry, the plastic products industry, and the non-metallic mineral products industry. Therefore, these highly polluting industries will largely affect the overall pollution emissions and their intensity. In the process of industrial upgrading, it is necessary to phase out pollution-intensive, high-energy-consuming and low-efficiency enterprises to effectively improve the environmental quality.

3.3 Spatial Relationship Between Industrial Geography and Industrial Pollution

3.3.1 Methods

We first apply the simple linear regression model, taking industrial employment, industrial pollutant emissions and emission intensity logarithmically, to draw scatter plots of the relevant variables, and then obtain linear regression results. In order to further explore the spatial relationship of variables, we use the bivariate spatial correlation method (Hu et al. 2018) to demonstrate the spatial correlation between the spatial distribution of industry and pollution. Meanwhile, the global bivariate spatial correlation and local bivariate spatial correlation method are used to analyse the relationship between the two variables in different regions. The analysis is realized using Geoda. The calculation formula for the bivariate spatial correlation is as follows:

$$I_{ab} = \frac{n \sum_{i=1}^{n} \sum_{j \neq i}^{n} W_{ij} Z_i^a Z_j^b}{(n-1) \sum_{i=1}^{n} \sum_{j \neq i}^{n} W_{ij}}, \ I_i^{ab} = Z_i^a \sum W_{ij} Z_j^b \qquad (3.4)$$

where I_{ab} represents the binary global Moran's I index of variables a and b, I_i^{ab} is the binary local Moran's I index of variables a and b, Z_i^a and Z_j^b represents the attribute value of variable a in region i and variable b in region j, and W_{ij} is the spatial weight matrix.

3.3.2 Linear Regression Analysis

Taking the pollutants of industrial wastewater as a case, we make scatter plots between industrial scale, polluted industrial scale, and industrial wastewater discharge, and displacement intensity in 2004 and 2013 using the logarithm of them (Fig. 3.15). As shown by the figures, there is a positive correlation between industrial scale and industrial wastewater discharge, and a negative correlation between industrial scale and industrial wastewater discharge intensity. This shows that although the total amount of industrial wastewater discharge increases with the expansion of industrial-scale, its emission efficiency is also improved due to the scale economies. In addition, the scale of the polluting industries is more explanatory than the overall industrial scale, and the interpretation of its influence on industrial wastewater discharge intensity is weaker. This result indicates that the expansion of the polluting industries affects industrial wastewater discharge more directly, and the environmental efficiency improvement generated by polluting industrial agglomeration is lower than the overall industrial level. From another perspective, the increase in industrial wastewater discharge efficiency caused by the overall industrial agglomeration is partly due

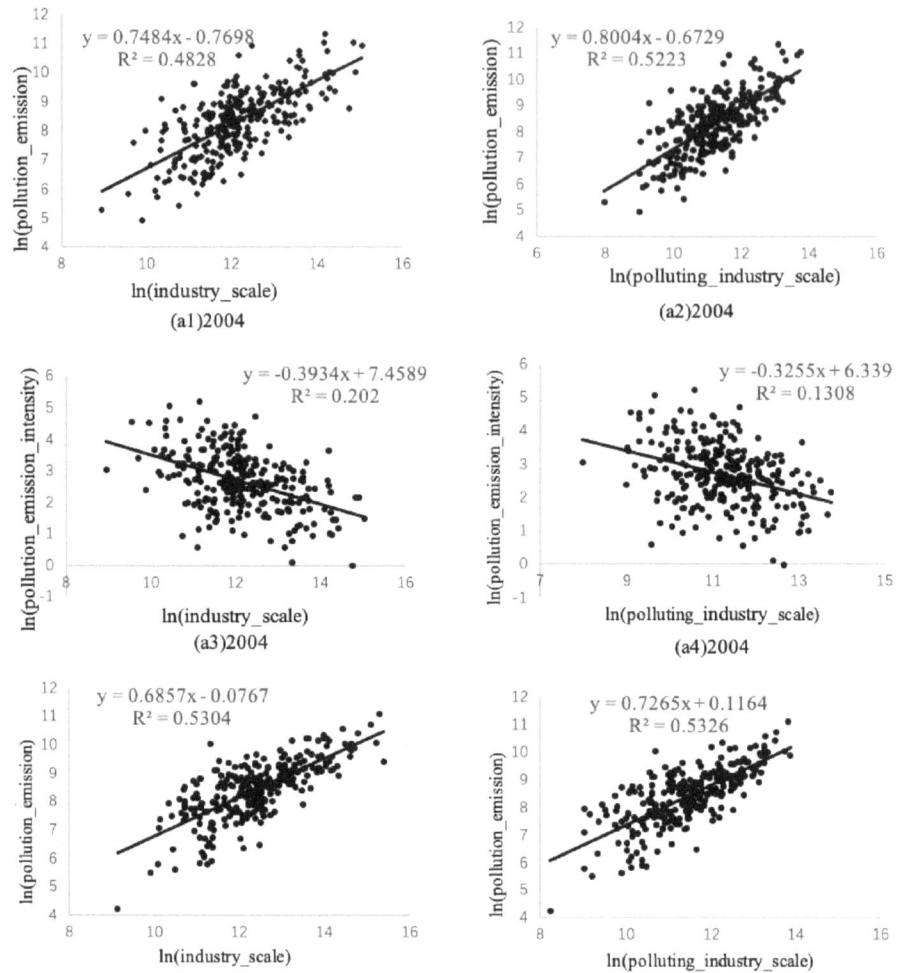

Fig. 3.15 Scatter plots of the relationships between industrial wastewater and industry

to the structural effect. Thus, the decline in the proportion of polluting industries is one of the important factors for the improvement of environmental quality.

Based on the regression results, comparing the two pollutant emissions, the industrial scale, and the scale of the polluting industries (Table 3.5 and Table 3.6), we can see that the overall pattern of industrial sulphur dioxide is similar to that of industrial wastewater. The discharge of pollutants is positively correlated with the industrial scale, especially the scale of polluting industries. The intensity of pollutant emissions is negatively correlated with both. Compared with 2004, the positive correlation between the emissions of the two pollutants and the industrial scale decreases significantly in 2013. Besides, the negative correlation between the discharge intensity of industrial wastewater and the industrial-scale is weakened, different from that

Table 3.5 Linear regression results for industry scale and pollution emission

x	y	2004		2013	
		a	R^2	a	R^2
ln(industry_scale)	ln(industrial_wastewater_emission)	0.748	0.483	0.686	0.530
ln(polluting_industry_scale)	ln(industrial_wastewater_emission)	0.800	0.522	0.727	0.533
ln(industry_scale)	ln(industrial_sulfur_dioxide_emission)	0.763	0.368	0.464	0.212
ln(polluting_industry_scale)	ln(industrial_sulfur_dioxide_emission)	0.796	0.379	0.506	0.226

of the other pollutants. Consequently, the statistical results suggest that for industrial wastewater, the environmental benefits of scale effects are weakened, while for industrial sulphur dioxide, the positive influence is enhanced.

3.3.3 Spatial Correlation Analysis

3.3.3.1 Global Bivariate Spatial Correlation Analysis

In order to further explore the spatial relationship of industrial scale and industrial pollution emissions, we calculate the global bivariate spatial correlation coefficient between two pollutant emissions and industrial scale (Table 3.7). The results show that there is a significant spatial positive correlation between the emissions of industrial wastewater and the industrial scale, same as industrial sulphur dioxide. As for the results of pollutants emission intensity, we can see a significant negative correlation between the emission intensity of these two pollutions and the industrial scale. For industrial wastewater, this spatial negative correlation decreases significantly in 2013, while industrial sulphur dioxide is even more negatively correlated with the industrial scale. Therefore, it confirms that from a spatial perspective, scale economies and industrial agglomeration can increase environmental efficiency and exert positive environmental externalities. In areas with higher levels of agglomeration, the efficiency of industrial pollution emissions is often better than that of backward areas, which should be attributed to the scale effect, technical effect, and structural effect in the process of industrial development.

3.3.3.2 Local Bivariate Spatial Correlation Analysis

In order to further explore the spatial relationship between pollutant emission efficiency and industrial distribution, we separately calculate the local bivariate spatial correlation coefficients of two pollutant emission intensity and industrial scale. Based on the local bivariate spatial correlation results, we plot four different correlation types on the maps, including high industry-high pollution, low industry-low pollution, low industry-high pollution, and high industry-low pollution.

Table 3.6 Linear regression results for industry scale and pollution emission intensity

x	*y*	2004		2013	
		a	R^2	a	R^2
ln(industry_scale)	ln(industrial_wastewater_emission_intensity)	−0.393	0.202	−0.264	0.131
ln(polluting_industry_scale)	ln(industrial_wastewater_emission_intensity)	−0.326	0.131	−0.266	0.120
ln(industry_scale)	ln(industrial_sulfur_dioxide_emission_intensity)	−0.379	0.124	−0.483	0.227
ln(polluting_industry_scale)	ln(industrial_sulfur_dioxide_emission_intensity)	−0.330	0.089	−0.484	0.204

Table 3.7 Bivariate spatial auto-correlation index for industry scale and pollution emission

a	b	2004	p	2013	p
Industry	Industrial wastewater emission	0.225	0.001	0.399	0.001
Polluting industry	Industrial wastewater emission	0.238	0.001	0.400	0.001
Industry	Industrial sulphur dioxide emission	0.159	0.001	0.034	0.001
Polluting industry	Industrial sulphur dioxide emission	0.177	0.001	0.038	0.001
Industry	Industrial wastewater emission intensity	−0.187	0.001	−0.102	0.001
Polluting industry	Industrial wastewater emission intensity	−0.210	0.001	−0.117	0.001
Industry	Industrial sulphur dioxide emission intensity	−0.194	0.001	−0.216	0.001
Polluting industry	Industrial sulphur dioxide emission intensity	−0.212	0.001	−0.251	0.001

First of all, we discuss the relationship between industrial wastewater discharge intensity and industrial-scale distribution (Fig. 3.16). In 2004, the high–high-type agglomeration occurred only in Xiamen, which did not happen in 2013. This shows that most cities with a high industrial agglomeration enjoy the high efficiency of industrial wastewater discharge. The situation of low–low type is concentrated in Shanxi and Shaanxi in 2004, and also distributed in some cities in Shandong, Jilin, Fujian and Guangdong. By 2013, the low–low type mainly occurs in northeast China. The low–high type areas concentrate in Guangxi and Hunan in 2004, and also partly spread in several cities in Yunnan and Heilongjiang provinces, which is further

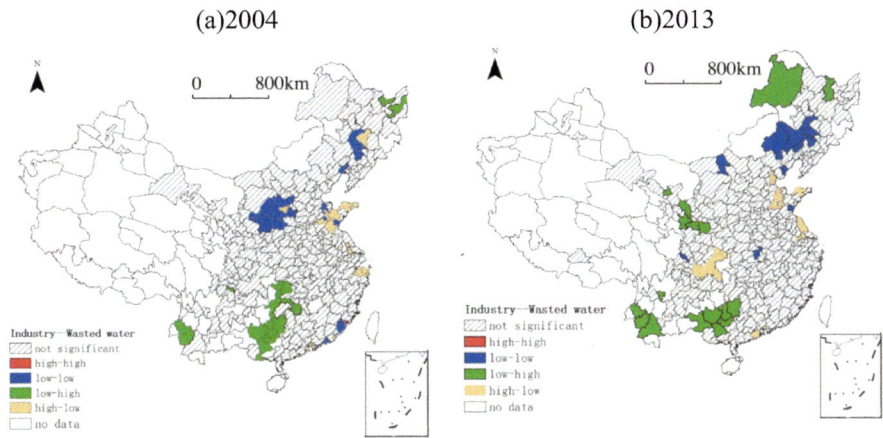

Fig. 3.16 Local bivariate spatial auto-correlation results for industry and industrial wastewater emission intensity

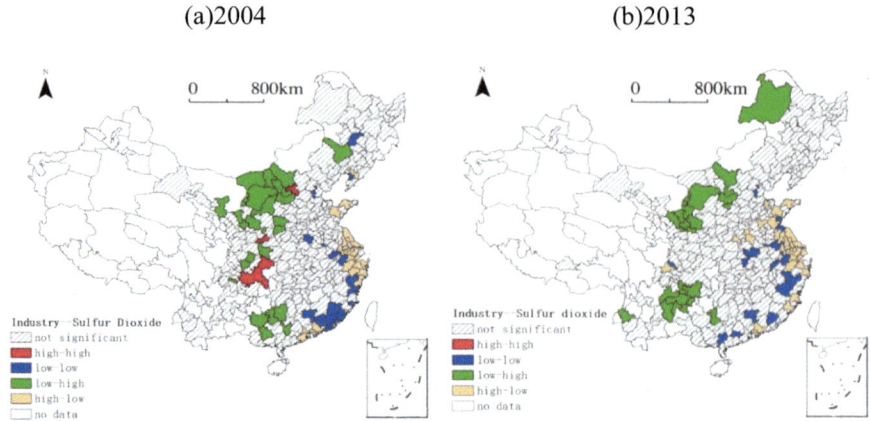

Fig. 3.17 Local bivariate spatial auto-correlation results for industry scale and industrial sulphur dioxide emission intensity

strengthened in 2013. These cities have a low degree of industrial agglomeration and high industrial wastewater discharge intensity, indicating more polluting industries in their industrial structures or lower levels of environmental regulation for industrial pollutant emissions. I n 2004, high–low agglomeration areas are concentrated in Shandong, Zhejiang, and Guangdong. By 2013, Chongqing, Tianjin, and Jiangsu Province also belong to the type of high industry-low pollution. Compared with other parts of the country, these cities not only occupy a prominent position in industrial development but also ensure the low intensity of industrial wastewater discharge.

Next, we focus on the spatial correlation between industrial sulphur dioxide emission intensity and industrial scale. From Fig. 3.17, it is obvious that the overall pattern is significantly different from that of industrial wastewater. Similarly, the type of high–high is relatively rare, only distributed in Chongqing, Xi'an, and Datong in 2004, and none in 2013. The low–low areas concentrate in Guangdong, Fujian, and Anhui in 2004, and more in Guangxi, Zhejiang, Henan, and Sichuan in 2013. Low–high areas are distributed more in provinces in the northwest and central regions. For example, in 2004, low–high type mainly occurred in Gansu, Ningxia, Inner Mongolia, Shaanxi, Sichuan, and Guangxi Provinces. By 2013, the situation has shifted to Guizhou and Yunnan, while the distribution in the Northwest has narrowed. The high–low areas are highly concentrated in the eastern coastal areas, mainly including the Shandong Peninsula, Jiangsu and Zhejiang, and the Pearl River Delta. Besides, some cities in Henan and Fujian also reach the level of high–low agglomeration in 2013.

3.4 Discussion

According to the existing empirical studies, the intensity of industrial pollution emission mainly depends on the industry scale, industrial structure, sewage technology, and environmental policies, which combine to play different roles depending on specific contexts. The industrial-scale effect has a certain threshold effect on the intensity of pollution discharge. In other words, the nascent industry agglomeration tends to intensify pollution, but its positive externalities will reduce the emission intensity when it comes to a certain stage (Yang 2015). The industrial scale in the eastern region is much developed than that in the central and western regions. For more advanced regions, industrial growth gradually exceeds that of pollution discharge, while the central and western regions are still at a relatively low level of industrialization, which then confront pollution aggravation.

The impact of industrial structure on pollution emissions is also critical, as it largely determines the type and intensity of pollutants. When the economy of the eastern region develops to a relatively advanced level, industrial upgrading becomes an unstoppable trend and energy-intensive and pollution-intensive industries are gradually fading out. To speed up their economic development, the inland regions take some of the polluting industries from the eastern region, resulting in a higher proportion of the polluting industries in their industrial structure. Moreover, the inland regions enjoy the resource advantages to develop pollution-intensive industries. In this way, some under-developed regions in China gradually become "pollution havens" in inter-regional industrial transfers (Li 2016).

With industrialization, technology advancement becomes increasingly crucial in pollution reduction. Knowledge spill-over effects, resource sharing and remarkable talent advantages in the eastern region are conducive to improving the technical level of enterprises, and thus improving the efficiency of environment protection. Enterprises in the eastern region have to improve the level of sewage technology and reduce the rate of pollution discharge in response to threats of withdrawal, because they are facing more stringent pollution emission requirements.

The importance of environmental regulation cannot be ignored since it has largely affected the location choice and technological innovation of industrial enterprises in China. Strict environmental regulations can also stimulate polluting enterprises to conduct innovations while forcing the exit and migration of some polluting enterprises, so as to achieve emission reduction. Of course, the performance of the environmental policy is dependent on the implementation and regional conditions. Higher quality requirements and stricter environmental regulations in the eastern region have triggered the migration of pollution-intensive industries to the central and western regions. Meanwhile, local governments of some backward regions tend to lower the threshold for polluting enterprises to attract more investments. Therefore, various types of environmental regulations in different parts are also significant forces shaping the geography of industries and pollution emission in China.

3.5 Conclusion

Based on national economic census data in 2004 and 2013, we analyse the geographical pattern of industries and industrial pollution emissions at the prefecture-level and region level, and their spatial relationship. China's industry has a ladder-shaped structure of "high east and low west". A large number of industries highly agglomerate in the coastal provinces. After ten years' development, although some industries have shown a trend of spreading to the central and western regions, the dominant position of the coastal areas has not changed significantly. Areas with the highest pollution-intensive industries are also in the eastern coast. However, the proportion of pollution industries in the central and western regions is much higher than that of the eastern region. During the rapid industrial development, the overall environmental efficiency has improved greatly although the total amount of industrial pollutants has not significantly decreased. Overall, the pollution discharge problems in North China, Southwest China and some cities in Northeast China are more serious, and the efficiency of sewage discharge in coastal industrialized areas is better.

Compared with the central and western regions, the coastal region has higher levels of industrial agglomeration and lower environmental pollution intensity. Some studies argue that the coastal region is already in the decline phase of the EKC and the central and western regions are in a relatively backward stage of industrial development. With industrial upgrading in the coastal region, the central and western regions have taken some polluting enterprises from more developed provinces, and to some extent acted as so-called "pollution havens". In the process of industrialization in the coastal region, factors such as scale effect, structural effect, technology effect, and strict environmental systems have declined the intensity of industrial pollution emissions.

In the past few decades, China's industrial development has been at the expense of environmental deterioration, and the contradiction between ecological improvement and economic development has become increasingly prominent. In the call for ecological civilization and green industrialization, it is necessary to solve problems from the source of industrial development. On the one hand, China needs to guide the rational layout of industrial enterprises in space, forming a more efficient geography pattern and controlling the "pollution haven" effect of backward areas. On the other hand, the government should improve the framework of environmental regulation and ensure its enforcement, employing corporate pollution emission charges, technical subsidies, and other means to stimulate enterprises to carry out technological innovation or adopt new technologies to reduce pollution discharges. It is not a one-off event to eventually solve the problem of environmental pollution in China. The coordinated development of ecological environmental and economic growth is a time-honoured mission of China.

References

Chen, L., & Li, Q. (2018). Study on timing change and spatial differentiation of industrial pollution in the Yangtze River Economic Belt. *Journal of Shanxi Normal University, 32*(3), 81–86. (In Chinese).

Chen, X., Xi, Q., & Li, G. (2015). Urbanization level and spatial distribution of manufacturing industry: An empirical research based on provincial panel data. *Scientia Geographica Sinica, 35*(3), 259–267. (In Chinese).

Fan, J., & Li, F. (2011). Effect of spatial concentration of manufacturing in China: A review. *South China Journal of Economics, 6,* 53–66.

He, C., Guo, Q., & Ye, X. (2016). Geographical agglomeration and co-agglomeration of exporters and non-exporters in China. *GeoJournal, 81*(6), 947–964.

He, C., Huang, Z., & Ye, X. (2014). Spatial heterogeneity of economic development and industrial pollution in urban China. *Stochastic Environmental Research and Risk Assessment, 28*(4), 767–781.

He, C., Pan, F., & Sun, L. (2007). Geography concentration of manufacturing industries in China. *Acta Geographica Sinica, 62*(12), 1253–1264. (In Chinese).

He, C., & Wang, J. (2012). Regional and sectoral differences in the spatial restructuring of Chinese manufacturing industries during the post-WTO period. *GeoJournal, 77*(3), 361–381.

He, C., Xie, X., & Pan, F. (2008). Locational studies of Chinese manufacturing industries. *Geographical Research, 27*(3), 623–634. (In Chinese).

He, C., Zhu, Y., & Zhu, S. (2010). Industrial attributes, provincial characteristics and industrial agglomeration in China. *Acta Geographica Sinica, 65*(10), 1218–1228. (In Chinese).

Hong, J., Liu, Z., & Huang, W. (2014). Regional revitailization strategies and industrial spatial strcuture changes in China-evidence based on China Industrial Survey Database. *Economic Research Journal, 8,* 28–40. (In Chinese).

Hu, W., & Zhang, Y. (2015). The evolution path of industrial spatial pattern in China. *Economic Geography, 35*(7), 105–112. (In Chinese).

Hu, Z., Miao, J., & Miao, C. (2016). Agglomeration characteristics of industrial pollution and their influencing factors on the scale of cities in China. *Geographical Research, 35*(8), 1470–1482. (In Chinese).

Hu, Z., Miao, J., & Miao, C. (2018). Spatial characteristics and econometric test of industrial agglomeration and pollutant emissions in China. *Scientia Geographica Sinica, 38*(2), 168–176. (In Chinese).

Ito, A. (2014). Industrial agglomeration and dispersion in China: Spatial reformation of the "workshop of the world". *RIETI Discussion Paper Series, 14*(68), 1–27.

Lemoine, F., Poncet, S., & Ünal, D. (2015). Spatial rebalancing and industrial convergence in China. *China Economic Review, 34,* 39–63.

Li, D. (2016). On the evolution of the spatial patterns of industrial pollution in China in the background of industrial transfer. *Economy and Management, 30*(1), 49–53. (In Chinese).

Li, X., Bai, Y., Zhou, P., et al. (2016). Pollution-intensive industry spatial relocation and driving mechanism in China-based on the statistical panel data from 2004-2014. *Resource Development & Market, 32*(11), 1286–1290. (In Chinese).

Liu, Q., Wang, Q., & Li, P. (2012). Regional distribution changes of pollution-intensive industries in China. *Ecological Economy, 1,* 107–112. (In Chinese).

Liu, Y., & Zeng, X. (2016). China's industrial spatial pattern evolution and the concentration difference-based on the EDSA and urban panel data of spatial econometrics. *Regional Economic Review, 1,* 80–88. (In Chinese).

Liu, Z., Ma, Z., & Liu, T. (2010). Analysis of the spatial and temporal distribution characteristics of major industrial pollutants in China over the past 20 years. *Environmental Pollution & Control, 32*(3), 94–97. (In Chinese).

Pei, Y., Zhu, Y., Liu, S., et al. (2020). Industrial agglomeration and environmental pollution: based on the specialised and diversified agglomeration in the Yangtze River Delta. *Environment, Development and Sustainability*. https://doi.org/10.1007/s10668-020-00756-4.

Ren, J., & Ma, Y. (2018). Studies on the spatiotemporal dynamics of industrial pollution in Northeast China. *Acta Scientiae Circumstantiae, 38*(5), 2108–2118. (In Chinese).

Shi, M., Yang, J., & Long, W. (2013). Changes in geographical distribution of Chinese manufacturing sectors and its driving forces. *Geographical Research, 32*(9), 1708–1720. (In Chinese).

Song, Z., & Liu, W. (2013). Spatial distribution of small and medium-sized enterprises (SMEs) and its determinants in China. *Geographical Research, 32*(12), 2233–2243. (In Chinese).

Tian, Y., & Sun, C. (2018). A spatial differentiation study on comprehensive carrying capacity of the urban agglomeration in the Yangtze River Economic Belt. *Regional Science and Urban Economics, 68,* 11–22.

Wang, G., & Wang, D. (2011). Porter Hypothesis, environmental regulation and enterprises' technological innovation-the comparative analysis between Central China and Eastern China. *China Soft Science, 1,* 100–112. (In Chinese).

Wei, W., & Ye, Y. (2013). Research on development of spatial pattern of manufacturing industry in China. *Economic Geography, 33*(3), 118–124. (In Chinese).

Xia, Y. (1999). A study on current situation, consequences and countermeasures of foreign investment in pollution-intensive industries in China. *Management World, 3,* 109–123. (In Chinese).

Yang, R. (2015). Whether industrial agglomeration can reduce environmental pollution or not. *China Population, Resources and Environment, 25*(2), 23–29.

Zhang, J., Wang, Y., Huang, B., et al. (2015). Spatial agglomeration and regional shift of pollution-intensive industries in China. *Journal of Industrial Technological Economics, 8,* 3–11. (In Chinese).

Zhou, K. (2016). Spatial-temporal differences and cluster features of environmental pollution in China. *Scientia Geographica Sinica, 36*(7), 989–997. (In Chinese).

Zhou, Y., He, C. F., & Liu, Y. (2015). An empirical study on the geographical distribution of pollution-intensive industries in China. *Journal of Natural Resources, 26*(1), 66–72.

Zhu, X., He, C., Li, Q., et al. (2018). Influence of local government competition and environmental regulations on Chinese urban air quality. *China Population, Resources and Environment, 28*(6), 103–110. (In Chinese).

Chapter 4
Do Polluting Firms Favour the Borders of Jurisdictions?

In Chap. 3, we provide an overview of the geography of industries and the geography of industrial pollution in China. The gap between these two geographies is largely ascribed to the location of polluting firms. As revealed in Chap. 2, one of the basic questions for Environmental Economic Geography can be "where the polluting firms are and why they are there". Regarding this issue, there have long been concerns about the "pollution-thy-neighbours" phenomenon (or a free-rider phenomenon). That is, polluting firms may take advantage of geographical conditions, such as the upstream along a river and a border region between two administrative units, to transfer their environmental burdens spatially so as to reduce costs for compliance with local regulation (Kahn et al. 2015; Chen et al. 2018).

As a result, the free-rider phenomenon increases the probability of transboundary pollution and simultaneously rises the difficulties of environmental governance. For instance, one local government is less likely to be authorized to regulate the polluting firms in places under the jurisdiction of the other local governments. It does not reap the benefit from the polluting production but still has to bear the impacts of pollution emissions. On the other hand, the local government for polluting firms has less incentive to enforce environmental regulation since there is no severe environmental conflicts within its jurisdiction. In China, the Environmental Protection Bureau (EPB) at the local level is responsible for the enforcement of environmental regulation. EPBs are supervised by the Ministry of Environmental Protection (MEP) vertically and simultaneously led by the local government. The enforcement of environmental regulation is thereby subject to the trade-offs between the national greening attempts and the local growth desires (Zhang et al. 2018).

From the perspective of economic geography, the location of firms is an equilibrium of multiple factors, such as the abundance of input factors, the spatial transport costs and transaction costs. Locating in one place also confronts firms with opportunity costs of not locating in others. In such a case, a free-rider effect is not sufficient to explain the preference of polluting firms to border location. In this chapter, we argue that firms can considerably benefit from agglomeration economies by locating in the regional core, which is less likely to achieve in the periphery. Locating in the

© Springer Nature Singapore Pte Ltd. 2020
C. He and X. Mao, *Environmental Economic Geography in China*,
Economic Geography, https://doi.org/10.1007/978-981-15-8991-1_4

border region, polluting firms face the trade-offs between the benefits of agglomeration effects in the core and the free-rider effects in the periphery. Based on this point, this chapter seeks to answer the following question: how does environmental regulation affect the trade-offs between agglomeration effects and free rider effects? How are the effects of environmental regulation manifested by the location of polluting firms?

To answer these questions, we start with a brief literature review. Next, we construct a stringency decision model of local governments to elaborate on the trade-off between economic growth and pollution damage within its jurisdiction. Then we use firm-level data of controlled polluting firms in China to test the "border-effect" hypothesis, which is derived from the theoretical model. We further depict the spatial distribution of polluting firms in China at the county level to present an intuitive judgment, followed by an econometric analysis to identify and explain the border effect of China. The final section concludes.

4.1 Locate Near the Border? Agglomeration Effects Versus Free-Rider Effects

4.1.1 Agglomeration Effects

As the top priority is assigned to economic development, local governments would like to direct investments to the most efficient places to generate high economic growth and local revenues. Economists have found a circular causation phenomenon in regional development. Once a new industry appears in one region, it will attract more industries by labour demand, skill upgrading, industrial linkage and the development of local services. Under the mechanism of positive feedbacks, industries begin to accumulate and grow in the space (Myrdal 1957). Keynes(1930) and Kahn(1931) propose a concept of the regional multiplier to measure the contribution of additional investment to regional growth. However, it is worth noting that regional multipliers are not equal across regions. Regions of bigger size, with a larger population and more specialized industrial structure, tend to have a larger multiplier (Shi 2013). Therefore, it is reasonable to assume that investments are most efficient when the existing industries and local population can provide sufficient supply endowments, production services, and qualified labour, which is usually absent in small towns and cities. Besides, the frequent adjustment of administrative regions at the prefecture-level since the 1970s also reduces the government's incentive to locate large investments near borders (Zhen et al. 2010; Li and Wu 2014).

From the perspective of entrepreneurs, firms benefit from agglomeration economies by locating in specialized clusters or big cities. The former effect, also known as localization externalities, is first addressed by Marshall (1890) to epitomize the external scale economy. Firms belonging to the similar sectors agglomerate to share skilled labour markets, suppliers, the intermediate goods market, specialized business service, and intra-industry knowledge spill-over. Saving transaction cost reduces average cost and attracts new firms to locate (Van Oort et al. 2012). Depending on the degree of local industrial development, the strength of local externalities is assumed to vary among sectors and regions (Duranton and Puga 2000). Urban-ization economies could benefit firms located in large cities. Large cities provide firms with access to large markets, strong research and development ability, and rich industrial diversity that allows for inter-industry knowledge spill-over. Other social, cultural and political factors in big cities, though hard to quantify, also support inno-vative behaviour and stimulate interregional growth (Harrison 1996). Agglomera-tion economies act as an important force in the location choice of firms. Empirical studies have verified that with other variables controlled, firms favour places with strong agglomeration economies, either specialized clusters or large cities (Head et al. 1995; Du et al. 2008; Chen et al. 2014). Local agglomeration is found to be effective in promoting greater economic growth (Martin and Ottaviano 1999; Davis et al. 2014; He and Pan 2010).

4.1.2 Free-Rider Effects

Environmental pollution is a typical common good with negative externality, causing substantive social cost. Internalizing negative externality would raise the private cost of firms, which is contradictory to the hypothesis of the rational man. When private cost does not reveal the real social cost, firms choose to produce more than the equilibrium level, resulting in the so-called "free-rider effect". The region, as a whole, fails to maximize its utility. Putting the free-rider effect in the context of inter-jurisdictional competition, a "border effect" of polluting firm location could be observed. At locations along jurisdiction borders, part of pollution emissions could disperse into neighbouring jurisdictions; hence the damage absorbed by local citizens within its jurisdiction is reduced. If local government endeavours to maximize the warfare of its jurisdiction, then the optimal location for polluting plants is at the border and the optimal amount of emission at bordering locations should be larger than at central locations. A series of research has identified the presence of "border effect". For instance, Helland and Whitford (2003) analyse the TRI data from 1987 to 1996 and found that counties at state borders tend to have a larger emission of pollutants into water and air than other in-state counties. Research of the US paper industry between 1985 and 1997 also confirms the "border effect" that paper plants near neighbouring states have a higher level of pollution (Gray and Shadbegian 2004). Governments have a set of regulatory tools to drive polluting firms away from central locations and towards border areas. For example, Konisky and Woods (2010) examine the county-level enforcement of the Clean Air Act during 1990 and 2000 and find that counties at national borders have a significantly weaker level of enforcement.

While economic growth remains the major concern of local governments, envi-ronmental and health issues are gaining increasing significance in the jurisdictional political tournament as countries become more developed. For the sake of residents, countries, and regions carry out various kinds of environmental standards and policies for firms and products. Environmental regulations bring substantial complying costs to firms, especially those from pollution-intensive industries, causing entrepreneurs to reconsider their decisions on firm location and production.

Under a decentralized system, there may be significant discrepancies between regulation stringencies of different jurisdictions. Then how would polluting firms react to different environmental regulations? This strand of research can be traced back to the "pollution haven hypothesis", presuming that economic activity would shift from regions with strict environmental standards to those with lenient environmental regulations when the transaction cost of relocation is low. Earlier works of literature have suggested that regional governments may lower its environmental standards to compete for dirty capital, resulting in a "race to bottom" competition (Cumberland 1979; Esty 1996; Barrett 1994; Esty and Geradin 1997). Other studies argue that decentralization may lead to a "not-in-my-backyard" phenomenon whereby local governments raise environmental regulation to prevent polluting firms from locating within their jurisdictions. Grossman and Kruger find the "reverse U shape" relationship between economic development and environmental pollution at the global level, which is further confirmed by Panayotou (1997). Despite the non-consensus on the consequences of environmental decentralization, there is no dispute that environmental regulation is a useful tool in constructing regional comparative advantage.

Agglomeration effect and free-riding effect constitute a pair of contradictory forces in the process of polluting firm location decision. On the one hand, local governments would like firms to invest in cities with large multipliers and firms themselves also incline to locate there to benefit from agglomeration externalities. On the other hand, when the responsibility of environmental protection is concerned, local governments would like polluting firms to locate at borders so that fewer pollution emissions stay in their jurisdictions. And polluting firms would favour border regions as well to avoid strict environmental regulation. In China, boundary areas are in most cases underdeveloped, where the agglomeration effect is weaker. Except for the disadvantage in locational conditions, one explanation could be the frequent adjustments of administrative division, which remains a concern of local government. For the probability of being shifted into neighbouring jurisdictions is higher for towns and counties near borders, local regulators are not as motivated to develop boundary areas (Zhen et al. 2010).

In this chapter, we try to verify if the "border effect" exists in the context of environmental decentralization and inter-jurisdictional competition in China. The political system of China consists of five layers, from national, provincial, municipal, county level to town. While the vertical linkage between layers is strong, the horizontal linkage is weak. Regional competition within the same layer is especially intensive (Cheung 2010). Here we choose the provincial border to investigate "border effect" as local legislative power. Previous studies have confirmed strong regional protectionism at the provincial level; while at the county level the impact of local protectionism is not significant (Bai et al. 2004; He et al. 2008).

4.2 A Decision Model of Regulation Stringency

In this section, we construct a regulation stringency decision model to illustrate the relative stringency of environmental regulation between border counties and central counties. Classic federalism models are based on the support of citizens, which is derived from democratic countries like the U.S. (Magat et al. 2013; Helland and Whitford 2003). However, within the top-down administrative structure, local governors in China are appointed by higher authorities according to the governor's performance in respective jurisdictions. Previously, economic growth was the sole goal. In recent years, however, environmental performance or green GDP has also been included in the evaluation of local governments. Therefore, we construct the jurisdiction model as a trade-off between economic growth G and pollution damage P with relative weight b (b > 0). T is the target for local governments.

$$T = G - b * P \tag{4.1}$$

Suppose a province with jurisdiction over z counties, each county contains n_i citizens and a typical polluting firm that chooses to produce at its optimal output m_i. Provincial governments have the right to set a pollution limit s_i for each county. A smaller s_i indicates a stricter environmental regulation, higher complying cost for firms and smaller damage for local citizens. The complying cost for a typical polluting firm is a function of the pollution limit per unit of output $t(s_i / m_i)$, which decreases as the total amount of pollution allowed to emit increases, $t'(s_i / m_i) > 0$ and $t''(s_i / m_i)t'' < 0$. The aggregate population, outputs, and pollutions can be derived:

$$N = \sum_{i}^{z} n_i \quad M = \sum_{i}^{z} m_i \quad S = \sum_{i}^{z} s_i \tag{4.2}$$

An additional firm and investment would stimulate local economic growth. The scale of growth is decided by the multiplier effect that relies on the level of local development. For cities with a large population and comprehensive industrial system, increasing investments could provide significant economic stimulation, while for small cities with a small population and weak industrial structure, such stimulation effect is small (Lin 2004). For each county, the multiplier e i s he function of the population n_i and its total industrial output y_i.

$$G = \sum_{i}^{z} G_i = \sum_{i}^{z} e(n_i, y_i) m_i \tag{4.3}$$

While lenient environmental regulations would attract more firms and create economic growth, it will also induce more pollution and impair the welfare of the local citizen. For each citizen, unhappiness $v(s_i)$ due to welfare loss from industrial pollution is a function of the total amount of pollution s_i and $v'(s_i) > 0$. To

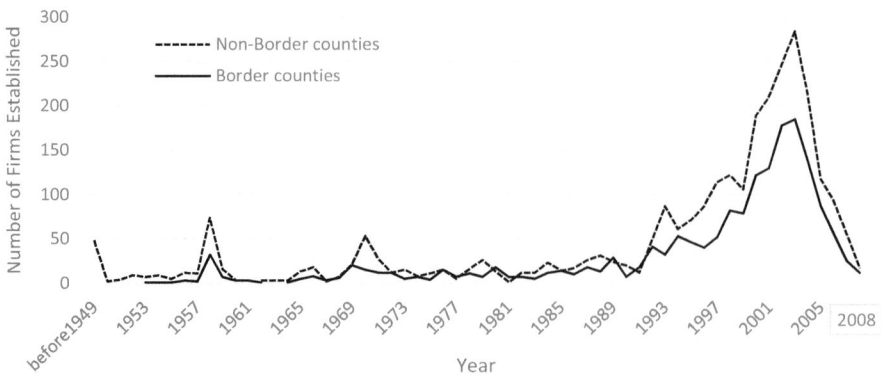

Fig. 4.1 Starting years of polluting firms in China

illustrate the free-riding effect at border counties, we follow Helland and Whitford's idea to include the variable $a_i \in (0, 1]$ to represent the proportion of each county's pollution emission that stays in the regulator's province. For counties at the center of a province, a_i approaches 1, while for those counties near borders, a_i is smaller because a good portion of pollution is exported to neighbour jurisdictions (Fig. 4.1). Only pollution stayed in the province would cause welfare loss and increase the unhappiness of residents.

$$P = \sum_{j=1}^{N} P_j P_j = \sum_{i=1}^{Z} v(a_i s_i) \tag{4.4}$$

Assume that all pollution, regardless of which county it is from, is evenly spread over the entire province, so that every citizen has the same level o f unhappiness. To simplify the calculation, we assume $v(s_i)$ to be linear-additive and homogeneous of degree one, $\sum v(s_i) = v(\sum s_i)$, $v(\lambda s_i) = \lambda v(s_i)$, which can be relaxed.

$$P = \text{N}* \sum_{i}^{z} v(a_i * s_i) = \sum_{j}^{z} n_j * \sum_{i}^{z} a_i * v(s_i) \tag{4.5}$$

Substitute G and P in the target function with the above expressions, we get the stringency decision model for regulators.

$$T = \sum_{j}^{z} e(n_j, y_j) * m^*(s_i) - b * \sum_{j}^{z} n_j * \sum_{i}^{z} a_i * v(s_i) \tag{4.6}$$

The solution of this model is a two-step process:

First, environmental regulation complying cost would affect a firm's location choice and production decision. Given the pollution limit s_i, a typical firm decides its optimal output m_i^* when profit is maximized:

$$\max_{s_i} \left[m_i - m_i * t \left(\frac{s_i}{m_i} \right) \right] \tag{4.7}$$

$$\text{First Order Condition } m_i^* = m_i^*(s_i) \tag{4.8}$$

Given the optimal output m_i^*, the target function of the provincial government is transformed to:

$$\max_{s_i} G\left(m_i^*(s_i), n_i, y_i\right) - b * P \tag{4.9}$$

$$\text{First Order Condition } s_i : G'\left(m_i^*(s_i), n_i, y_i\right) * m_i^*(s_i) - b * P'(s_i) = 0 \tag{4.10}$$

where $P'(s_i)$ is the differential of $P(s_i)$.

We are interested in regulation stringency variable s_i^* and location variable a_i and we derive the total differentials with respect to s_i^* using comparative statics.

The competitive static analysis is as follows:

$$\frac{ds_i}{db} = \frac{p'(s_i)}{\gamma} \tag{4.11}$$

$$\frac{ds_i}{dn_i} = -\frac{m_i'^*(s_i) * \frac{dG'\left(m_i^*(s_i), n_i, y_i\right)}{dn_i}}{\gamma} \tag{4.12}$$

$$\frac{ds_i}{dy_i} = -\frac{m_i'^*(s_i) * \frac{dG'\left(m_i^*(s_i), n_i, y_i\right)}{dy_i}}{\gamma} \tag{4.13}$$

$$\frac{ds_i}{da_i} = \frac{b * s_i * P'(a_i s_i)}{\gamma} \tag{4.14}$$

$$\gamma = \left(\frac{dG'\left(m_i^*(s_i), n_i, y_i\right)}{dm_i^*(s_i)} * m_i'^*(s_i) + G'\left(m_i^*(s_i), n_i, y_i\right) * m_i'^*(s_i) \right) - p''(s_i) < 0 \tag{4.15}$$

The relationship between other factors and s_i is given in Table 4.1.

Although the pollution limit s_i, which reflects both the regulation itself and the local enforcement stringency, can't be observed directly, we can take the number and aggregate output of polluting firms as a proxy of s_i. Bigger s_i indicates a more lenient regulation in county i, where more polluting firms are likely to locate and more polluting outputs may be produced.

From the above analysis, four testable hypotheses are developed:

Table 4.1 Relationship between variables

Parameter increased	Sign associated with change in s_i
n	Ambiguous
b	−
a	−
y	+

H1: For counties near borders, whose a_i is small, the free-riding effect is strong and environmental regulation stringency is weak (a larger s_i). Hence polluting firms are more likely to locate at border counties and produce more "dirty GDP" than in other counties.

H2: Although the relationship between n_i and s_i can be both ways in the model, we hypothsize that counties with a larger population have stricter regulation or stringency of enforcement, while counties with a smaller population tend to have more lenient regulations.

H3: Local industrial GDP y_i is positively related to s_i, implying that counties with greater industrial GDP may have more polluting firms and greater dirty output.

H4: b implies the relative importance assigned to economic development and environmental protection. The negative relationship between b and s_i indicates that for the province who give greater importance to economic development, environmental regulation would be more lenient.

4.3 Data Description

Since it is hard to measure the pollution limit of each county, we use polluting firms and "dirty output" (gross industrial output from polluting firms in each county) as proxies for environmental regulation stringency. However, despite the effort made by scholars in recent years, there is no widely accepted classification of pollution-intensive industries. Categorizing pollution intensity by SIC code is not feasible as there are great differences within the same two-digit or even three-digit industry. To define the pollution characteristics of a firm, a method based on the emission of every single firm is needed. Starting from the year 2006, the Ministry of Environmental Protection (MEP, SEPA before 2008) in China has published the list of Key National Monitoring Sources of Pollution Firms. Over 80,000 key firms are surveyed and sorted according to their major pollutant emissions in the previous year, with respect to air pollution emissions (measured by industrial sulfur dioxide, soot, industrial dust) and air pollution emissions (measured by industrial COD and ammonia emissions). Firms account for 65% of the total pollution emission of all key surveyed firms that are included in the list of key national monitoring sources. The list is modified annually.

To study the difference in environmental regulations and its consequences on firm behaviour, an ideal way is to see how the spatial pattern of pollution-intensive firm changes overtime in panel data. However, the industrial firm data with location and production details after 2008 is not published yet. Given the restraints on data, our research is based on a cross-sectional database, the core of which is the list of Key National Monitoring Sources of Pollution Firms (KNMSP) of the year 2009. The cross-sectional database is constructed in three steps.

First, we match the Key National Monitoring Sources of Pollution Firms list with the Annual Survey of Industrial Firms 2008 using legal person code to get detailed information about polluting firms. Those haven't been merged successfully are further matched using firm name. 5124 out of 7660 polluting firms are matched successfully whose characteristics are summarized by county. Those still haven't been matched are enterprises below the designated size of sales revenue of 5 million RMB. Since small firms are not the major concern to local governments, losing these samples will not bias our research. Among them, 921 firms belong to the power, heat, gas and water supply industry and 162 firms are extractive industries. Since supply firms are created to support urban lives and locations of extractive firms heavily depend on the availability of mineral resources, the mechanism for their location choices runs beyond our topic. Therefore, we exclude these 1083 firms and our sample size reduces to 4558 firms (Table 4.2).

Next, we match the list of polluting firms with counties' social-economic data from Yearbooks. Municipal districts of the same city are combined into one observation. Finally, we construct the border characteristic of each county on the map using ArcGIS 10.0 and match it with the county-level data from the first two steps. 2403 county-level observations are generated, among which 1134 are border counties and 1269 are non-border counties.

Three waves could be observed from the establishing time of these polluting firms. The establishments of polluting firms boomed since 1990, hand in hand with the taking off of the Chinese economy (Fig. 4.1). More than half of polluting firms were established after 1997, and peaked in 2003. One explanation for the decline of establishments since 2003 is the threshold of KNMSP data. While firms created after 2003 are still in their infancy and the production capacity hasn't expanded, their pollution emission is not large enough to enter the KNMSP list. Another reason could be the discrepancy between old and new production equipment. Newly established firms may use the updated production equipment with cleaner technology, whereas for older firms replacing equipment would be a difficult decision to make.

4.4 Spatial Pattern of Polluting Firms in China

Our sample covers 2403 county-level geographic units (counties, county cities, and municipal districts) of China, of which 1134 are at provincial borders. 4558 controlled polluting firms in manufacture industries are distributed across these counties, with 1834 firms having intensive air pollution emission and the other 2724 firms were heavy polluters of wastewater.

Table 4.2 Data description

	All Counties	Border		Regions		
		Border County	Non-border County	East	Middle	West
Number of counties	2403	1134	1269	825	612	966
Polluting characteristics						
Number of polluting firms	4558	1761	2796	1681	1591	1285
Waste gas	1834	701	1133	533	735	566
Waste water	2724	1060	1663	1148	856	719
Total output of polluting firms (Million Yuan)	10396.6	3242.2	7154.3	6303.4	2420.7	1672.5
Waste gas	5244.5	1664.5	3579.9	3158.7	1221.7	864.1
Waste water	5152.1	1577.7	3574.4	3144.7	1199.1	808.4
Tax income from polluting firms (Million Yuan)	19.0	5.2	13.8	10.0	6.0	3.0
Waste gas	9.7	2.3	6.8	4.9	3.1	1.6
Waste water	9.3	2.9	7.0	5.0	2.9	1.3
Employment from polluting firms (Million Yuan)	8.2	2.3	5.9	4.1	2.5	1.6
Waste gas	3.7	1.0	2.7	1.8	1.2	0.8
Waste water	4.5	1.3	3.2	2.4	1.3	0.8
Social-economic characteristics						
Average GDP(Billion Yuan)	12.464	11.296	13.507	22.406	9.960	5.559
Average population (Thousands)	500.99	490.14	520.64	620.21	580.26	360.80
Average consumption(Billion Yuan)	4.044	3.457	4.568	7.220	3.428	1.721
Average local fiscal revenue (Billion Yuan)	0.758	0.691	0.818	1.446	0.572	0.289
Average local fiscal expenditure(Billion Yuan)	1.344	1.254	1.425	2.019	1.184	0.870

Relatively speaking, border counties, in general, has a smaller number in terms of the number of polluting firms, industrial gross output, tax income and employment generated from polluting firms. Impressive gaps between border counties and non-border counties are found in social-economic characteristics. Non-border counties on average have a larger GDP, population, consumption, and local fiscal revenue and expenditure, which is consistent with our intuition. China's administrative jurisdictions are divided in accordance with geographical conditions, as a result, mountains and rivers are often used as borders. Hence, counties at provincial borders often face harsh natural environments, which hurdle their economic development.

Concerning the uneven development across the country, we further look into the issue from a regional perspective. Although the west region accounts for the biggest share of land area and has the highest number of counties, it has the smallest number of polluting firms, as well as gross outputs, tax revenue and employment generated by polluting firms. This could be explained by the fact that western China is rather underdeveloped with a smaller GDP, population, consumption and fiscal revenue compared to the coastal region. Industrial development still concentrates on China's coastal and central regions.

The 4558 polluting firms belong to 26 manufacturing industries (see Table 4.3). Smelting and Pressing of Ferrous Metals (32) and Processing of Petroleum, Coking, Processing of Nuclear Fuel (25) generate the dirtiest output, for both air and water pollution. These two industries are capital-intensive heavy industries, providing intermediate goods for other industries. Concerning their importance to local industrial development, they are regarded as pillar industries in many provinces. However, these two industries are pollution-intensive. In previous studies, they ranked the top five polluting industries, either measured by intensity or volume of pollution emission (Low and Yeat 1992; Lucas et al. 1993).

Manufacture of Non-metallic Mineral Products (31) has the largest number of polluters for air pollution, other industries with more than 100 polluters include Smelting and Pressing of Ferrous Metals (32), Manufacture of Chemical Materials and Chemical Products (26), Processing of Petroleum, Coking, Processing of Nuclear Fuel (25) and Manufacture of Paper and Paper Products (22). For enterprises controlled for wastewater, Manufacture of Paper and Paper Products (22) ranks the first in terms of number of polluters, followed by Manufacture of Chemical Materials and Chemical Products (26), Processing of Food from Agricultural Products (13), Manufacture of Textile (17), Manufacture of Beverages (15), and Manufacture of Foods (14). Those industries are key industries in many provinces in China.

While private-owned firms are the main force of the polluters, state-owned enterprises contribute the largest share of polluting enterprises' output (Fig. 4.2). Comparing with enterprises with other types of ownership, SOEs are generally larger and older. They concentrate in heavy chemical sectors like Processing of Petroleum, Coking, Processing of Nuclear Fuel (36.8% of all SOE output) and Smelting and Pressing of Ferrous Metals (42.2% of all SOE output). These two industries are the most pollution-intensive industries listed by MEP. This is consistent with the empirical studies that report a positive relationship between SOEs and industrial pollution (He et al. 2012a, b).

Table 4.3 Statistics by industries

SIC	Industry	Number of Polluting Firms		Gross Industrial Output (million yuan)	
		Air	Water	Air	Water
13	Processing of Food from Agricultural Products	39	325	64.1	153.7
14	Manufacture of Foods	28	114	25.9	94.9
15	Manufacture of Beverages	24	222	57.9	141.5
16	Tobacco Processing	5	2	73	4.6
17	Manufacture of Textile	30	301	135.5	283.4
18	Manufacture of Textile Wearing Apparel, Footwear and Caps	3	13	15.3	25.7
19	Manufacture of Leather, Fur, Feather and Related Products	0	47	0	20.4
20	Processing of Timber, Manufacture of Wood, Bamboo, Rattan, Palm and Straw Products	9	13	1.2	2
21	Manufacture of Furniture	0	0	0	0
22	Manufacture of Paper and Paper Products	104	625	122.8	232
23	Printing, Reproduction of Recording Media	1	1	1.7	1.7
24	Manufacture of Articles For Culture, Education and Sport Activities	0	0	0	0
25	Processing of Petroleum, Coking, Processing of Nuclear Fuel	181	92	1376	1375
26	Manufacture of Chemical Materials and Chemical Products	286	600	363.9	563.1
27	Manufacture of Medicines	12	111	8.8	62.1
28	Manufacture of Chemical Fibers	16	43	28.1	39.3
29	Manufacture of Rubber	8	6	25.1	27.4
30	Manufacture of Plastics	3	3	2.2	1.4
31	Manufacture of Non-metallic Mineral Products	728	14	179.1	7.3

(continued)

Table 4.3 (continued)

SIC	Industry	Number of Polluting Firms			Gross Industrial Output (million yuan)	
		Air	Water	Air	Air	Water
32	Smelting and Pressing of Ferrous Metals	249	97	2455	1628	
33	Smelting and Pressing of Non-ferrous Metals	77	17	211.5	108.3	
34	Manufacture of Metal Products	8	2	8.2	3.5	
35	Manufacture of General Purpose Machinery	1	6	0.1	17	
36	Manufacture of Special Purpose Machinery	9	8	48.9	43.6	
37	Manufacture of Transport Equipment	8	16	27.6	125.3	
39	Manufacture of Electrical Machinery and Equipment	3	6	11.5	22.5	
40	Manufacture of Communication Equipment, Computers and Other Electronic Equipment	1	36	1.1	168	
41	Manufacture of Measuring Instruments and Machinery for Cultural Activity and Office Work	0	0	0	0	
42	Manufacture of Artwork and Other Manufacturing	1	3	0.6	1	

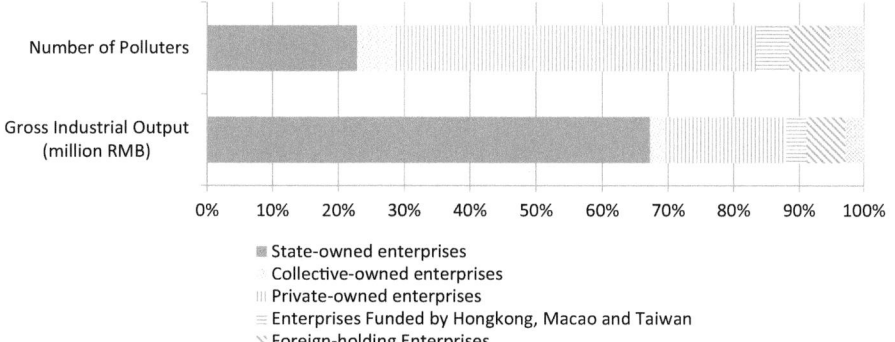

Fig. 4.2 Ownership structure of polluting firms in China

Interestingly, industries show different preferences for border locations (Fig. 4.3). While firms in Smelting and Pressing of Ferrous Metals (32), Textile (17) and Manufacture of Special Purpose Machinery (36) prefer locations in provincial centres, more outputs from metal products manufacturing industry (34) are generated in border counties. The "Border effect" phenomenon does not seem obvious. For air pollution, only 5 out of 26 industries have more gross output in border counties, and for water pollution, 6 out of 27 industries have more output in border counties. It appears that firms are more likely to take advantage of agglomeration effects rather than being free riders to divert pollution to other jurisdictions.

However, we find significant differences in the average firm size between the border and non-border counties (Fig. 4.4). Larger firms are more likely to locate in the border counties in industries like Processing of Petroleum, Coking, Processing (25), Manufacture of Rubber (29), Manufacture of Electrical Machinery and Equipment (39) and Manufacture of Transport Equipment (37). Large firms may enjoy scale economies, mitigating the importance of external economies and are more likely to survive in remote areas.

Statistically, the "dirty output" from the polluting firms is positively related to local GDP and population, though correlation to the latter is not as significant as that to the former (Fig. 4.5). Correlation between dirty output and ln(GDP) for non-

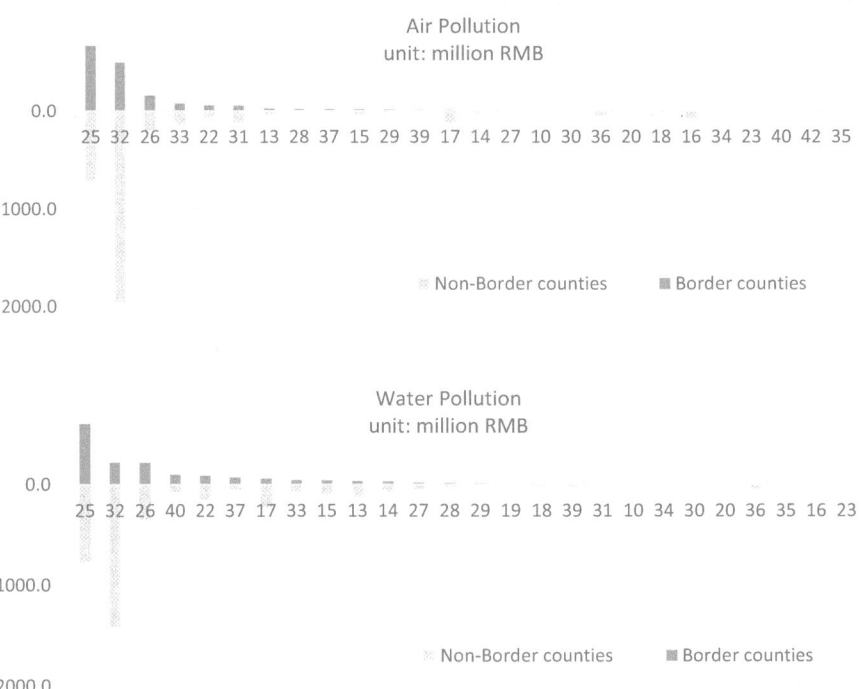

Fig. 4.3 Border and non-border locations of the gross industrial output of polluting firms

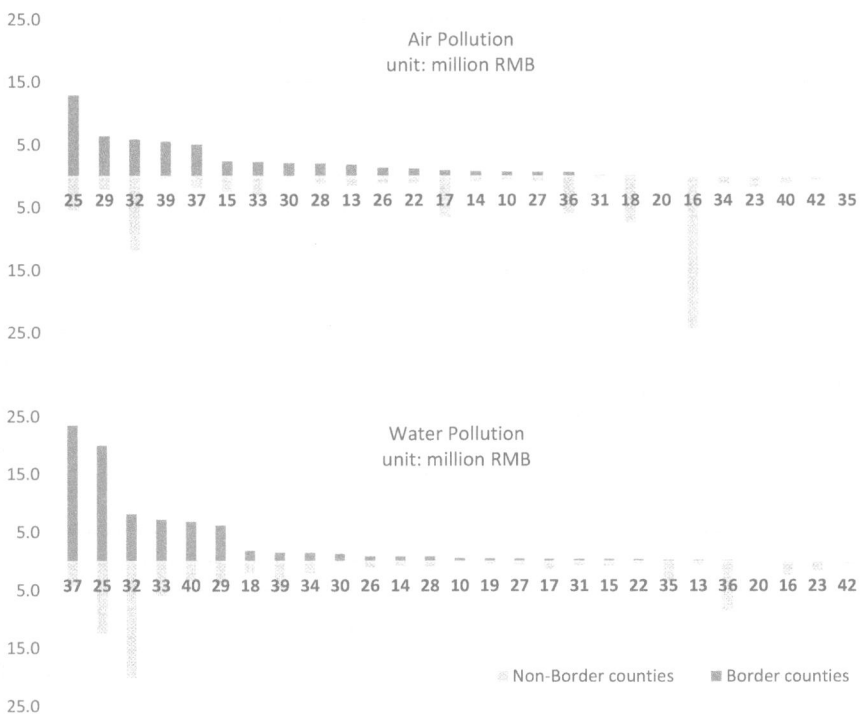

Fig. 4.4 Average size of polluting firms in the border and non-border locations

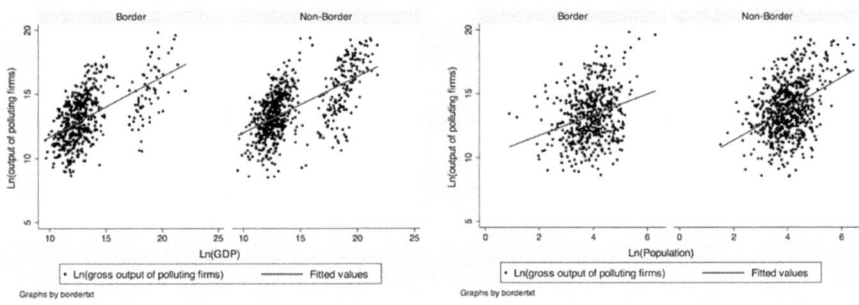

Fig. 4.5 Relationship between GDP, population and "dirty output"

border counties is 0.5390, while that for border counties is 0.4838. The same is for the population. Correlation between dirty output and GDP for non-border counties is 0.4531, while that for border counties is 0.3930. This implies a weaker agglomeration effect for border counties, which may result from the offset by the "border effect".

Meanwhile, the variation of dirty output for border counties is larger than that of non-border counties. Overall, this shows some evidence of border effects in the location of polluting firms in China although agglomeration effects may be stronger. The next part is to test the significance of border effect and agglomeration effects in the locational choices of polluting firms.

4.5   Empirical Strategy of Polluting Firm Locational Choices

We perform a set of regressions to test the presence of border effect and agglomeration effects with other factors controlled. Our interests focus on the four variables from the model in Sect. 4.2.

4.5.1   Variables

n_i is represented by the population of each county **ln(POP)**, a_i is represented by the location dummy variable **Border**. A series of variables may influence on b_i, the relative weight of economic growth to environmental protection, and y_i, the local industrial foundation and the agglomeration effects.

4.5.1.1   County-Level Explanatory Variables

As discussed above, the effect of **POP** is ambiguous. On one hand, the larger population provides sufficient labour market, makes greater local consumption, thus the local multiplier of investment is larger. On the other hand, a larger population implies greater welfare loss due to industrial pollution. The unsatisfactory environment leads to public complaints and protest, which potentially evolve into collective actions and bring negative impacts on local governments

Municipal districts (**MD**) of prefectural cities are county-level jurisdictions. Compared to counties, municipal districts are more urbanized, populous, and developed in the cultural and technological realm. Therefore, the multiplier effect may be larger in MD, which often plays the role of growth engine for the local economy. Firms in districts are more likely to benefit from urbanization economies. Local governments may prefer investments in municipal districts. We introduce a dummy to indicate the municipal districts (**MD**) in the model.

The number of local enterprises other than our sample firms (**ln(Firm)**), is one of the most widely used indicators of local industrial development and agglomeration economy. The co-location of firms reflects a better business environment, which is of particular attractiveness to polluters. We also include urbanization level, measured as the ratio of the agricultural population in total population (**Urbanization**), to quantify the urbanization economies.

4.5.1.2   Prefectural Level Explanatory Variables and Regional Dummies

Since county-level data is very limited, we include some prefectural level variables. The development of a county is embedded in its prefectural and provincial contexts. **Tax Revenue**, valued as the proportion of tax revenue (sum of VAT and Enterprise Income Tax) in local governmental fiscal revenue, reflects local governments' attitude towards polluting firms. The more dependent they are on tax revenues, the more likely that local governments would compromise to polluting firms. The same is with **Social Expenditure**, the proportion of social, health and education expenditure in governmental expenditure, which reflects local governments' care for social welfare. Since the development of tourism is contradictory to environmental pollution, we also include the proportion of tourism revenue accounts for local GDP (**Tour Ratio**).

Also, we should not overlook the unbalanced pattern of regional economic development in China. Marketization first took place in the eastern provinces, establishing significant first-mover advantages over central and western regions in attracting firms. The structure of regional specialization has further enlarged the regional gap. The densely distributed industrial clusters in East China enhance the agglomeration externalities while the relatively stringent environmental regulation drives polluting firms away (Fan 2006; He and Wang 2010). Taking WEST China as the base group, we introduce dummy variables of **EAST** and **Central** to test the differences of environmental regulation among regions.

Other control variables include the average wage (**ln(Wage)**), the length of road mileage per squared kilometer (**ln(Road)**) and the number of college students (**ln(EDU)**). Those variables are often used in vocational studies. All variables and their definitions are summarized in Table 4.4.

The model is specified as follows:

$$\text{Number of Polluters}_i = \beta_0 + \beta_1 \text{Border}_i + \gamma \text{County\_Variable}_i$$
$$+ \delta \text{Prefectural\_Variable}_i + \eta \text{Region}_i + \lambda Control_i + \varepsilon_i$$

$$\text{Gross Output}_i \text{ of Polluters} = \beta_0 + \beta_1 \text{Border}_i + \gamma \text{County\_Variable}_i$$
$$+ \delta \text{Prefectural\_Variable}_i + \eta \text{Region}_i + \lambda Control_i + \varepsilon_i$$

$$(i = \text{County}_1, \text{County}_2, \text{County}_3, \dots \text{County}_{2086})$$

We note that the data are hierarchically organized. The variance of dependent variables results not only from the difference between counties but also from that between prefectural cities. In our case, the fraction of variance due to prefectural

Table 4.4 Definitions of variables

Variable Name	Definition
Border	Dummy variable, 1 for border county, 0 for non-border county
Ln(POP)	Ln value of local population for each county
Municipal District	Dummy variable, 1 for border municipal district, 0 for ordinary county
Ln (Firms)	Ln value of number of local enterprises for each county
Urbanization	Urbanization rate for each county calculated by the portion of urban population
East China	Dummy variable, 1 for counties East Region, 0 for others
Central China	Dummy variable, 1 for counties Central Region, 0 for others
Tax Revenue	Proportion of tax revenue (sum of VAT and Enterprise Income Tax) in local governmental fiscal revenue
Social Expenditure	Proportion of social, health, and education expenditure in governmental expenditure
Tour ratio	Proportion tourism revenue in local GDP
Ln (wage)	Ln value of local average wage (RMB) in each prefecture
Ln (Road Density)	Ln value of road density (km/km^2) of each prefecture
Ln (Education)	Ln value of number of college graduates in each prefecture

difference is 34.2% (dependent variable is the number of firms) and 28.0% (dependent variable is the gross output of firms), which are significant at 1% level. While with traditional approaches prefectural variables can be included to explain the variances in dependent variables, heteroscedasticity can't be solved. Hence we perform the multilevel regression, which is created to deal with hierarchical data (Fazio and Piacentino 2010). With county being the first level, provinces are set as the second level variable. However, regional dummies can't be included in the two-level analysis, therefore we conduct Poisson regression and Tobit regression in model 3 and model 6. For Beijing, Tianjin, Shanghai, and Chongqing, which are municipalities directly under the central government, the area is very small and border effect can't be distinguished at the county level. Therefore, we exclude counties and districts from these four provinces in our regressions.

4.5.2   Empirical Results

Table 4.5 presents the correlation coefficients among the explanatory variables. As expected, border counties have a smaller population, fewer industrial enterprises, and less urbanized. The correlation coefficient between **ln(POP)** and **ln(Firm)** is 0.7714, **MD** and **Urbanization** have a correlation coefficient of 0.5986. Other correlations are relatively weak. To avoid collinearity in the model, we put **ln(POP)** and **ln(POP)** into different model specifications. We have two dependent variables, including the

Table 4.5 Correlation matrix of explanatory variables

	Border	Ln(POP)	MD	Ln(Firm)	Urban	Tax	Expenditure	Tour	Ln(Road)	Ln(Wage)	Ln(Edu)
Border	1										
Ln(POP)	−0.088	1									
MD	−0.121	0.344	1								
Ln(Firm)	−0.063	0.766	0.375	1							
Urbanization	−0.064	0.084	0.596	0.246	1						
Tax	−0.036	0.021	0.028	0.120	0.098	1					
Expenditure	0.085	0.107	−0.073	−0.198	−0.132	−0.101	1				
Tour	0.042	−0.088	0.014	−0.004	0.078	−0.039	−0.136	1			
Ln(Road)	−0.054	0.650	0.113	0.633	−0.097	0.088	0.037	−0.068	1		
Ln(Wage)	−0.002	−0.318	0.012	−0.099	−0.023	0.034	−0.544	0.125	−0.308	1	
Ln(Edu)	−0.144	0.423	0.049	0.533	−0.010	0.004	−0.308	0.000	0.454	0.076	1

number of firms and industrial output. We apply the Poisson model to the first dependent variable and Tobit model to industrial output due to the zero values in many counties. We also divide the sample into firms controlled for air pollution and water pollution, since they may differ in locational behaviors. The regression results are reported in Table 4.6 and Table 4.7. The statistics indicate that both models have significant explanatory power.

Table 4.6 Regression results for firms controlled for air pollution

Dependent Variable	Number of Polluting Firms			Gross Industrial Output of Polluting Firms		
	Multilevel regression		Poisson	Multilevel regression		Tobit
Model	1	2	3	4	5	6
County-level variables						
Border	−0.0365	−0.107	−0.143***	−0.414	−0.492*	−1.231
Ln(POP)	0.595***		0.619***	0.824***		3.525***
Municipal District	1.216***		0.776***	1.594***		4.610***
Ln(Firms)		0.479***			0.743***	
urbanization		2.767***			8.013***	
urbanization^2		−1.334**			−6.479***	
Regional dummy variables						
East China			−0.133*			−3.531**
Central China			0.496***			4.406***
Prefectural level Variables						
Tax Revenue	1.001	−0.374	1.818***	2.847	0.562	17.22***
Social Expenditure	−1.619	1.296	−1.031**	−4.837*	−0.296	−20.93**
Tour Ratio	−0.00273	−0.00361	−0.0121***	−0.0116*	−0.0135**	−0.142**
Ln(Wage)	0.635**	0.804**	0.614***	0.0591	0.509	−0.735
Ln(Road Density)	−0.174**	−0.139	−0.140***	−0.0317	−0.0198	−0.323
Ln(Edu)	−0.0279	−0.128	−0.0358	0.0302	−0.141	0.728
Constant	−6.497*	−9.428**	−8.267***	1.491	−5.586	-10.28
sigma_u	0.851***	0.930***		2.031***	2.081***	
sigma_e	1.368***	1.342***		5.256***	5.186***	
sigma						17.72***
Observations	2,086	2,067	2,086	2,086	2,067	2,086
Number of groups	315	314		315	314	

Note *** $p < 0.01$, ** $p < 0.05$, * $p < 0.1$

Table 4.7 Regression result for firms controlled for water pollution

Dependent Variable	Number of Polluting Firms			Gross Industrial Output of Polluting Firms		
	Multilevel regression		Poisson	Multilevel regression		Tobit
Model	1	2	3	6	5	6
County-level variables						
Border	−0.0659	−0.194*	−0.139***	0.00265	−0.240	0.392
Ln(POP)	0.716***		0.570***	1.397***		5.278***
Municipal District	1.899***		0.717***	2.188***		3.725***
	(0.164)		(0.0521)	(0.403)		(1.187)
Ln(Firms)		0.708***			1.235***	
Urbanization		1.210			3.260*	
Urbanization^2		0.659			−1.550	
Regional dummy variables						
East China			0.128**			2.379**
Central China			0.0844			−0.149
Prefectural level Variables						
Tax Revenue	1.774*	0.522	1.551***	4.310**	2.199	14.30***
Social Expenditure	−2.769*	1.164	−1.470***	−4.238	2.511	−13.73*
Tour Ratio	−2.95E-06	−0.00107	0.000712	0.00119	−0.00105	0.0154
Ln(Wage)	1.096**	1.237**	0.0640*	−1.809*	−1.532	−11.09***
Ln(Road Density)	0.0237	−0.185	0.126	−0.0449	−0.128	0.370
Ln(Edu)	0.231**	0.0968	0.0845***	0.458*	0.169	0.834
Constant	−10.92*	−14.31**	−3.552**	18.01	13.05	86.09***
sigma_u	1.421***	1.409***		2.453***	2.329***	
sigma_e	2.184***	2.214***		5.407***	5.413***	
sigma						14.39***
Observations	2,086	2,067	2,086	2,086	2,067	2,086
Number of groups	315	314		315	314	

Note *** $p < 0.01$, ** $p < 0.05$, * $p < 0.1$

Surprisingly, there is no significant border effect in the location of polluting firms in China. For the coefficients of Border are either negative or insignificant in all models, we reject the hypothesis H1. Statistically speaking, polluting firms intend to locate away from provincial borders. Locating at provincial borders places jurisdictions at a disadvantage situation in the competition for industrial investments, because borders are typically less developed and less populated, landlocked, and more isolated. This indicates that the cost of complying with environmental regulations in China is not high yet, thus free-riding benefits in border counties do not seem attractive to firms. Local governments currently emphasize more about economic growth and effectiveness rather than environmental protection. Moreover, local governments face strong resistance to implementing stringent environmental regulation from large firms and state-owned enterprises (He et al. 2012a). Social pressure for environmental protection has not been effective in forcing local governments to care about the local environment (He et al. 2012a). Furthermore, to secure their economic benefits, local governments are also unwilling to locate large firms in the borders (An 2004; Han et al. 2011).

Meanwhile, **MD** and **ln (Firm)** have positive and significant coefficients, which suggests that polluting firms are more likely to locate in municipal districts and counties with a larger number of industrial enterprises. This indicates that polluting firms are stimulated to take advantage of agglomeration effects. Previous studies already find significant agglomeration effects in work in liberalized and globalized industries and regions to promote innovation, growth and productivity in China (Pan and Zhang 2002; He and Pan 2010; Ke 2010). Polluting firms co-locate with other firms to improve their performance. Previous research on plant location in Japan suggests a similar result, that co-location benefits far outweigh the grants like subsidies or lenient regulations in a firm's location choice (Devereux et al. 2007). In line with this finding is the significant positive effect of population-**ln(POP)**. Despite the increasing attention to the environmental problem, the influence of public opposition on regulation is still weak. This is consistent with He et al. (2012b), who find that public pressure has not helped to clean China's urban sky. Districts and counties with more population and industries are typically those favoured by local governments, which in turn encourage large firms to locate in their favoured regions. Designated area-based policies are often reported to attract industrial firms and businesses in China. Wu (1999) and Wu and Radbone (2005) report that foreign investments are significantly attractive to industrial development zones in Guangzhou and Shanghai, respectively. Local governments purposely lead industrial firms to their favourable locations through area-based policies in China. Overall, both firms controlled for air and water pollution avoid border location and favour good locations that generate positive externalities.

Urbanization explains a significant part of the models. There is a positive relationship between urbanization level and firms controlled for air and water pollution, and a significant inverse U shape between urbanization and firms controlled for air pollution. Urbanization economies are certainly a key factor to draw polluting firms. Air polluting firms are also less likely to concentrate in the coastal region but obviously favour the inland regions. This may be the result of industrial relocation from the coastal region to the inland region in the process of industrial upgrading (Fan 2013). Moreover, coastal regions often face increasingly severe punishments for pollution violations than inland China (van Rooij and Lo 2010). On the contrary, water-polluting firms favour the coastal and central regions more than the west region. Water polluting firms are typically located closer to waters such as seas, rivers, and lakes. Coastal and central regions have better natural conditions to host water-polluting firms.

For firms controlled for air pollution, Tax Revenue has positive and significant coefficients in model 3 and model 6, which is consistent with **H4** that if the regulator is more dependent on tax revenues from industrial firms, it would be more lenient to polluters. Endogeneity is not a problem since Tax Revenue is a prefectural variable. We have to understand this positive relationship under the context of regional fiscal decentralization in China. Under fiscal decentralization, local governments often face restrictive budgets. And to promote revenue growth, local officials have become increasingly entrepreneurial. The decentralization of economic decision-making to local governments, combined with calls for rapid economic growth, have created further incentives for local governments to pursue economic growth at the expense of environmental degradation (Jahiel 1997). Existing studies report that local governments in China have consistently undermined pollution enforcement to protect local economy interests (Swanson et al. 2001; Tang et al. 2003). Taguchi and Murofushi (2010) even found evidence on the inter-jurisdictional competition for polluting industries in China.

While the structure of local government revenue is not a direct proxy of b_i, local expenditure on social welfare reflects the regulator's value of environmental problems more directly. The negative and significant coefficients of Social Expenditure in different models confirm our hypothesis that the more local governments spend on social welfare, the stricter the environmental regulation would be, and there would be fewer polluters and dirty outputs, which is particularly true for air polluters.

Among other controlling variables, the Tour Ratio has a negative and significant coefficient in the model for air pollution, indicating that the development of tourism would discourage the entry of air polluters. It has no significant impact on the location of water polluters. Tourism does need better air quality to attract tourists. In the Poisson models, Wage has positive coefficients, which suggests that both air and water polluters are attracted to regions with higher salaries. Polluters may see the higher wage rate as complementary. In the Tobit model, Wage has negative coefficients for output of water polluters, implying more output of water polluters in the regions with cheap labour. Ln (Road Density) and Ln (EDU) are barely significant in the model of air polluters. Water polluters, however, are drawn to regions with more college students. Water polluters may need more skilled labour or more human capital input.

4.6 Conclusion and Implications

Regional environmental pollution is rooted in the externalities of environment and jurisdictional competition. This research hypothesizes that under the fiscal decentralization and decentralized environmental management system, the "border effect" would emerge. Pollution Haven Hypothesis implies that jurisdictions, especially those undeveloped ones, would compete for "dirty investments". The free-riding effect then directs these "dirty investments" towards locations near jurisdiction borders. By constructing an environmental regulation stringency decision model for provincial governments, we illustrate that border, population, local industrial foundation and the relative importance of environment protection given by local governments influence the stringency of environmental regulation. The other force aside from the "border effect" is the agglomeration effect, taking the form of localization and urbanization economies. Jurisdictions with better industrial foundation, more enterprises, larger labour pool, and better market environment could provide larger positive externalities, thus attract more firms and yields larger output.

A key finding in this chapter is that polluting firms do not always prefer to become a free rider. Instead, they still exhibit the desire to locate in the core and benefit from agglomeration economies. Empirical results demonstrate that an inverted U-shaped relationship between the level of pollution and urbanization, which is especially explicit between urbanization and gross industrial output of polluting firms. This suggests that as the jurisdiction becomes more urbanized and industrialized, people might give more attention to the environmental problem, forcing the regulator to either drive polluters out or reduce their production through environmental regulation. The regional effect in polluting firms shows that there are more air-polluting firms located in the central region. While the coastal region is escalating industrial structure, the central region is attracting more polluting firms transferred from the east.

All evidence points to the fact that economic growth tends to be prioritized in most regions. The importance of environmental protection appears poor in comparison. Despite a number of environmental regulations that have been established, the stringency of enforcement is far from enough. Under such a paradigm of development, polluting firms are barely restricted by environmental regulation. As a result, polluting firms tend to locate in places with a larger population and more industrial firms. This goes against the idea of urban environmental planning that polluting firms should avoid densely populated places to minimize the number of people at the risk of pollution exposure. However, the solution to pollution control does not lie in changing its location, but in reducing the amount of pollution emission. The multi-level governments are confronted with difficulties in the enforcement of environmental regulations to guarantee the reallocation of resources for green development rather than the relocation of industries.

Overall, findings in this chapter reveal that the border location of polluting firms is not only determined by the free-rider effect but also subject to the interactions between centrifugal and centripetal forces in the core location. After all, polluting firms can also benefit from agglomeration economies in the core, which is less likely to be achieved in the periphery. In other words, the border locations confront polluting firms with considerable opportunity costs, relating to the benefit of leaving agglomeration centers and becoming a free rider. Both of them are closely related to the spatial governance in environmental terms, which will be further discussed in next chapter.

References

An, S. (2004). *On economic development in the borders of administrative regions.* Beijing: China Economic Publishing Press.

Bai, C. E., Du, Y., Tao, Z., et al. (2004). Local protectionism and regional specialization: Evidence from China's industries. *Journal of International Economics, 63*(2), 397–417.

Barrett, S. (1994). Strategic environmental policy and intrenational trade. *Journal of public economics, 54*(3), 325–338.

Chen, K., Bao, S., Mai, Y., et al. (2014). Agglomeration and location choice of foreign financial institutions in China. *GeoJournal, 79*(2), 255–266.

Chen, Z., Kahn, M. E., Liu, Y., et al. (2018). The consequences of spatially differentiated water pollution regulation in China. *Journal of Environmental Economics and Management, 88,* 468–485.

Cheung, S. N. S. (2010). China in transition: Where is she heading now? *Contemporary Economic Policy, 4*(4), 1–11.

Cumberland, J. H. (1979). *Interregional pollution spillovers and consistency of environmental policy.* New York: NYU Press.

Davis, M. A., Fisher, J. D., & Whited, T. M. (2014). Macroeconomic implications of agglomeration. *Econometrica, 82*(2), 731–764.

Devereux, M. P., Griffith, R., & Simpson, H. (2007). Firm location decisions, regional grants and agglomeration externalities. *Journal of Public Economics, 91*(3), 413–435.

Du, J., Lu, Y., & Tao, Z. (2008). FDI location choice: agglomeration vs institutions. *International Journal of Finance & Economics, 13*(1), 92–107.

Duranton, G., & Puga, D. (2000). Diversity and specialisation in cities: Why, where and when does it matter? *Urban studies, 37*(3), 533–555.

Esty, D. C. (1996). Revitalizing environmental federalism. *Michigan Law Review, 95*(3), 570–653.

Esty, D. C., & Geradin, D. (1997). Market access, competitiveness, and harmonization: Environmental protection in regional trade agreements. *Social Science Electronic Publishing, 21*(2), 265–336.

Fan, J. (2006). Industrial agglomeration and difference of regional labor productivity: Chinese evidence with international comparison. *Economic Research Journal, 11,* 72–81.

Fazio, G., & Piacentino, D. (2010). A spatial multilevel analysis of Italian SMEs' productivity. *Spatial Economic Analysis, 5*(3), 299–316.

Gray, W. B., & Shadbegian, R. J. (2004). 'Optimal'pollution abatement-whose benefits matter, and how much? *Journal of Environmental Economics and Management, 47*(3), 510–534.

Han, Y., Jiao, H., & Li, J. (2011). Research process and prospect of study on provincial border-regions of China since the Reform and Opening-up. *Areal Research and Development, 30*(2), 1–5. (In Chinese).

Harrison, A. (1996). Openness and growth: A time-series, cross-country analysis for developing countries. *Journal of Development Economics, 48*(2), 419–447.

He, C., & Pan, F. (2010). Economic transition, dynamic externalities and city-industry growth in china. *Urban Studies, 47*(1), 121–144.

He, C., & Wang, J. (2010). Geographical agglomeration and co-agglomeration of foreign and domestic enterprises: a case study of Chinese manufacturing industries. *Post-Communist Economies, 22*(3), 323–343.

He, C., Pan, F., & Yan, Y. (2012a). Is economic transition harmful to China's urban environment? Evidence from industrial air pollution in Chinese cities. *Urban Studies, 49*(8), 1767–1790.

He, C., Wei, Y. D., & Xie, X. (2008). Globalization, institutional change, and industrial location: Economic transition and industrial concentration in China. *Regional Studies, 42*(7), 923–945.

He, C., Zhang, T., & Wang, R. (2012b). Air quality in urban China. *Eurasian Geography and Economics, 53*(6), 750–771.

Head, K., Ries, J., & Swenson, D. (1995). Agglomeration benefits and location choice: Evidence from japanese manufacturing investments in the United States. *Journal of International Economics, 38*(3), 223–247.

Helland, E., & Whitford, A. B. (2003). Pollution incidence and political jurisdiction: Evidence from the TRI. *Journal of Environmental Economics and Management, 46*(3), 403–424.

Jahiel, A. (1997). The contradictory impact of reform on environmental protection in China. *China Quarterly, 149,* 81–103.

Kahn, M., Li, P., & Zhao, D. (2015). Water pollution progress at borders: The role of changes in China's political promotion incentives. *American Economic Journal: Economic Policy, 7*(4), 223–242.

Kahn, R. F. (1931). The relation of home investment to unemployment. *The Economic Journal, 41*(162), 173–198.

Ke, S. (2010). Agglomeration productivity and spatial spillovers across Chinese cities. *The Annals of Regional Science, 45*(1), 157–179.

Keynes, J. M. (1930). *Treatise on money: Pure theory of money.* London: Macmillan.

Konisky, D. M., & Woods, N. D. (2010). Exporting air pollution? Regulatory enforcement and environmental free riding in the United States. *Political Research Quarterly, 63*(4), 771–782.

Li, Y., & Wu, F. (2014). Reconstructing urban scale: New experiments with the "provincial administration of counties" reform in China. *The China Review, 14*(1), 147–173.

Lin, J. Y. (2004). Is China's growth real and sustainable? *Asian Perspective, 28*(3), 5–29.

Low, P., Yeat, A. (1992) Do dirty industries migrate? International Trade and the Environment. Word Bank Discussion Papers: 159.

Lucas, R., Wheeler, D., & Hettige, H. (1993). Economic development, environmental regulation and the international migration of toxic industrial pollution. *International Trade and the Environment, 159,* 1960–1988.

Magat, W., Krupnick, A. J., & Harrington, W. (2013). *Rules in the making: A statistical analysis of regulatory agency behavior.* London: Routledge.

Marshall, A. (1890). *Principles of Economics.* London: Macmillan.

Martin, P., & Ottaviano, G. I. (1999). Growing locations: Industry location in a model of endogenous growth. *European Economic Review, 43*(2), 281–302.

Myrdal, G. (1957). *Economic theory and under-developed regions.* London: Gerald Duckworth & Co., Ltd.

Pan, Z., & Zhang, F. (2002). Urban productivity in china. *Urban Studies, 39*(12), 2267–2281.

Panayotou, T. (1997). Demystifying the environmental Kuznets curve: Turning a black box into a policy tool. *Environment and development economics, 2*(4), 465–484.

Shi, M. (2013). *Regional Economics.* Beijing: Science Press.

Swanson, K. E., Kuhn, R. G., & Xu, W. (2001). Environmental policy implementation in rural China: A case study of Yuhang. *Zhejiang. Environment Management, 27*(4), 481–491.

Taguchi, H., & Murofushi, H. (2010). Evidence on the interjurisdictional competition for polluted industries within China. *Environment and Development Economics, 15*(3), 363–378.

Tang, S., Lo, C. W. H., & Fryxell, G. E. (2003). Enforcement styles, organizational commitment and enforcement effectiveness: An empirical study of local environmental protection officials in urban China. *Environment and Planning A, 35*(1), 75–94.

Van Oort, F. G., Burger, M. J., Knoben, J., et al. (2012). Multilevel approaches and the firm-agglomeration ambiguity in economic growth studies. *Journal of Economic Surveys, 26*(3), 468–491.

Van Rooij, B., & Lo, C. W. (2010). Fragile convergence: Understanding variation in the enforcement of China's industrial pollution law. *Law and Policy, 32*(1), 14–37.

Wu, F. (1999). Intrametropolitan FDI firm location in Guangzhou. *China. Annals of Regional Science, 33*(3), 535–555.

Wu, J., & Radbone, I. (2005). Global integration and the intraurban determinants of foreign direct investment in Shanghai. *Cities, 22*(4), 275–286.

Zhang, B., Chen, X., & Guo, H. (2018). Does central supervision enhance local environmental enforcement? Quasi-experimental evidence from China. *Journal of Public Economics, 164,* 70–90.

Zhen, F., Shen, Q., Jian, B., et al. (2010). Regional governance, local fragmentation and administrative division adjustment: Spatial integration in Changzhou. *The China Review, 10*(1), 95–128.

Chapter 5
Do Environmental Regulations Affect Air Quality and SO$_2$ Emissions?

Along with the great economic achievements through rapid industrialization and urbanization in China, serious challenges of resources, energy and environment have emerged from the extensive development model. Among these challenges, environmental pollution, especially air pollution, has received increasing attention from both the public and academia. Theoretically, the Environmental Kuznets Curve indicates that environment pollution would arise with industrialization and urbanization at the initial development stages. Practically, interregional competition in economic development has further triggered a race to the bottom competition to relax environmental standards. And some localities even compete for profitable and high-value-added polluting industries such as chemical, papermaking, and smelting industries at the expense of environmental quality (Wheeler 2001; Zhou et al. 2015). Recently, to control environmental pollution, a series of environmental regulations have been introduced to constrain firms from emitting pollution. Firms respond to environmental regulations differently. Some firms develop or adopt environmental-friendly production technologies to achieve the dual goals of economic efficiency and environmental protection (Chen et al. 2013; Milliman and Prince 1989). Others are forced to relocate to avoid the constraints of environmental regulations (Christmann and Taylor 2001).

In China, the central government comes up with environmental regulations. It is the responsibility of local governments to implement environmental regulations. Local governments, however, differ in their ability, incentive, pressure, and resistance to enforcing environmental regulations. As a result, it is assumed that the spatial variation of environment pollution in China is related to the implementation of environmental regulations, together with economic development, industrial structure, and technological progress. Moreover, different types of environmental regulations may exert different effects on environmental pollution.

Sulphur dioxide (SO$_2$) is one of the major air pollutants in China. There are two types of environmental regulations for SO$_2$ reduction: command and control regulations and economic incentive regulations (Conrad and Morrison 1985). The state council approved a typical command and control type environment regulation

© Springer Nature Singapore Pte Ltd. 2020
C. He and X. Mao, *Environmental Economic Geography in China*,
Economic Geography, https://doi.org/10.1007/978-981-15-8991-1_5

"Acid rain control zone and sulfur dioxide control zone division scheme" ("two control zones") in 1998. The "two control zones" include 175 cities, accounting for 11.4% of China's territory. The SO_2 emissions in the "two control zone" account for nearly 60% of the total emissions. Besides the policies of "two control zones", a series of other policies were further introduced and strengthened to complete the SO_2 emission reduction task (Shi et al. 2017). The economic incentives include subsidies for reducing emissions, taxes, and charges of pollution discharge, emissions trading and so on (Tang and Liang 2012). In 2003, *"The regulations on the collection and use of the sewage charges"* was implemented to stipulate the standards of the SO_2 emission charges. Since then, the standards of SO_2 emission charges have effectively contributed to the reduction of SO_2 emissions (Zhao et al. 2009). In addition, a series of specific policies and standards, such as *"The notice on strengthening the prevention and control of SO_2 pollution in coal-fired power plants"* and *"Standards of air pollutants emission in thermal power plants"*, have been constantly updated and implemented to control SO_2 emissions.

This chapter will first identify the influencing factors of air pollution in China, taking local government competition, pollution industry dynamics, environmental regulations, and technological progress into consideration. After confirming the significant role of environmental regulation in reducing environmental pollution, this chapter further examines the impacts of different types of environmental regulations on SO_2 emissions.

5.1 Environmental Regulations and Air Quality in China

5.1.1 Literature Review and Hypotheses Development

5.1.1.1 Local Government Competition, Polluting Industries, and Environmental Pollution

Local government competition is a behaviour of different regional governments within the country to compete for capital and business, using taxation waiver, environmental policies, and other favourable policies (Wittman 1998). In China, fiscal revenues, economic growth, and political promotion are the key incentives for local government competition. The tax sharing reform in 1994 provided localities with self-governing financial power, which formed a development model of "pro-growth government". In the context of free flows of resources and factors of production, fiscal decentralization can stimulate fierce competition among local governments to increase fiscal revenues (Oi 1992). Studies also argue that the rise of China's economy in the past 40 years is not only related to resource endowment, capital accumulation, and technological innovation but also due to the tournament competition" (Zhou and Tao 2011; Qiao et al. 2014). As political promotion is a zero-sum game, there is limited space for cooperation but considerable room for competition among officials at the same levels (Zhou 2004). Whether for the purpose of fiscal revenue or political promotion, local governments intend to develop industries with high economic returns and large scales. As polluting industries are often resources and capital intensive with significant scale effects, they could contribute to the local economic growth and revenues (Xia and Yan 2009; Wang 1994). The introduction of pollution-intensive industries would contribute to environmental degradation and air pollution.

Environmental policy has often been used as an effective tool in the local government competition. Fredriksson et al. (2003) apply a panel data from the U.S. to test the "Pollution Haven Hypothesis" (PHH), and shows that the state government would lower the environmental standards to attract foreign investment and protect local competitive advantages. Fredriksson et al. (2003), using data at the county level in New York, also report a "Race to bottom" competition among local governments. Empirical studies in China indicate that local government competition has a negative environmental impact (He et al. 2011). Local government competition could promote the development of polluting industries and increase the discharge of industrial wastewater, waste gas, and waste (Liu and Li 2013). Local governments would also attract pollution-intensive enterprises by increasing the supply of industrial land and sacrificing environmental quality (Liu and Zhang 2012). Overall, empirical studies show that local government competition would promote the development of polluting industries and cause damage to the local environment.

Local government competition means that neighbours affect local decision-makings. When neighbours relax their environmental standards to attract polluting industries, local government will intend to adopt similar strategies. Therefore, local competition would attract polluting industries to enter the local and surrounding areas, resulting in more pollution at the regional level. Thus, we propose the research hypothesis 1 as follows,

The economic development model under local government competition will exacerbate air pollution in local and surrounding areas.

5.1.1.2   Environmental Regulation, Technical Progress, and Environmental Pollution

Environmental regulation is the force of tangible rules or intangible consciousness aimed at protecting the environment (Zhao et al. 2009). There are explicit regulations and implicit regulations. Among explicit regulations, there are command and control types and economic incentive type. Implicit regulations refer to agreements, commitments, or plans proposed by firms or other entities to participate in protecting the environment and individual invisible environmental protection thought, idea, consciousness, attitude, and cognition (Table 5.1).

Empirical studies have shown that environmental regulations can effectively prohibit environmental pollutions by changing the mode or location of firm production (Li et al. 2014; Zhu et al. 2014). With the increasingly strict environmental regulation and constantly upgraded industrial structure, environmental regulations have

Table 5.1 Types and characteristics of environmental regulations

Environmental regulation	Type	Main characteristics
Explicit regulation	Command and control type	Laws, regulations, policies, and systems that directly influence polluters to act environmentally friendly
	Economic incentive type	Market incentives for enterprises to adopt cheaper and better pollution control technologies
	Voluntary type	Agreements, commitments, or plans proposed by enterprise itself or other entities that the enterprise voluntarily participates in protecting environment
Implicit regulation		Individual invisible environmental protection thought, idea, consciousness, attitude, and cognition

become progressively effective in discouraging pollution emissions (Liu et al. 2012). Yang et al. (2008) argue that there is regional environmental regulation competition in China, and the level of environmental regulation is easy to be aligned with areas with loose regulations. Zhao (2014a, b) find the environmental regulations have different spatial spill-over effects that are stronger in the coastal and northeastern regions. Following the literature, we propose the second hypothesis as follows,

Environmental regulations can effectively reduce air pollution, and its impact has a spatial spill-over effect.

The "cost hypothesis" proposes that the increase in production costs caused by pollution control will result in a decrease in production efficiency (Jaffe et al. 1995; Christainsen and Haveman 1981). The "Porter Hypothesis" however believes that an appropriate level of environmental regulation can promote firms to optimize resource allocation and innovate to offset cost increases. The "innovation offset effect" can neutralize the "compliance cost" of firms and ultimately deliver both environmental and economic benefits (Porter 1991; Ambec et al. 2013). The two hypotheses differ in the impact of environmental regulations on the production efficiency of firms, but both emphasize the role of technological progress in pollution reduction. Empirically, Levinson (2009) find that R&D investments and the knowledge innovation accumulated from practices can reduce CO_2 emissions. Hence technological progress plays an important role in reducing pollution emissions. We therefore propose the third hypothesis as follows,

Technological progress is a key factor in reducing air pollution in Chinese cities, and its effects would be affected by other factors.

5.1.2 Air Pollution and Polluting Industries

Data are collected from 245 prefecture-level cities during 2003 and 2013. Data for air quality include the annual average monitoring concentration of urban SO_2, NO_2, and PM_{10}, derived from the Chinese Environmental Quality Report. Data for polluting industries are from the Chinese Industrial Enterprises Database. The Urban Statistical Yearbook of China provides data on pollution emission and socio-economic indicators.

A comprehensive pollution index is used to assess air pollution, which is weighted by different pollutants and can better fully reflect the pollution level in a region. Thus, it is appropriate for comparison among regions (Lv et al. 2012). The calculation formula is as follows,

$$CAI_{it} = \sum_{j=1}^{1} \left(C_{ijt} / S_j \right) \tag{5.1}$$

where CAI_{it} is the comprehensive pollution index of city i in period t. C_{ijt} is the measured concentration of the pollution j in city i in period t. S_j is the National Secondary Standard Value (GB3 095-2012) of the air quality of the pollutant j. The results of the comprehensive pollution index evaluation are shown in Table 5.2.

As shown in Table 5.2, the number of prefecture-level cities with a national air pollution index that is greater than 3 has decreased year by year since 2003, and the number of cities in the range of 1.5–3.0 has increased. This indicates that urban air pollution in China has been decreasing during the study period on average. Table 5.3 shows the top five two-digit industries with the highest energy consumption among all industries. They consume 92% of the total industrial energy consumption in China. We consider the five industries as the major air polluters.

Table 5.2 Proportion of cities by comprehensive pollution index (%)

Year	Comprehensive pollution index					Total
	0.0–1.5	1.5–3.0	3.0–4.5	4.5–6.0	>6.0	
2003	6.8	46.2	30.9	11.6	4.4	100
2004	4.8	44.6	37.3	9.6	3.6	100
2005	4.8	55.8	34.1	3.6	1.6	100
2006	6.4	55.4	34.5	2.8	0.8	100
2007	5.6	59.0	34.9	0.4	0.0	100
2008	5.6	67.5	26.9	0.0	0.0	100
2009	6.4	70.7	22.5	0.4	0.0	100
2010	5.6	73.9	20.5	0.0	0.0	100
2011	5.2	74.3	20.5	0.0	0.0	100
2012	9.6	75.5	14.9	0.0	0.0	100
2013	6.4	57.0	27.7	6.8	2.0	100

Table 5.3 Two-digits heavy-chemical industry

Industry code	Industrial divisions	Energy consumption as proportion of total industrial energy consumption(%)
25	Petroleum, coking and nuclear fuel processing industries	6.94
26	Chemical raw materials and chemical products manufacturing industries	14.16
31	Nonmetal mineral products industries	12.19
32	Ferrous metal smelting and rolling processing industries	23.81
33	Non-ferrous metal smelting and rolling processing industries	5.27

Note The percentage of energy consumption is the mean value of the research period, and the data are obtained from the National Bureau of Statistics of China

The scale of polluting industries is represented by the total industrial output of the above five two-digit industries. The annual average air pollution index and the scale of polluting industries at the provincial level in 2003–2008 and 2009–2013 are calculated and shown in Table 5.4. Output has been calculated at the 2003 constant price.

Table 5.4 shows that air pollution in North China is the most serious before 2008, including Shanxi, Beijing, Henan, Tianjin, Hebei, and so on. South China is less polluted except Chongqing and the Yangtze River Delta. Since 2008, the degree of air pollution in most provinces has decreased, while the air pollution in the Shandong Peninsula, Beijing-Tianjin-Hebei Region, and Yangtze River Delta remain serious. Besides, air pollution shows a trend of migration to the east. The scale of polluting industries in all provinces has increased. During 2003 and 2008, the polluting industries are concentrated in the Yangtze River Delta, the Pearl River Delta, and the Shandong Peninsula. From 2009 to 2013, the scale of the polluting industry in Beijing-Tianjin-Hebei regions has risen considerably and became the most important concentrated areas for polluting industries. Overall, the polluting industries have heavily concentrated in the coastal region.

5.1.3   Model Specification and Variables

To identify the factors influencing the spatial patterns of air quality, we construct a panel data regression model of prefecture-level cities during 2003 and 2013. The spatial effect of variables is analyzed using the Spatial Panel Durbin Model (Lesage 2008). The model is as follows:

$$CPI = \alpha_1 + X\beta_1 + \lambda_1 + \mu_1 + \varepsilon_1 \tag{5.2}$$

Table 5.4 The dynamics of air pollution and polluting industries

Region	Annual average air pollution index			Annual average output of heavy-chemical industries(Billion Yuan)		
	2003–2008	2009–2013	Change	2003–2008	2009–2013	Change
Eastern Region	2.92	2.64	−0.29	3319.77	4656.22	1336.44
Beijing	**4.52**	**3.5**	**−1.02**	115.86	97.79	**−18.07**
Tianjin	**3.86**	**3.52**	−0.35	156.38	261.62	**105.23**
Hebei	**3.63**	3.09	−0.54	418.11	680.66	**262.54**
Shanghai	3.61	2.84	−0.77	289.44	297.56	8.12
Jiangsu	3.15	2.91	−0.24	704.29	1030.90	**326.62**
Zhejiang	2.88	2.7	−0.18	326.28	426.48	100.20
Fujian	2.4	2.06	−0.34	116.06	183.79	67.73
Shandong	2.9	3.47	0.58	679.61	1019.93	**340.32**
Guangdong	2.1	1.79	−0.31	430.29	491.09	60.80
Guangxi	2.16	2.18	0.01	73.05	139.75	66.70
Hainan	0.94	0.96	0.01	10.41	26.65	16.24
CentralRegion	3.32	2.67	−0.65	1118.25	1786.15	667.90
Shanxi	**4.89**	2.56	**−2.32**	184.61	244.23	59.62
Neimenggu	**3.82**	3.08	−0.74	95.40	187.89	92.49
Anhui	2.42	2.32	−0.1	121.93	212.52	90.59
Jiangxi	2.42	2.22	−0.2	118.16	240.76	**122.60**
Henan	**3.97**	3.35	−0.62	296.65	438.97	**142.32**
Hubei	2.62	2.56	−0.06	154.38	252.13	97.75
Hunan	3.1	2.62	−0.48	147.13	209.65	62.52
Western Region	3.12	2.69	−0.42	612.53	876.36	263.83
Chongqing	**4.37**	3.01	**−1.36**	504.99	77.07	26.57
Sichuan	3.14	2.58	−0.57	152.70	231.35	78.65
Guizhou	2.63	2.44	−0.19	44.71	56.87	12.16
Yunan	2.22	1.97	−0.25	91.25	109.52	18.27
Shaanxi	3.16	2.69	−0.46	72.21	128.20	55.98
Gansu	2.63	2.41	−0.22	94.47	110.27	15.80
Qinghai	3.03	3.27	0.24	17.55	28.29	10.74
Ningxia	3.42	2.59	−0.82	23.43	35.16	11.73
Xinjiang	3.45	3.28	−0.16	65.13	99.07	33.93
Northeast Region	2.63	2.32	−0.31	529.60	836.78	307.18

(continued)

Table 5.4 (continued)

Region	Annual average air pollution index			Annual average output of heavy-chemical industries(Billion Yuan)		
	2003–2008	2009–2013	Change	2003–2008	2009–2013	Change
Liaoning	3.07	2.64	−0.43	378.51	598.42	**219.91**
Jilin	2.54	2.32	−0.22	75.21	133.28	58.08
Heilongjiang	2.28	2.01	−0.27	75.88	105.07	29.19

Note Industrial output has been calculated at the 2003 constant price

$$CPI = \alpha_2 + X\beta_2 + X'\beta_2' + \lambda_2 + \mu_2 + \varepsilon_2 \qquad (5.3)$$

$$CPI = \alpha_3 + X\beta_3 + WX + \lambda_3 + \mu_3 + \varepsilon_3 \qquad (5.4)$$

where *CPI* is the air pollution index. α_1, α_2, and α_3 are constant terms. *X* is the explanatory variable matrix (*HCI, GHCI, ER, TEC, GGDP, LOAN, PARK, PGDP*). *X'X* represents the interaction term. β_1, β_2, β'_2, β_3 are coefficient matrices. λ_1, λ_2, λ_3 are individual fixed effects. μ_1, μ_2, μ_3 are time fixed effects. ε_1, ε_2, ε_3 are error terms.

W is a spatial weight matrix, which is characterized by the nearest four spatial units. If region *i* is one of the four spatial units closest to the region *m*, then W_{im} is 1, otherwise, W_{im} is 0. The spatial weight matrix was normalized before being introduced into the model. Model 1 (Formula 2.2) is the base model, Model 2 (Formula 2.3) analyzes the effects of the cross-terms and Model 3 (Formula 2.4) adds the spatial effects of the variables. The variables are described in Table 5.5.

Local government competition is represented by GDP growth rate of the prefecture-level cities. Practically, local governments compete for growth while local officials compete for promotion through economic growth. Local government competition is also reflected by favourable policies that set up for businesses, including the proportion of the firms with banking loans and the proportion of firms in industrial parks. We apply the proportion of the output value of the new products in the regional polluting industries to represent technological progress. Higher proportion of output value of new products indicates better technological progress.

There are two methods to measure environmental regulations. The first one measures the number of environmental regulations, the number of sewage charges, the proportion of pollution control investments to the total costs of enterprises and so on (Berman and Bui 2001; Lanoie et al. 2008). The second method uses different pollutant emission densities to measure the intensity of environmental regulations (Cole and Elliott 2003). The results of pollution control to measure environment regulation is more comprehensive (Salamon and Lund 1989). Following Zhao (2014a), we apply the pollution density method to measure environmental regulations.

We define the intensity of pollution emission in city *i* as follows:

Table 5.5 Definitions and Description of Variables

Variables	Mark	Variable description	Sample size	Mean	Standard deviation	Minimum value	Maximum value
Air pollution index	CPI	The sum of pollution index of the three pollutants	2 646	2.72	0.97	0.59	9.03
Scale of polluting industry (million yuan)	HCI	The output sum of five two-digits manufacturing industries	2 646	71	129	0	2357
Growth rate of polluting industry (%)	GHCI	Annual growth rate of output value of polluting industries	2 646	0.46	1.47	-0.93	24.30
Environmental regulation	ER	The reciprocal of weighted emission intensity of the three pollutants	2 646	4.05	4.50	0.04	54.04
Technical progress (%)	TEC	The proportion of the output value of new products in the regional polluting industries	2 646	0.06	0.09	0.00	0.88
The growth of GDP (%)	GGDP	The growth rate of prefectural GDP	2 646	17.99	26.95	−19.40	394.87
Enterprise loans (%)	LOAN	The proportion of enterprises that loan in prefectures	2 646	0.67	0.14	0.10	1.00
Park (%)	PARK	The proportion of local enterprises that entry into industrial park in the prefecture	2 646	0.15	0.14	0.00	0.83
Economic development level (yuan)	PGDP	Per capita GDP of prefecture-level city	2 646	27 589	22 480	2 379	156 356

$$P_{ijt} = \left(T_{ijt}/Y_{it}\right) \Big/ \frac{1}{245} \sum\nolimits_{i=1}^{245} \frac{T_{ijt}}{Y_{it}} \tag{5.5}$$

where P_{ijt} is the pollution emission intensity of the pollutant j of city i in year t. T_{ijt} is the total discharge of pollutant j of city i in year t. Y_{it} is the total industrial output value of city i in year t. The higher the value of P_{ijt}, the higher the emission intensity of the pollutant j in year t of the city and the weaker the environmental regulation level.

We then calculate the intensity of environmental regulation as follows:

$$ER_{it} = 3 \Big/ \sum\nolimits_{j=1}^{3} P_{ijt} \tag{5.6}$$

The average value of the three different pollutant emission P_{ijt} is used to indicate the regional emission intensity. ER_{it} is the intensity of environmental regulation in the year t of city i. The greater the ER_{it} is, the better the environment pollution effect of the local government would be, which means the environmental regulation is stricter, and vice versa.

5.1.4   Empirical Results

The Hausman test shows that there are both the time fixed effects and individual fixed effects in the model, thus the Two-way Fixed Effect Model should be used to estimate the parameters. The results are shown in Table 5.6.

From model 1, the coefficients of *HCI* and *GHCI* are significantly positive, indicating that the stocks and growth of polluting industries exacerbate air pollution. The coefficients of *ER* and *TEC* are significantly negative, meaning that environmental regulations and technological advancement have an inhibitory effect on air pollution, which are consistent with our research hypothesis. Among the variables that represent local government competition (*GGDP, LOAN, PARK*), the coefficient of *PARK* is significantly negative, which means that the strategies of local government competition, such as investment policy and industry park policy, have increased air pollution. Such a result is consistent with Xiong et al. (2016), which report that local government competition under the context of decentralization in China has impeded pollution control. The coefficient of *PGDP* is significantly positive, which is similar to the results of He et al. (2002), which find that the air pollution is more severe in the areas where the economy is more developed during the study period.

From Model 2, the cross-term of *TEC* and *HCI* is not significant, which indicates that the technological advancement and the scale of the polluting industry would work against each other, confirming the Hypothesis 3. The cross-terms of *ER* and *HCI*, *ER* and *PARK*, and *TEC* and *PARK* are all significantly negative, while the coefficients of *HCI* and *PARK* in Model 1 are significantly positive. It means that

Table 5.6 Results of regression models

Variables	Model 1: base model	Model 2: cross-term		Model 3: spatial Durbin model		
				Spatial	Spatial	Spill-over
HCI	0.0702**	0.0939**	0.0662**		−0.025 3	−0.347***
ER	−0.0164**	−0.0965***	−0.0106		−0.0725*	0.167**
GGDP	0.0499	0.0568	0.0497		0.000330	0.000210
LOAN	0.329	0.343	0.328		0.515***	1.018**
PARK	0.487***	0.504***	0.861***		0.884***	0.737*
TEC	−0.441*	−0.472**	−0.442*		−0.0301*	−0.0121
PGDP	7.61e−10***	7.94e−10***	7.69e−10***		−0.00667	−0.0120
GHCI	0.00885*	0.0130**	0.0120**		0.00766	−0.000407
ER*HCI		−0.00141**				
TEC*HCI		−0.0995				
ER*PARK			−0.0508**			
TEC*PARK			−0.0374**			
rho				0.0784*		
Constant	2.866***	2.852***	2.710***			
Observations	2646	2646	2646	2646	2646	2646
R-squared	0.218	0.221	0.225	0.201	0.201	0.201
Number of cities	245	245	245	245	245	245

Note $^{*}p < 0.10$, $^{**}p < 0.05$, $^{***}p < 0.01$

the effects of the variables to suppress air pollution (*ER*, *TEC*) are stronger than the effects of variables to increase air pollution (*PARK*, *HCI*).

In the Spatial Panel Durbin Model, *HCI*, *ER*, *PARK*, and *LOAN* have significant spatial spill-over effects, while other explanatory variables are not significant. The *HCI* of surrounding areas has a negative impact on local air, indicating that the concentration of polluting industries in the surrounding area would discourage the development of local polluting industries, contributing to the reduction of local air pollution. The direct effect and the spill-over effect of local government competition variables (*LOAN* and *PARK*) are both significantly positive, indicating that when local governments provide economic incentives for the development of polluting industries, surrounding areas will follow and thereby exacerbate regional air pollution, which supports Hypothesis 1. The direct effect of *ER* is negative while the spill-over effect of *ER* is positive, confirming Hypothesis 2, indicating that the stronger environmental regulation in the surrounding areas would encourage polluting enterprises to concentrate in the local area, increasing of local air pollution.

5.2   Environmental Regulations and Industrial SO_2 Emission Reduction

5.2.1   Environmental Regulations and SO_2 Emission

Different types of environmental regulations work differently to control pollution (Kellenberg 2009; Lanoie et al. 2008; Zeng 2010). Cofala et al. (2004) analyse the SO_2 emission reduction in Asian countries and find that the initial SO_2 abatement cost is only 500–1000$/t while the later cost rise to 3000$/t. When firms' enthusiasm for SO_2 emission reduction is reduced due to the increasing cost, the effect of command and control environmental regulations would be restrained. Kanada et al. (2013) also report similar findings from a study on key cities of China. In developed cities, such as Beijing and Shanghai, the increased abatement costs would increase the difficulty of achieving the targets of total emission reduction. Compared with the command and control regulations, economic incentive regulations provide greater freedom for firms, who can trade the excess emission reduction target for profits, stimulating technological innovation, and adopting new technologies to promote sustainable emissions reduction (Yao et al. 2017; Zhang et al. 2016). It can be seen that in the initial stage of emission reduction, relatively simple means such as command and control regulations could remarkably reduce emissions. With the increase of the emission reduction rate, the rise of the marginal cost would increase the difficulty of emission reduction.

Environmental regulation achieves the goal of SO_2 emission reduction through scale effect, structural effect and technological effect (Chen et al. 2016; Grossman and Krueger 1992). The "Pollution Haven Hypothesis (PHH)" focuses on the relationship between environmental regulation and industrial scale, while "Porter Hypothesis (PH)" discusses the relationship between environmental regulation and technological level. PHH believes that firms tend to shift their dirty production to areas with low environmental standards to reduce the costs (Tang 2014). Correspondingly, the actions of relaxing environmental regulations to attract polluting enterprises are called "Race to the Bottom" competition (Cole 2004). PH considers that felicitous environmental regulations can force firms to conduct technological innovation so that innovation compensation could make up for environmental costs and even gain additional revenues (Porter 1991; Porter and Van 1995).

Another type of studies suggests that the development speed of polluting and non-polluting industries differs under environmental regulation, and environmental regulation can reduce emissions by optimizing industrial structure (Yuan and Xie 2014; Li 2013; He 2015). Huang et al. (2015) report that environmental regulations in eastern region of China can significantly improve the technical level of polluting industries while regulations in the central and western regions of China could lead to "Race to the Bottom" competition and increase air pollution. Zhao (2014a) also find a similar phenomenon that environmental regulations have a positive growth effect in eastern and northeastern regions of China, but a negative effect in central and western regions. Existing studies empirically test the "Porter Hypothesis" by analyzing the

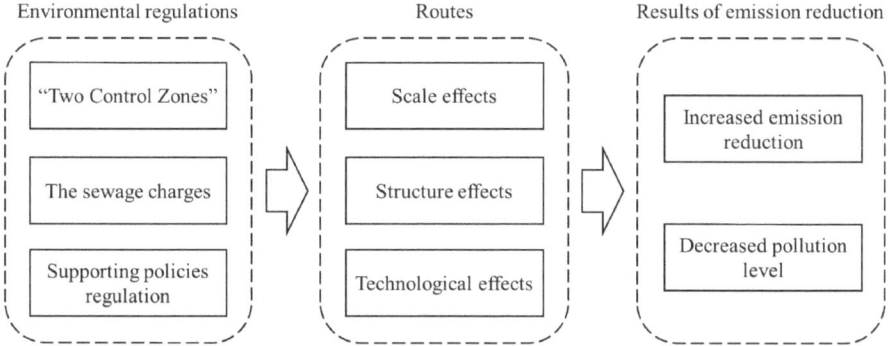

Fig. 5.1 Environmental regulations, emission reduction routines and reduction effects

role of environmental regulation on technological development, explore the "Pollution Haven Hypothesis" based on the scale effect, and analyze the relationship between environmental regulations and local industrial structure (He et al. 2013; Zhou et al. 2017). But they ignore that the three effects are the three parallel paths to contribute to pollution emission reduction.

We decompose the SO_2 emission reduction effects at the city level using the Laspeyres decomposition method and explore the relationship between different types of environmental regulations and emission reduction effects (Fig. 5.1).

5.2.2 Spatial-Temporal Pattern of SO₂ Emission Reduction

5.2.2.1 Data and Methods

The study samples include 261 prefecture-level cities in China in 2004–2008 and 273 prefecture-level cities in 2008–2013. SO_2 emission data are collected from the "*China City Statistical Yearbook*" over the years. The production value data of each industry comes from "China Industry Business Performance Data". The other socio-economic data are from relevant statistical yearbooks, the website of the Chinese Ministry of Environmental Protection, and the website of the Chinese National Bureau. The socio-economic data are adjusted to a comparable price in 2011.

The index decomposition analysis (IDA) is originated from the efficiency research in the energy field in the 1970 s. Due to its simple and flexible characteristics, IDA is expansively applied to researches on pollution discharge, material flow and other aspects (Gonzalez et al. 2014; Zhang 2000; Fan et al. 2007). Compared to the input-output analysis, IDA can obtain more sufficient data and has advantageous aspects for cross-regional and sectoral comparisons (Ang and Zhang 2000). There are two main methods of IDA, Laspeyres and Divisa. The core difference is that the Divisia uses logarithmic data and increases or decreases the same amount in different years to facilitate the discussion of the percentage change, while the Laspeyres uses raw data and increases or decreases to the normal values for easy visual understanding (Ang 2004). Grossman et al. (1992) first introduce IDA into the environmental study field. Since then, pollution decomposition has been widely used in environmental field (Shrestha and Timilsina 1997).

Due to data limitation, the existing industrial SO$_2$ decomposition in China has not reached to the prefecture-level scale, making it difficult to conduct spatial analysis. In combination with the "China Industry Business Performance Data", this chapter would use Laspeyres Decomposition Method to decompose industrial SO$_2$ emission reductions in prefecture-level cities from 2004 to 2013 and identify the emission reduction routines by simplifying industrial types. Considering the impact of the 2008 financial crisis, we divide the research period into two periods: 2004–2008 and 2008–2013. The formula of Laspeyres Decomposition is as follows:

$$S_i = Y_i \times \sum_{j=0}^{1} \left(\frac{Y_{j,i}}{Y_i} \times \frac{S_{j,i}}{Y_{j,i}} \right) \tag{5.7}$$

where S_i is the industrial SO$_2$ emission of city i and Y_i is the total output value of city i. $Y_{j,i}$ is the total output value of j industry in i city; $S_{j,t}$ is the industrial SO$_2$ emission of j industry in i city. $j = 0$ represents the non-polluting industry and $j = 1$ represents the polluting industry. Y_i is the scale effect, which suggests the impact of industrial scale change on emissions. $Y_{j,i}/Y_i$ is the structure effect, reflecting the impact of industrial structure changes on emissions. $S_{j,i}/Y_{j,i}$ is the technical effect, which indicates the impact of changes of production technology on emission. Due to data limitations, it is difficult to obtain the structural and technical effects of prefecture-level cities on SO$_2$ abatement effects. Therefore, we divide all two-digit industries into polluting and non-polluting industries based on the SO$_2$ emission intensity of each industry's output value (Lucas et al. 1992; Qiu et al. 2013). We also assume that the distribution of technology between industries is similar across the country, which means that the ratio of the technical level of polluting industries is the same as that of non-polluting industries.

Combined with the industrial output value and industrial SO$_2$ emissions data, we could obtain the structure effect and technical effect. Furthermore, the formula (5.1) is transformed to get the predicted value of the total amount of pollution in the prefecture-level cities when only scale changes in order to investigate the three effects. Then the scale effect of SO$_2$ emission reduction and the proportion of SO$_2$ emission reduction can be calculated. The formula is as follows:

$$R_{s,i} = \frac{P_{s,i} - S_i}{S_i} \tag{5.8}$$

where $R_{s,i}$ is the proportion of SO$_2$ emission reduction of scale effect in city i. $P_{s,i}$ is the predicted value of industrial SO$_2$ emissions when the only scale is changed in city i. S_i is the actual industrial SO$_2$ emission of city i. In the same way, the emission

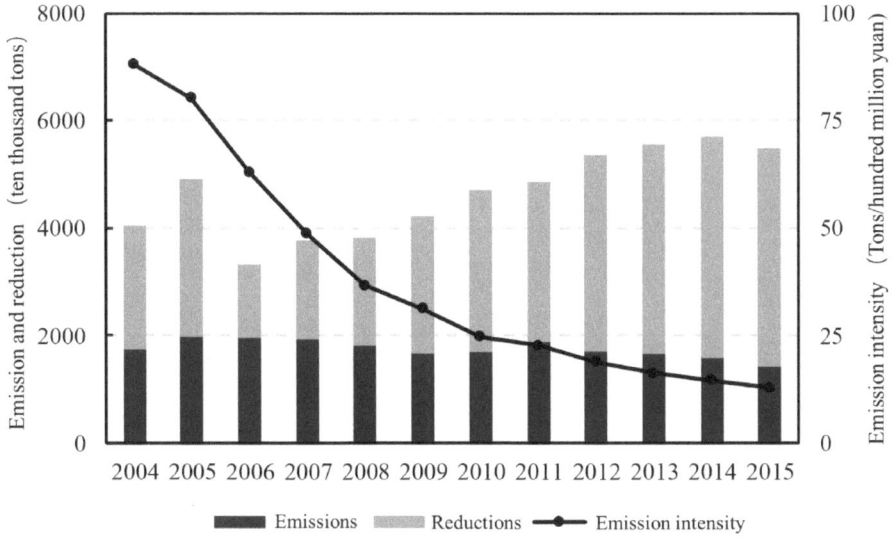

Fig. 5.2 Total emission, reduction, and emission intensity of SO$_2$ in China from 2004 to 2015

reduction and emission reduction ratio of structure effects and technical effects can be calculated.

5.2.2.2 China's Industrial SO$_2$ Emissions

Industrial SO$_2$ emission is the dominant SO$_2$ emission, accounting for more than 80% of total SO$_2$ emissions in China (Li et al. 2011; Qiu et al. 2011). Figure 5.2 shows the total emission, reduction and emission intensity of SO$_2$ in China from 2004 to 2015. The total emissions of SO$_2$ formed "M"-shaped fluctuations. There are two emission peaks, with 19.8 million tons and 18.8 million tons in 2005 and 2011. While the low emissions appear in 2009 and 2015 with 16.7 million tons and 16.5 million tons. Compared with the top in 2005, emissions decrease by 28.7% in 2015. SO$_2$ emission reduction shows an increasing trend during the study period. The drop in emission reduction from 2005 to 2006 may be due to the changes in statistics. The SO$_2$ emission intensity shows a monotonous type decline, from 7.10 t/billion yuan in 2004 to 1.28 t/billion yuan in 2015, with a drop of 82.0%.

5.2.2.3 Spatial Pattern of SO$_2$ Emission Reduction Effect

The Laspeyres decomposition method is used to identify the proportion of SO$_2$ emission reduction effects. The temporal and spatial patterns of the proportions of the emission reduction effects are shown in Figs. 5.3, 5.4 and 5.5.

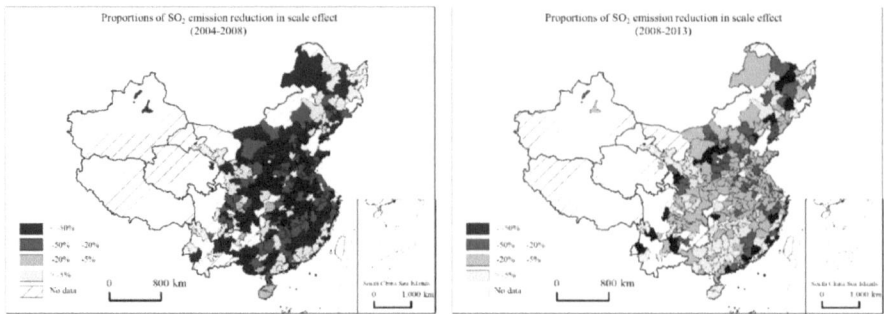

Fig. 5.3 Temporal and spatial variation of the proportion of SO$_2$ emission reduction through scale effects

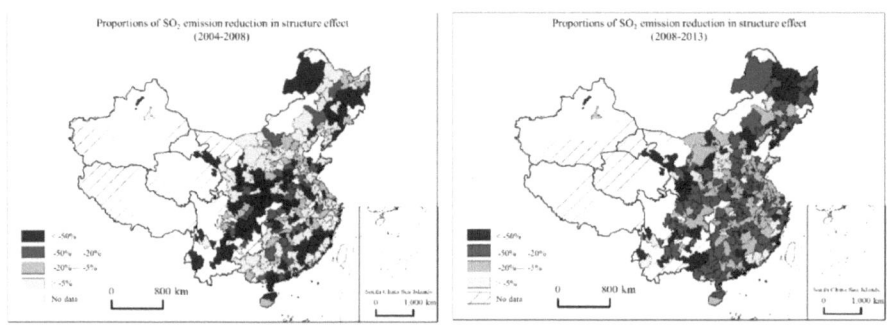

Fig. 5.4 Temporal and spatial variation of the proportion of SO$_2$ emission reduction through structure effects

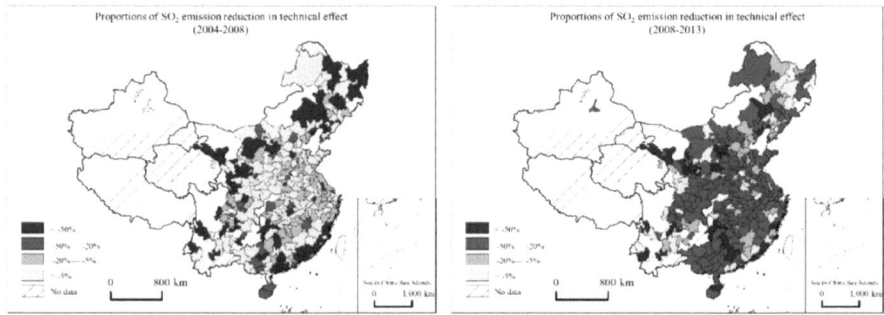

Fig. 5.5 Temporal and spatial variation of the proportion of SO$_2$ emission reduction through technical effects

Scale effects are the biggest obstacles for industrial SO$_2$ emission reduction. The proportion of SO$_2$ emission reduction caused by the scale effect in most prefecture-level cities is reduced by more than 50% during 2004 and 2008. And during 2008 and 2013, on average, the proportion of reduction is decreased by 20%. The impact of structure effects is limited. The proportion of emission reduction in structure effects fluctuates within ±10% in most cities. The technical effect is an important path for SO$_2$ emission reduction. In the first period, the emission reduction of technical effects accounted for more than 140% of the total emission reduction in most regions. Although the impact of technical effects has weakened in the second period, the emission reduction still reached 100%–120% of the total reduction in major regions of the country.

5.2.3 Empirical Analysis of Factors Influencing Industrial SO$_2$ Emission

5.2.3.1 Model Specification and Variables

The objective is to explore the effects of different types of environmental regulations on SO$_2$ emission reduction. The dependent variable is the industrial SO$_2$ emission reduction at the prefecture-level cities. The independent variables include command and control type environmental regulation, economic incentive type environmental regulation and the scale effects, structure effects, and technical effects of emission reduction.

$$Y = f_1(S, C, T, Z, M) + f_2((Z, M) \times (S, T)) + f_3 + \varepsilon \qquad (5.9)$$

where $f_1(S,C,T,Z,M)$ is the function of the independent variables, including the scale effects (S), structure effects (C) and technology effects (T) of emission reduction, and the policies of sewage charges (M) and "Two Control Zones" (Z). $f_2((Z, M) \times (S,T))$ is the cross-term of environmental regulations and scale and technical effects. f_3 is the control variable and ε is an idiosyncratic error term. Table 5.7 reports the definitions of the variables.

The dependent variable is the difference in annual actual SO$_2$ emissions during the study period. To deal with potential endogeneity problems of Y and the number of sewage charges, the unit discharges are used to represent sewage charges (M) and is assigned the value 1–5 from low to high according to the quartile. Following He et al. (2011) and Zhou et al. (2015), our control variables include the removal rate of SO$_2$ (RESO), local economic development level (GDPPC), the effects of globalization (EXP), marketization (SOE) and decentralization (PARK).

Table 5.7 Variables and definitions

Variables		Definition
Dependent variable	SO_2 emission reduction (Y)	Difference in SO_2 emission between year t_0 and year t_1
Environmental regulations	Sewage charges level (M)	Sewage charges level (1~5)
	"Two Control Zones" (Z)	If it belongs to the "Two Control Zones", the value is 1; otherwise it is 0
Effects of emission reduction	Scale effect (S)	Emission reduction of scale effect from year t_0 to year t_1
	Structure effect (C)	Emission reduction of structure effect from year t_0 to year t_1
	Technical effect (T)	Emission reduction of technical effect from year t_0 to year t_1
Control variable	Removal rate of SO_2 (RESO)	Implementation results of supporting policies and regulations
	Economic level (GDPPC)	Per capita GDP of prefecture-level cities
	Export proportion (EXP)	Exports proportion of output value in prefecture-level city
	Marketization factors (SOE)	Proportion of output value of state-owned enterprises in prefecture-level city
	Proportion of park (PARK)	Proportion of output value of enterprises located in the industrial park in prefecture-level city

5.2.3.2 Empirical Results

Based on the Hausmann test results, the fixed-effect model is used for estimation and the standard error is corrected by the White Heteroskedasticity Test. The regression results for SO_2 emission reduction are presented in Table 5.8, where model I is the base model and models II, III, IV, and V include cross-terms.

In the period of 2004–2008, S (scale effects) and T (technical effects) are the main factors of influencing SO_2 emissions, thus the compensation of technical effects on scale effects is an efficient way to reduce SO_2 emissions. M (sewage charge policy) and Z ("Two Control Zones" policy) can both significantly reduce pollution emissions, of which M is more significant, indicating that the sewage charge policy has a closer relationship with SO_2 emissions. The cross-term of M and S is significantly positive, which suggests that economic incentive regulations can limit the emission reduction caused by scale effects. However, the cross-term of Z and S is not significant, showing that the "Two Control Zones" policy has a limited influence on scale effects. The cross-term of M and T is also significantly positive. It is believed that

Table 5.8 Regression results

Variables	Industrial SO$_2$ emission reduction from 2004 to 2008					Industrial SO$_2$ emission reduction from 2004 to 2013				
	Model I	Model II	Model III	Model IV	Model V	Model I	Model II	Model III	Model IV	Model V
S	−0.117**	−0.081**	−0.154**	−0.088***	−0.114**	−0.127*	−0.06	−0.156*	−0.133**	−0.130**
T	0.501***	0.380***	0.500***	0.386***	0.600***	0.127**	0.126**	0.126**	0.097**	0.105**
C	−0.053	−0.048	−0.045	−0.048	−0.049	−0.044	−0.043	−0.047	−0.0564	−0.052
Z	0.040*	0.014	0.108	0.015	0.086	−0.159***	−0.158***	−0.233	−0.136**	−0.229***
M	0.181***	0.168***	0.181***	0.182***	0.180***	0.277***	0.202***	0.275***	0.296***	0.255***
M×S	–	0.396***	–	–	–	–	0.206	–	–	–
Z×S	–	–	0.152	–	–	–	–	0.077	–	–
M×T	–	–	–	–	–	–	–	–	1.036***	–
Z×T	–	–	–	0.372***	0.143*	–	–	–	–	0.149
RESO	0.112**	0.137***	0.118***	0.139***	0.119***	0.108*	0.113**	0.105*	0.126**	0.116**
GDPPC	0.088	0.083	0.091	0.089	0.09	−0.125*	−0.125*	−0.122*	−0.0907	−0.121*
EXP	−0.188***	−0.103*	−0.181***	−0.114**	−0.195***	0.179***	0.180***	0.183***	0.124**	0.153**
SOE	−0.123**	−0.111**	−0.129**	−0.112**	−0.124**	−0.165**	−0.159**	−0.160**	−0.0845	−0.159**
PARK	0.008	0.056	0.005	0.055	0.009	0.053	0.069	0.053	0.047	0.049
Constant	−0.544***	−0.508***	−0.546***	−0.548***	−0.539***	−0.850***	−0.632***	−0.843***	−0.851***	−0.785***
R^2	0.512	0.666	0.514	0.654	0.519	0.537	0.553	0.537	0.545	0.551

Note * $p < 0.10$, ** $p < 0.05$, *** $p < 0.01$

there is a synergy between economic incentive regulations and technical effects, supporting the "Porter Hypothesis". The cross-term of Z and T is also significantly positive, which indicates that command and control type regulations also play a role in promoting technological innovation. The cross-term of M and T is significantly higher than that of Z and T, showing that command and control type regulations can promote firms' compliance but could hardly encourage technological innovation. However, under economic incentive regulations, firms can benefit from trading excess emission reductions, which could encourage technological innovation and introducing new technologies (Jorgenson and Wilcoxen 1990).

In the second period of 2008–2013, S and T are still the main factors of influencing SO₂ emissions, but the significance of S has declined, which further supports the reduction of scale effects. C (Structure effects) is not significant, indicating that the changes in industrial structure have not significantly reduced pollution reduction. Z is significantly negative, showing that SO₂ emissions are gradually dispersed so that there is no significant difference between the key control area and non-key control areas. Even the emission reductions are fewer in the key control areas than those in the non-key control areas. M is still significant, highlighting the robust role of economic incentive regulations. The cross-terms of M-S and Z-S are both not significant, showing that at this period, environmental regulations could no longer reduce pollution emission by controlling the industrial scale. The cross-term of M and T is positive, confirming again that economic incentive regulations can improve the technical level of industries. The cross-term of Z and T is no longer significant, indicating that command and control type regulations have lower impacts on technical effects. As the policy of "Two Control Zones" has a negative influence at the second period and all provinces have cities outside the "Two Control Zones", the command and control type regulations could cause the transference of pollution industries to non-control zones, which further lead to the increase of pollution emissions in non-control areas (Cai et al. 2016). Consistent with this argument, the cross-term of Z and S is not significant, illustrating that "Two Control Zones" fails to reduce emission through scale effects.

5.3 Conclusion and Discussion

Air pollution in China is not only an environmental issue but also an economic and institutional matter. Interregional competition, environmental regulations, industrial development, and technological advancement work together to influence air pollution. In particular, environmental regulations impact air pollution through scale effect, structural effect and technical effect. Moreover, different types of environment regulations exert different effects at different development stages.

It is interesting to see that both polluting industries and air pollution show a trend of shifting to the eastern region during the study period. Both growth and stocks of polluting industries can effectively aggravate air quality. Panel data regression indicates that local government competition can significantly increase air pollution in local and surrounding areas. Environmental regulations are important ways to suppress air pollution. However, the strict regulations in local areas will increase the pollution in surrounding areas. Technological progress can effectively reduce air pollution, and it does not show significant spatial spill-over effects. It is worthwhile to point out that factors that inhibit air pollution, such as environmental regulations, and technological progress, are stronger than factors that could increase air pollution, like local government competition and the scale of polluting industries. This, to some extent, explains the phenomena that air pollution in China is gradually reduced during the study period.

We further explore the spatial pattern of industrial SO_2 emission reduction in China by highlighting the role of different types of environmental regulations. We find that scale and technical effects are the main factors to influence SO_2 emission during the study period, while the structural effect has limited impact. The compensation of technical effects to scale effect is the main way of SO_2 emission reduction. During 2004–2008, both command and control type and economic incentive type environmental regulations significantly contribute to SO_2 emission reduction, which inhibits the increase of emissions caused by scale effects and improves the effect of emission reduction through technological effects. While over time, the difficulty of SO_2 emission reduction has risen, and the effectiveness of command and control type environmental regulations has declined. The effects of emission reduction by economic incentive type environmental regulations are constant over the time, at the same time, the cross-term of sewage charges and technical effects are significantly positive. This indicates that the economic incentive type environmental regulations are the key method to promote technological progress and realize the "Porter Hypothesis".

To conclude, local government competition would aggravate environmental pollution while environmental regulations and technological progress are important ways to reduce air pollution. Based on the empirical findings, we provide the following suggestions. (1) It is urgent to systematically revise the evaluation of local officials to avoid the unhealthy competition of local governments to achieve environmental protection. Specifically, changing the "results evaluation" to "process evaluation" to balance the economic and ecological performance to avoid the excessive weight of GDP growth in the evaluation. (2) Technological progress is an important way for firms to reduce pollution emissions. Policies should be directed to punish polluting enterprises and reward enterprises that follow the regulations at the same time, guiding enterprises to develop or adopt environment-friendly technologies. (3) Environmental regulations should be enforced more effectively. More resources should be directed to local environmental bureaus to implement environmental regulations. (4) Economic incentive type environmental regulations can provide firms with full flexibility and autonomy to achieve better emission reduction.

References

Ambec, S., Cohen, M. A., Elgie, S., et al. (2013). The Porter hypothesis at 20: Can environmental regulation enhance innovation and competitiveness? *Review of Environmental Economics and Policy, 7*(1), 2–22.

Ang, B. W. (2004). Decomposition analysis for policymaking in energy. *Energy Policy, 32*(9), 1131–1139.

Ang, B. W., & Zhang, F. Q. (2000). A survey of index decomposition analysis in energy and environmental studies. *Energy, 25*(12), 1149–1176.

Berman, E., & Bui, L. T. (2001). Environmental regulation and productivity: Evidence from oil refineries. *Review of Economics and Statistics, 83*(3), 498–510.

Cai, X., Lu, Y., Wu, M., et al. (2016). Does environmental regulation drive away inbound foreign direct investment? Evidence from a quasi-natural experiment in China. *Journal of Development Economics, 123,* 73–85.

Chen, Y., Jin, G. Z., Kumar, N., et al. (2013). The promise of Beijing: Evaluating the impact of the 2008 Olympic Games on air quality. *Journal of environmental Economics and Management, 66*(3), 424–443.

Chen, Y., Ren, J., Chen, Y., et al. (2016). Spatial evolution and driving mechanism of china's environmental regulation efficiency. *Geographical Research, 39*(1), 123–136.

Christainsen, G. B., & Haveman, R. H. (1981). Public regulations and the slowdown in productivity growth. *The American Economic Review, 71*(2), 320–325.

Christmann, P., & Taylor, G. (2001). Globalization and the environment: Determinants of firm self-regulation in China. *Journal of International Business Studies, 32*(3), 439–458.

Cofala, J., Amann, M., Gyarfas, F., et al. (2004). Cost-effective control of SO_2 emissions in Asia. *Journal of Environmental Management, 72*(3), 149–161.

Cole, M. A. (2004). Trade, the pollution haven hypothesis and the environmental Kuznets curve: Examining the linkages. *Ecological Economics, 48*(1), 71–81.

Cole, M. A., & Elliott, R. J. (2003). Determining the trade-environment composition effect: The role of capital, labor and environmental regulations. *Journal of Environmental Economics and Management, 46*(3), 363–383.

Conrad, K., & Morrison, C. J. (1985) The impact of pollution abatement investment on productivity change: An empirical comparison of the U. S., Germany, and Canada. *Southern Economic Journal, 55*(3), 684–698.

Fredriksson, P. G., List, J. A., & Millimet, D. L. (2003). Bureaucratic corruption, environmental policy and inbound US FDI: Theory and evidence. *Journal of Public Economics, 87*(7), 1407–1430.

Fan, Y., Liu, L. C., Wu, G., et al. (2007). Changes in carbon intensity in China: Empirical findings from 1980–2003. *Ecological Economics, 62*(3), 683–691.

Gonzalez, P. F., Landajo, M., & Presno, M. J. (2014). Tracking European Union CO_2 emissions through LMDI (Logarithmic-Mean Divisia Index) decomposition, the activity revaluation approach. *Energy, 73*(7), 41–50.

Grossman, G. M., & Krueger, A. B. (1992). Environmental impacts of a North American free trade agreement. *Social Science Electronic Publishing, 8*(2), 223–250.

He, C. F., Pan, F. H., & Yan, Y. (2011). Is economic transition harmful to China's urban environment? Evidence from industrial air pollution in Chinese Cities. *Urban Studies, 49*(8), 1767–1790.

He, C. F., Huang, Z., & Ye, X. (2013). Spatial heterogeneity of economic development and industrial pollution in urban China. *Stochastic Environmental Research and Risk Assessment, 28*(4), 767–781.

He, H. (2015). Environmental quality, environmental regulation and industrial structure optimization: Based on the panel data empirical analysis of the Esatern, Central and Western area of China. *Areal Research and Development, 34*(1), 105–110.

He, K., Huo, H., & Zhang, Q. (2002). Urban air pollution in China: Current status, characteristics, and progress. *Annual Review of Energy and the Environment, 27*(1), 397–431.

Huang, Z., He, C., Yang, F., et al. (2015). Environmental regulation, geographic location and growth of firms' productivity in China. *Acta Geographica Sinica, 70*(10), 1581–1591.

Jaffe, A. B., Peterson, S. R., Portney, P. R., et al. (1995). Environmental regulation and the competitiveness of US manufacturing: What does the evidence tell us? *Journal of Economic Literature, 33*(1), 132–163.

Jorgenson, D. W., & Wilcoxen, P. J. (1990) Environmental regulation and U.S. economic growth. *Rand Journal of Economics, 21*(2), 314–340.

Kanada, M., Dong, L., Fujita, T., et al. (2013). Regional disparity and cost-effective SO_2 pollution control in China: A case study in 5 mega cities. *Energy Policy, 61,* 1322–1331.

Kellenberg, D. K. (2009). An empirical investigation of the pollution haven effect with strategic environment and trade policy. *Journal of International Economics, 78*(2), 242–255.

Lanoie, P., Patry, M., & Lajeunesse, R. (2008). Environmental regulation and productivity: Testing the Porter Hypothesis. *Journal of Productivity Analysis, 30*(2), 121–128.

Lesage, J. P. (2008). An introduction to spatial econometrics. *Revue D'économie Industrielle, 123,* 19–44.

Levinson, A. (2009). Technology, international trade, and pollution from US manufacturing. *The American Economic Review, 99*(5), 2177–2192.

Li, M., Zhang, J., Luo, H., et al. (2011). Sulphur dioxide reduction and potential in China. *Scientia Geographica Sinica, 31*(9), 1065–1071.

Li, Q. (2013). Environmental regulation and industrial structure adjustment: The theoretical analysis and empirical study based on Baumol Model. *Economic Review, 5,* 100–107.

Li, S., Chu, S., & Shen, C. (2014). Local governments competition, environmental regulation and regional eco-efficienc. *The Journal of World Economy, 37*(4), 88–110.

Liu, G., & Zhang, C. (2012). The spatial organization form evolution of heavy chemical industry in china. *On Economic Problems, 4,* 40–44.

Liu, J., & Li, W. (2013). Environmental pollution and intergovernmental tax competition in China: Based on spatial panel data model. *Chinese Population. Resources and Environment, 23*(4), 81–88.

Liu, Q., Wang, Q., & Li, P. (2012). Regional distribution changes of pollution-intensive industries in China. *Ecological Economy, 28*(1), 107–112.

Lucas, R. E. B., Wheeler, D. R., & Hettige, H. (1992) Economic development, environmental regulation, and the international migration of toxic industrial pollution: 1960–1988. *Policy Research Working Paper, 2007*(4), 13–18.

Lv, J., Zhang, Z., Liu, Y., et al. (2012). Sources identification and hazardous risk delineation of heavy metals contamination in Rizhao city. *Acta Geographica Sinica, 67*(7), 971–984.

Milliman, S. R., & Prince, R. (1989). Firm incentives to promote technological change in pollution control. *Journal of environmental Economics and Management, 17*(3), 247–265.

Oi, J. C. (1992). Fiscal reform and the economic foundations of local state corporatism in China. *World Politics, 45*(1), 99–126.

Porter, M. E. (1991). America green strategy. *Scientific American, 264*(4), 168–170.

Porter, M. E., & Van, D. L. (1995). Toward a new conception of the environment-competitiveness relationship. *The Journal of Economic Perspectives, 9*(4), 97–118.

Qiao, K., Zhou, L., & Liu, C. (2014). Interim rank, promotion incentives and current performance: Evidence from the dynamic tournament among Chinese local officials. *China Journal of Economics, 1*(3), 84–106.

Qiu, F., Shen, Z., Zhang, J., et al. (2011). Research on the quantitative relationship between urbanization level and urban resource stress: A case study on Baoji City. *Areal Research and Development, 30*(3), 67–72.

Qiu, F., Jiang, T., Zhang, C., et al. (2013). Spatial relocation and mechanism of pollution-intensive industries in Jiangsu Province. *Scientia Geographica Sinica, 33*(7), 789–796.

Salamon, L. M., & Lund, M. S. (1989). *Beyond privatization: The tools of government action.* Washington: The Urban Insitute.

Shi, B., Chen, F., Yan, Z., et al. (2017). Increasing marginal effect of environmental regulation dividend. *China Industrial Economics, 12,* 42–60.

Shrestha, R. M., & Timilsina, G. R. (1997). SO_2 Emission intensities of the power sector in Asia: effects of generation-mix and guel-intensity changes. *Energy Economics, 19*(3), 355–362.

Tang, J. (2014). Testing the pollution haven effect: Does the type of FDI matter? *Environmental & Resource Economics, 60*(4), 549–578.

Tang, Y., & Liang, R. (2012). TCZ policy and SO_2 abatement: An empirical study based on did. *Journal of Shanxi Finance and Economics University, 6,* 9–16.

Wang, J. (1994). *Modern industrial geography.* Beijing: Science and Technology of China Press.

Wheeler, D. (2001). Racing to the bottom? Foreign investment and air pollution in developing countries. *The Journal of Environment and Development, 10*(3), 225–245.

Wittman, D. (1998). Competitive governments: An economic theory of politics and public finance. *Southern Economic Journal, 64*(4), 1011–1013.

Xia, L., & Yan, X. (2009). A study on the evolution of industrial spatial structure of pearl river delta based on the heavy industry development. *Human geography, 24*(6), 68–72.

Xiong, B., Chen, W., & Liu, P., et al. (2016) Fiscal policy, local governments competition and air pollution control quality. *Journal of China University of Geosciences (Social Sciences Edition), 16*(1), 20–33.

Yao, L., Yang, H., & Wang, X. (2017). Analysis on the influence of different environmental regulation tools on enterprise' performance. *Collected Essays on Finance and Economics, 12,* 109–115.

Yang, H., Chen, S., & Zhou, Y. (2008). Local government competition and environmental policy: Empirical evidence from province's governments in China. *South China Journal of Economics, 6,* 15–30.

Yuan, Y., & Xie, R. (2014). Research on environmental effect of industrial restructuring about coastal ocean of Yancheng. *Jiangsu province, China Environmental Science, 8,* 57–69.

Zeng, X. (2010). Environmental regulation, foreign direct investment and "pollution haven" hypothesis. *Economic Theory and Business Management, 11,* 65–71.

Zhang, Z. (2000). Decoupling China's carbon emissions increase from economic growth: An economic analysis and policy implications. *World Development, 28*(4), 739–752.

Zhang, P., Zhang, P., & Gai, G. (2016). Comparative study on impacts of different types of environmental regulation on enterprise technological innovation. *China Population, Resources and Environment, 26*(4), 8–13.

Zhao, X. (2014a). Environmental regulation, environmental regulation competition and regional industrial economic growth: An empirical study cased on spatial panel Durbin Model. *Journal of International Trade, 7,* 82–92.

Zhao, X. (2014b). Inter-local government strategies of environmental regulation competition and its economic growth effect. *Finance and Trade Economics, 10,* 105–113.

Zhao, Y., Zhu, F., & He, L. (2009). Definition, classification and evolution of environmental regulations. *Chinese Population, Resources and Environment, 19*(6), 85–90.

Zhou, L. (2004). The incentive and cooperation of government officials in the political tournaments: An interpretation of the prolonged local protectionism and duplicative investments in China. *Economic Research Journal, 39*(6), 33–40.

Zhou, L., & Tao, J. (2011). Political tournament and border effect: A case study of economic development in the border regions of China's provinces. *Journal of Financial Research, 3*(3), 15–26.

Zhou, Y., He, C., & Liu, Y. (2015). An empirical study on the geographical distribution of pollution-intensive industries in China. *Journal of Natural Resource, 30*(7), 183–1196.

Zhou, Y., Zhu, S., & He, C. (2017). How do environmental regulations affect industrial dynamics? Evidence from China's pollution-intensive industries. *Habitat International, 60,* 10–18.

Zhu, S., He, C., & Liu, Y. (2014). Going green or going away: Environmental regulation, economic geography and firms' strategies in China's pollution-intensive industries. *Geoforum, 55,* 53–65.

Chapter 6
How Does China's Economic Transition Contribute to Air Pollution?

Dealing with the determinants that shape the geography of environmental performance faces the issue that these determinants are never static, particularly the socio-economic determinants. In China, the dynamic of socio-economic determinants is largely manifested by the remarkable economic transition during last decades. Since the late 1970s, China has experienced a triple process of economic transition, namely, marketization, globalization, and decentralization. The economic transition, however, has posed a serious contradiction for environmental protection efforts in China (Jahiel 1997). The contradiction has arisen from their different environmental effects. Thus, this chapter is to investigate the different environmental effects of the triple process of economic transition. Particularly, has marketization and regional decentralization increased industrial pollution in Chinese cities? What is the possible relationship between globalization and air pollution in China? Those questions deserve further and thorough empirical investigations.

As introduced in 2, there are rich studies concerning the pollution emissions and resource depletion across Chinese provinces based on the Environmental Kuznets Curve Hypothesis, using provincial and industrial level data (Poon et al. 2006; He 2010; Wang et al. 2017a, b; Xu 2018; Xie et al. 2019). Another popular way is using micro-level data to investigate the environmental performance of firms (Dasgupta et al. 2001; Wang and Jin 2007; Long et al. 2017; Liang and Liu 2017). Those studies certainly offer insights into the severe environmental challenges in China, tracing both the sources and causes. However, there lacks a systematic investigation about the relationship between economic reform and industrial pollution. China's economic transition includes a two-pronged decentralization process: power and fiscal decentralization from the central to local governments; and decentralization of decision-making from governments to firms and households (Qian and Weingast 1997; He 2006). China has increasingly integrated into the world economy through international trade and the utilization of foreign direct investment since the late 1970s. The triple process of marketization, globalization, and decentralization has different impacts on the environmental performance of industrial enterprises through scale, composition, and technique effects.

© Springer Nature Singapore Pte Ltd. 2020
C. He and X. Mao, *Environmental Economic Geography in China*,
Economic Geography, https://doi.org/10.1007/978-981-15-8991-1_6

Marketization has gradually introduced market forces to allocate resources across sectors and cities and allows various ownerships of enterprises. Unlike the private enterprises, state-owned enterprises (SOEs) are usually considered more powerful in bargaining with the local governments so that they perform worse in pollution reduction. As a consequence, less economically liberalized cities may host more pollution-intensive industries. However, SOEs may also respond to social responsibility in a more active way than private enterprises so that their environmental performance could be theoretically better.

The impacts of globalization on pollution emissions in China are also ambiguous. Pollution haven thesis proposes that international trade and foreign investment are harmful to a country's environment, especially in developing countries where environmental regulations are relatively lax. However, the international fragmentation of production theoretically allows China to develop cleaner production (Zhang et al. 2017; Wang et al. 2018). Both technological contents embodied in trade and foreign direct investment (FDI) would help clean the air in China (Dean and Lovely 2008; Xu et al. 2019). Associated with these flows, the interregional/intergovernmental cooperation is also expected to make a contribution (Cherniwchan 2017; Morin et al. 2018).

Decentralization results in insufficient authority and a lack of coordination between institutional actors in China's environmental protection apparatus (Jahiel 1998). Economic decentralization has given local officials the means and incentives to develop local economies. The pervasive emphasis on development, consumerism, and profit in government proclamations has further made local governments reluctant to enforce environmental regulation, which is deemed to reduce the competitiveness for growth (Oi 1995). Fiscal decentralization would trigger "race to the bottom competition", in which cities lower their environmental standards to compete for capital and even attract pollution-intensive industries (Liu et al. 2019). Thus, regional decentralization would be responsible for the environmental degradation in Chinese cities. Besides, facing fierce interregional competition, some developed regions may have strong incentives to improve environmental quality to attract high-value added industries or high-tech industries. As a consequence, how and whether economic reform should be responsible for China's environmental degradation remains an empirical question.

These changes provide several important facets to understand the effects of China's economic transition on the environment pollution. In the following section, we first provide in-depth discussion about how the trio of globalization, marketization, and decentralization, representing the prominent features of China's economic transition, will affect its environment. Next, we further conduct descriptive analysis to show environmental performance by sectors and regions. In the third section, we perform a systematic empirical investigation to identify the specific effects of economic transition on the environment. Such an investigation will also provide evidence to either support or reject various theoretical propositions that are also mentioned in previous chapters.

6.1 Air Pollution in China: A Perspective of Economic Transition

Economic development is nonlinearly associated with industrial pollution, relating to a wide range of determinants (See the discussion of Environmental Kuznets Curve in Chapter 2). Panayotou (1997) argues that policies and institutions can significantly reduce environmental degradation at lower income levels and speed up improvement at higher income levels, thereby flattening the EKC and reducing the environmental price of economic growth. Better policies such as more secure property rights and better enforcement of contracts and effective environmental regulations can help flatten the EKC and reduce the environmental price of economic growth.

Economic transition is the institutional environment for industrial pollution in China. There is no doubt that pollution demand and supply is affected by the institutional changes associated with economic transition, which has been conceptualized as a triple process of marketization, globalization, and decentralization. The triple process has been found to shape the economic geography and industrial restructuring of Chinese manufacturing industries (He and Zhu 2007; He et al. 2008). Marketization, globalization, and decentralization would exert different direct and indirect impacts on the environmental behaviour of industrial enterprises and the local government's attitudes towards environmental regulations. The triple process may contribute to industrial pollution in Chinese cities by moderating the scale, composition, and technique effects.

6.1.1 Marketization and Industrial Air Pollution

Institutionally, the Chinese economic transition is to transfer from a command economy to a market-oriented economy. In the command economy, governments distribute resources and firms are executors of state-orders. There is literally no well-functioning markets of goods and factors. As economic transition proceeds, market forces and competition are progressively introduced and markets have played an increasingly important role in resource allocation. The establishment of market systems has encouraged the development of non-state-owned enterprises, including privately owned, collectively owned, and foreign-owned enterprises and other types of enterprises.

Meanwhile, facing fierce competition from non-state-owned enterprises, state-owned enterprises have gradually withdrawn from certain industries such as labour-intensive industries. Currently, state-owned enterprises (SOEs) largely remain in the heavy and capital-intensive industries, such as the energy sector, machinery and equipment manufacturing, ferrous and non-ferrous metal smelting and pressing, petroleum refining and coking, and chemical industries. Less liberalized regions and cities are dominated or controlled by SOEs while privately owned enterprises agglomerate in regions that have significantly benefited from economic liberalization.

Private enterprises can utilize resources more efficiently and produce less pollution with same resources. They may also have lower bargaining power with local environmental authorities with respect to the enforcement of pollution charges and regulations (Wang et al. 2007). A better environmental quality could be achieved with the greater existence of the private sector. For instance, using annual data for 44 developing countries from 1987 to 1995, Talukdar and Meisner (2001) disclose a significantly negative relationship between the degree of private sector involvement in terms of its investment in the total domestic investment, national GDP, or its value of output share in the national GDP and CO_2 emission levels, suggesting that an increased role by the private sector in an economy is more likely to help the economy environment. However, although private-owned enterprises may have higher efficiency in resource utilization, they are solely profit-oriented and may not seek to internalize environmental externalities and may compromise the environment to avoid the potential cost of environmental investments (Eiser et al. 1996). China's non-state-owned sector has been the most rapidly growing, which implies that it will become more and more difficult to enforce the policy as institutional channels of state control over industry become weaker and weaker (Jahiel 1997).

On the contrary, SOEs have their incentives to internalize the environmental costs resulting from pollution discharge to obtain higher national or local social welfare. SOEs normally take more social impacts into account for decision-making and their environmental performance could be theoretically better than private sectors. While responding to the market, SOEs are supposed to internalize their environmental externalities for the local communities and the whole nation. Furthermore, SOEs are more likely to be furnished with advanced technological capabilities to deal with air pollution. The dominance of SOEs may provide opportunities to clean the air in Chinese cities controlling for the structural effects of SOEs.

On the other hand, Wang et al. (2003) argue that SOEs have stronger bargaining power with central and local governments regarding environmental regulations in contrast to the privately owned enterprises. SOEs in China have strong connections with the governments and some managers of SOEs have higher political status than the local environmental authorities. As a result, SOEs are able to elicit a lower pollution payment or punishment and have less incentive to decrease their pollution and reduce pollution intensity. Existing studies do support the argument that the dominance of SOEs in a city may deteriorate its environmental quality and fuel environmental degradation. Using plant-level data, Wang and Wheeler (2000) find that SOEs are more likely to pollute than private enterprises. Wang and Jin (2007) claim that foreign-owned enterprises and collectively owned enterprises have better environmental performance in terms of water pollution discharge intensity while SOEs and privately owned enterprises in China are the worst performers. Less liberalized cities in China are thereby more polluted due to structural and institutional effects. How economic liberalization affects industrial pollution indeed depends on other economic and institutional conditions.

6.1.2 Economic Globalization and Industrial Air Pollution

China has participated in the globalization process through the utilization of foreign investment and international trade. Since the 1990s, China has been the largest FDI host among developing countries and the second most attractive location for multinational corporations. The net environmental effect of globalization in developing economies is debatable. Assuming that regions are identical except for exogenous differences in pollution policy, the pollution haven hypothesis proposes that it is cheaper to produce dirty goods in regions with weaker environmental regulation. Trade and investment induced by environmental regulation differences create a pollution haven i n the poor regions. Pollution-intensive industries would migrate to regions or cities that have weaker environmental regulations to save production costs.

Meanwhile, governments in developing regions may hold their lax environmental policies to give the domestic producers an advantage in the competitive international market, generating the so-called eco-dumping hypothesis (Christmann and Taylor 2001). Both pollution haven hypothesis and eco-dumping hypothesis argue that globalization is harmful to the environment of developing countries, whose environmental regulations are weak. However, the simple factor endowment hypothesis suggests that dirty capital-intensive processes would relocate to the relatively capital-abundant developed regions (Antweiler et al. 2001). Cole (2003) further argues that the impact of trade liberalization on the environment depends on whether or not they have a comparative advantage in pollution-intensive production. It depends on a region's relative factor endowments and/or its relative environmental regulations. However, trade and investment policies and environmental policies are not separate. Many factors drive trade and investment policies and environmental policies simultaneously. Developed regions are likely to have both abundant capitals and stricter pollution policy while poor regions are likely to be the opposite.

However, globalization is likely to help the environment in developing regions through technique effects. Openness would encourage technological and managerial innovation, which are beneficial to environmental improvement as well as economic progress and could facilitate an international ratcheting up of environmental standards. Increasing trade and investment liberalization can provide favourable conditions for the diffusion of global environmental norms and standards by creating opportunities and necessities for environmental institution-building and policy processes (Shin 2004). Opportunities that more globalized economies can bring about include (a) foreign advanced technology transfer, (b) exposure to environmental practices of advanced economies, and (c) learning and adopting new policy instruments, eco-labelling, and cleaner production. Trade liberalization may push governments to ensure that their export products comply with the environmental quality standards of importing countries and international standards. FDI may induce policy changes as a responding strategy of developing countries to cope with possible environmental damage. In addition, multilateral trade agreements such as WTO require environmental commitments for exporting countries. Advanced technology and international environmental standards brought through international trade and foreign investment provide globalized regions an improved environment and less pollution.

Some studies provide supportive evidence for a pollution haven (Hettige et al. 1992; Birdsall and Wheeler 1993; Suri and Chapman 1998; Xing and Kolstad 2002; Cole 2004a, b; Jorgenson 2007). More studies prove insignificant results or opposite impacts (Jaffe et al. 1995; Smarzynska and Wei 2001; Antweiler et al. 2001; Wheeler 2001; Frankel and Rose 2005; Kellenberg 2008). Free trade and openness appear to be good for the environment even in developing countries. The existing literature has generated mixed results regarding the relationship between globalization and pollution emission in China. Wang and Jin (2007) find that foreign-invested firms have better environmental performances than state-owned and privately owned firms. They suggest that foreign firms pollute less because they use superior technology in production and are more energy-efficient. Shin (2004) examines the effects of trade and investment liberalization on the environment in two Chinese cities—Shenyang and Dalian, and found that economic openness positively affected domestic environmental policy by providing the necessity and opportunities for strengthening environmental institutions. However, Dean (2002) estimates the effect of trade liberalization on water pollution in China's provinces using a simultaneous equation system and suggested that trade aggravates pollution via the improved terms of trade but mitigates it via income growth. He (2006, 2009) provide convincing supportive evidence for the pollution haven hypothesis.

6.1.3 Regional Decentralization and Industrial Air Pollution

Economic transition in China has resulted in considerable decentralization of power from the central government to a more regional locus. As a result, local governments have primary responsibility and great autonomy for economic development in their jurisdictions (He et al. 2008). Fiscal decentralization, in particular, has enhanced the importance of local revenues (Zhao and Zhang 1999; Young 2000). Fiscal decentralization inherently and explicitly emphasized autarchic development because the localities had to self-finance their budgets and their development (Zhao and Zhang 1999). Fiscal decentralization and designated local tax structures generate strong incentives and pressure to develop manufacturing industries to enhance local revenues.

Since the late 1980s, the Chinese central government has also given much autonomy and responsibility in environmental policy to local authorities so that each provincial and municipal government has to compete with each other in environmental as well as economic performance (Jahiel 1997, 1998). Environmental protection agencies now report to the administratively higher levels of the national environmental protection apparatus and the local governments where they reside. Decentralization has removed central government's guarantees of financial resources for floundering localities and has deprived localities of the financial security they once had. It is the local government that provides environmental agencies with their annual budgetary funds, approves institutional advancements in rank, and determines increases in personnel, and even the allocation of such resources as cars and office buildings. Increasingly hard budgets particularly make the local environment protection agencies difficult.

As a consequence, China's environmental protection apparatus has suffered from two lingering problems: insufficient authority and lack of coordination between institutional actors (Jahiel 1998). Fragmented authority structure undermines the effective enforcement of environmental policies in China. Van Rooij and Lo (2010) find considerable regional variations in the enforcement of environmental pollution violations. Coastal areas have more and higher punishments than inland in China and factors as central government support, community pressure, local government commitment, enforcement capacity, regulated firm characteristics, and general economic conditions are responsible for the variations. The root of enforcement problems of environmental policies lies with the institutional arrangement in which the local governments pay and directly manage China's main local environmental enforcement authorities (Jahiel 1998; Van Rooij and Lo 2010).

The effect of fiscal and environmental decentralization has been that many local officials have become entrepreneurial trying to promote growth in their particular locality. While this entrepreneurship by local leaders has translated into an economic boom for many localities, its effect on regulations has been far less beneficial. As Oi (1995) observed, "with the transformation from administrators to entrepreneurs, local governments are shifting from regulators to advocates of their local enterprises". Meanwhile, fiscal decentralization has induced fierce interregional competition and may trigger the "race to the bottom competition" in which regions lower environmental standards to compete for investments and firms.

Decentralization of economic decision-making to local governments and factory managers, combined with calls for rapid economic growth and production for profit, has created further incentives for local governments and managers to pursue economic growth and profitability at the expense of environmental degradation (Jahiel 1997). Local governments have strong incentives to circumvent those policies that might constrain local growth, such as environmental regulations (Lieberthal 1995). Reform incentives thus "have actually distorted the role of local governments as agents of the central state", making the local authorities laxer with the enforcement of environmental regulations (Jahiel 1997). Destructive regulatory competition in the form of a race to the bottom would lower environmental quality with decentralization. Existing studies have reported that local governments consistently undermine pollution enforcement in order to protect local economic interests (Ma and Ortolano 2000; Jahiel 1998, 1997; Sinkule and Ortolano 1995; Tang et al. 1997; Swanson et al. 2001; Tang et al. 2003; Van Rooij 2006). With decentralization, pollution-intensive industries gain opportunities to grow in regions facing hard budgets with limited local revenues. Regional decentralization thereby may be harmful to the environment through scale and composition effects in China.

Overall, the economic transition in China has gradually introduced market forces and global forces and resulted in economic decentralization, which shapes the scale, composition, and technique effects of economic development on industrial pollution. Marketization has significantly stimulated economic growth and changed the ownership structure of the Chinese economy with a growing share of non-state-owned enterprises and decreasing the status of SOEs. Globalization has been a key factor contributing to China's rapid economic expansion, dramatic industrial restructuring, and substantial technological progress while introducing global environmental norms. Decentralization has stimulated industrial growth and triggered a race to the bottom competition regarding environmental regulations. The net effects of marketization, globalization, and regional decentralization on industrial pollution in Chinese cities, however, remain empirical questions, which deserve a thorough investigation. The following section is first to describe the structural and spatial patterns of industrial pollution and investigate the driving forces of industrial pollution in Chinese cities, with special attention to the impacts of economic transition.

6.2   Structural and Spatial Patterns of Industrial Air Pollution in China

6.2.1   Industrial Patterns of Pollution in Chinese Cities

There have been some structural changes in industrial air pollution in the last decade (Fig. 6.1). The total industrial SO_2 emissions decreased from 15.87 million tons in

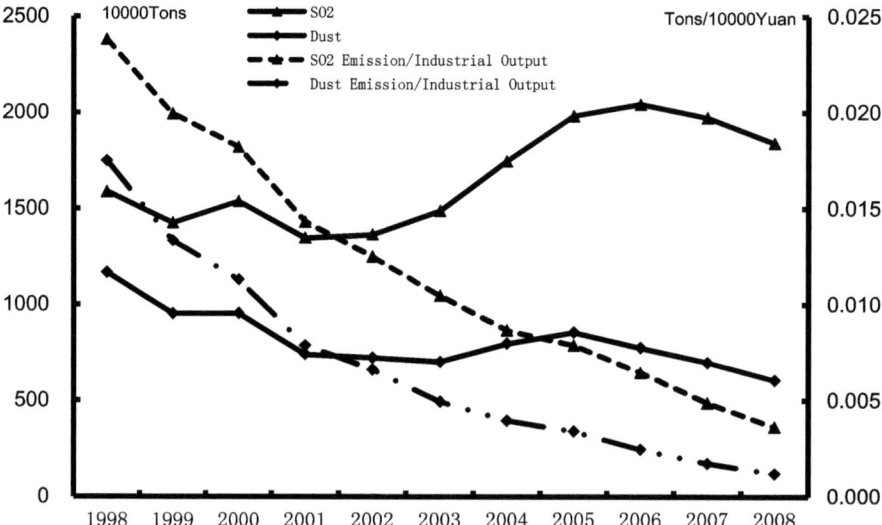

Fig. 6.1 Industrial SO_2 and dust emissions and their intensities in China. *Note* Emission refers to the left vertical axis and intensity to the right vertical axis

1998 to 13.46 million tons in 2001 but went up to 20.42 million tons in 2006. The rising trend has been reversed recently, with 18.39 million tons of SO_2 in 2008. Industrial dust showed a clear decreasing trend, dropping from 11.64 million tons in 1998 to 60.44 million tons in 2008. Both pollutant intensities had been reducing during 1998–2008, indicating that industrial production in China had been cleaner. SO_2 emissions per industrial output dropped from 2.38 to 0.36 tons/Million Yuan. Industrial dust per industrial output decreased from 1.75 to 0.12 tons/Million Yuan. The decreasing trend of SO_2 and dust emissions and reducing intensities are highly expected in the future.

Industrial pollution in China, however, concentrated in several pollution-intensive industries (Tables 6.1 and 6.2). Power, natural gas, and water production is the most pollution-intensive industry. In 1998 and 2008, this industry is responsible for 6.9679 million tons and 10.6279 million tons of SO_2 emissions, respectively, accounting for 43.92 and 57.79% of total industrial SO_2 emissions in China. Its SO_2 emission intensity topped all industries, reaching 17.46 tons/million Yuan in 1998 and 3.27 tons/million Yuan in 2008. Other important SO_2 emitters and SO_2 intensive industries include non-metal mineral products, chemical materials, and products, ferrous metal smelting and rolling processing, non-ferrous metal smelting and rolling processing, petroleum refining and coking, mining, papermaking and paper products, food, drinks, and tobacco manufacturing. Those industries are resource processing and heavy industries, demanding a large number of energy inputs. During 1998–2008, all industries had been much less SO_2 intensive. Petroleum refining and coking, ferrous metal smelting and rolling processing industry, plastic products, papermaking, and paper products experienced significant growth of SO_2 emissions. The annual growth rate of SO_2 emissions is 8.38% for petroleum refining and coking, and 7.36% for ferrous metal smelting and processing. Fortunately, many more industries have successfully cut their SO_2 emissions. Machinery, electrical equipment and electronics, printing and copying, rubber plastics, fabricated metal products, and non-metal mineral products saw the fastest decrease in SO_2 emissions, with an annual decreasing rate of greater than 3%.

The structural pattern of industrial dust is similar to that of SO_2 emissions. Power, natural and water production has dominated industrial dust emission, accounting for 29.36% in 1998 and 41.79% in 2008, followed by non-metal mineral products (22.92%, 16.30%), mining (8.52%, 3.07%), chemical materials and products (4.44%, 7.77%), and ferrous metal smelting and processing (3.04%, 9.44%). Except for petroleum refining and coking, plastic products, and ferrous metal smelting and processing, all industries witnessed a significant drop in industrial dust during 1998–2008. Mining, printing, and copying, non-metal mineral products experienced the fastest decrease.

There have been remarkable changes in the intensities of SO_2 and dust emissions in China during 1998–2008. This chapter compares the pollution-intensive and less pollution-intensive industries. Figure 6.2 shows the changes in the sectoral SO_2 intensity. Most pollution-intensive industries have been increasingly and significantly cleaner and more pollution-intensive industries underwent considerable decreases in SO_2 intensity. For instance, non-metal mineral products decreased its intensity

Table 6.1 Industrial SO_2 emissions by sector

Sector	1998		2008		Annual growth rate (%)	Emission intensity	
	Volume (tons)	%	Volume (tons)	%		1998 (tons/billion yuan)	2008 (tons/billion yuan)
Mining	409,470	2.58	451,975	2.46	0.99	10,004.7	134.5
Food, drinks, and tobacco manufacturing	464,426	2.93	411,698	2.24	−1.20	604.4	97.2
Textiles and garments making	286,415	1.81	263,827	1.43	−0.82	654.5	123.3
Leather fur and down products	17,587	0.11	17,347	0.09	−0.14	147.6	29.5
Paper making and paper products	359,459	2.27	462,959	2.52	2.56	2889.6	588.0
Printing and copying	6953	0.04	3727	0.02	−6.05	127.8	13.9
Petroleum refining and coking	281,262	1.77	629,183	3.42	8.38	1207.4	278.0
Chemical materials and products	940,762	5.93	1,035,423	5.63	0.96	2032.8	304.9
Pharmaceutical industry	80,620	0.51	76,271	0.41	−0.55	587.3	96.9
Chemical fibre	141,916	0.89	117,031	0.64	−1.91	1717.0	294.8
Rubber products	64,986	0.41	38,203	0.21	−5.17	848.8	90.3
Plastic products	14,886	0.09	24,196	0.13	4.98	99.4	24.4
Non-metal mineral products	2,281,547	14.38	1,680,618	9.14	−3.01	7119.9	802.5
Ferrous metal smelting and processing	790,319	4.98	1,607,470	8.74	7.36	2035.2	359.4
Non-ferrous metal smelting and processing	700,462	4.42	668,786	3.64	−0.46	4300.7	319.2
Fabricated metal products	64,382	0.41	42,277	0.23	−4.12	299.4	28.1

(continued)

Table 6.1 (continued)

Sector	1998		2008		Annual growth rate (%)	Emission intensity	
	Volume (tons)	%	Volume (tons)	%		1998 (tons/billion yuan)	2008 (tons/billion yuan)
Machinery, electrical equipment, and electronics	262,635	1.66	138,107	0.75	−6.23	146.5	9.1
Power, natural, and water production	6,967,926	43.92	10,627,957	57.79	4.31	17,464.6	3272.2
Other industries	1,727,253	10.89	94,725	0.52	−25.20	5144.6	37.8
Total	15,863,266	100	18,391,780	100	1.49	2379.1	362.4

from 7.11 to 0.80 tons/million Yuan, and non-ferrous metal smelting and processing from 4.30 to 0.32 tons/million Yuan. For less pollution-intensive industries, it is also true that more pollution-intensive industries experienced a more considerable reduction in SO_2 emission intensity. Production in rubber products, textiles, food, and drinks manufacturing, the medical and pharmaceutical industry has significantly been cleaner in the last decade. Other less pollution-intensive industries kept much stable SO_2 emission intensity. Large sectoral variations in SO_2 intensity still exist.

As a dust intensive industry, non-metal mineral products experienced the most dramatic reduction in dust intensity during 1998–2008, decreasing from 8.33 to 0.47 tons/million Yuan. The decline largely occurred from 1998 to 2001. All other dust intensive industries also made significant progress in reducing dust emissions, especially mining and papermaking and paper products, realizing more significant achievements. By 2008, all dust intensive industries bear dust intensities lower than 0.50 tons/million Yuan. Among the relative less dust intensive industries, food, drinks, and tobacco manufacturing, medical and pharmaceutical industry, textiles and garment making, rubber products and fabricated metal products have significantly brought down their dust intensities during 1998–2008. For instance, food, drinks, and tobacco manufacturing reduced its dust intensity from 0.40 to 0.059 tons/million Yuan, rubber products from 0.37 to 0.043 tons/million Yuan, and medical and pharmaceutical industry from 0.36 to 0.058 tons/million Yuan. By 2008, all less dust intensive industries have intensities lower than 0.10 tons/million Yuan. Similarly, industries still differ significantly in their dust intensities (Fig. 6.3).

Table 6.2 Industrial dust by sector

Sector	1998		2008		Annual growth rate (%)	Intensity	
	Volume (tons)	%	Volume (tons)	%		1998 (tons/billion yuan)	2008 (tons/billion yuan)
Mining	992,537	8.52	185,388	3.067	−15.45	**2435.4**	**55.2**
Food, drinks, and tobacco manufacturing	304,290	2.61	251,282	4.16	−1.90	396.0	59.3
Textiles and garments making	148,575	1.28	128,294	2.12	−1.46	339.5	60.0
Leather fur and down products	11,305	0.10	10,298	0.17	−0.93	94.8	17.5
Paper making and paper products	259,543	2.23	239,860	3.97	−0.79	**2086.4**	**304.6**
Printing and copying	6844	0.06	1827	0.03	−12.37	125.8	6.8
Petroleum refining and coking	187,479	1.61	255,154	4.22	3.13	804.8	112.8
Chemical materials and products	517,487	4.44	469,443	7.77	−0.97	**1118.2**	**138.3**
Pharmaceutical industry	48,536	0.42	45,732	0.76	−0.59	353.6	58.1
Chemical fibre	58,541	0.50	27,070	0.45	−7.42	**708.3**	**68.2**
Rubber products	28,403	0.24	18,168	0.30	−4.37	371.0	43.0
Plastic products	7116	0.06	11,414	0.19	4.84	47.5	11.5
Non-metal mineral products	2,669,604	22.92	985,294	16.30	−9.49	**8330.8**	**470.5**
Ferrous metal smelting and rolling processing	354,375	3.04	570,353	9.44	4.87	**912.6**	**127.5**
Noferrous metal smelting and rolling processing	145,747	1.25	134,907	2.23	−0.77	**894.9**	**64.4**
Fabricated metal products	42,778	0.37	23,020	0.38	−6.01	198.9	15.3

(continued)

Table 6.2 (continued)

Sector	1998		2008		Annual growth rate (%)	Intensity	
	Volume (tons)	%	Volume (tons)	%		1998 (tons/billion yuan)	2008 (tons/billion yuan)
Machinery, electrical equipment and electronics	177,597	1.52	88,601	1.47	−6.72	99.1	5.8
Power, natural and water production	3,419,778	29.36	2,525,716	41.79	−2.99	**8571.4**	**777.6**
Other industries	2,268,428	19.47	72,691	1.20	−29.11	6756.5	29.0
Total	11,648,963	100	6,044,512	100	−6.35	1747.1	119.1

6.2.2 Industrial Air Pollution in Chinese Cities

This chapter covers 287 cities, including prefecture cities, sub-provincial cities, and centrally administered cities of Beijing, Tianjin, Shanghai, and Chongqing. The China Urban Statistical Yearbook started to report data on industrial air pollution for Chinese cities from 2003. However, data is missing in 25 cities in the western provinces including Shanxi, Gansu, Qinghai, Ningxia, Xinjiang, and Tibet in 2003. Only Lasha did not report data since 2004. This chapter compares the spatial distribution of two typical industrial pollutants of SO_2 and Dust and their intensities in 2004 and 2007.

Figure 6.4 shows the distribution of the total SO_2 emissions in Chinese cities. In 2004, Chongqing topped the city list, emitting 0.64 million tons of SO_2, followed by Shanghai (0.35 million tons) and Tangshan (Hebei) (0.28 million tons). Other heavily polluted cities generating more than 0.20 million tons of SO_2 include Weinan (Shaanxi), Ordos (Inner Mongolia), Suzhou (Jiangsu), Guiyang (Guizhou), Luoyang (Henan), Tianjin and Ningbo (Zhejiang). Those are large cities with a large industrial production or heavy industrial structure. The least SO_2 emitter is Ningde (Fujian), with SO_2 emissions of 64 tons, followed by Sanya (305 tons) and Haikou (454 tons) in Hainan. Guyuan (Ningxia) and Bazhong (Sichuan) emitted less than 1000 tons of SO_2 in 2004. Those cities are small and have very few industrial activities. In 2007, Chongqing and Shanghai remained the top two SO_2 emitters, with 0.68 and 0.36 million tons, respectively. The other heavy SO_2 polluters with more than 0.20 million tons include Weinan (Shaaxi), Tangshan (Hebei), Luoyang (Henan), Ordos (Inner Mongolia), Laibing (Guangxi), Suzhou (Jiangsu), Tianjin and Chifeng (Inner Mongolia). Laibing and Chifeng replaced Guiyang and Ningbo in 2007 and were added to the top list. The least SO_2 emitters in 2007 basically remained the same.

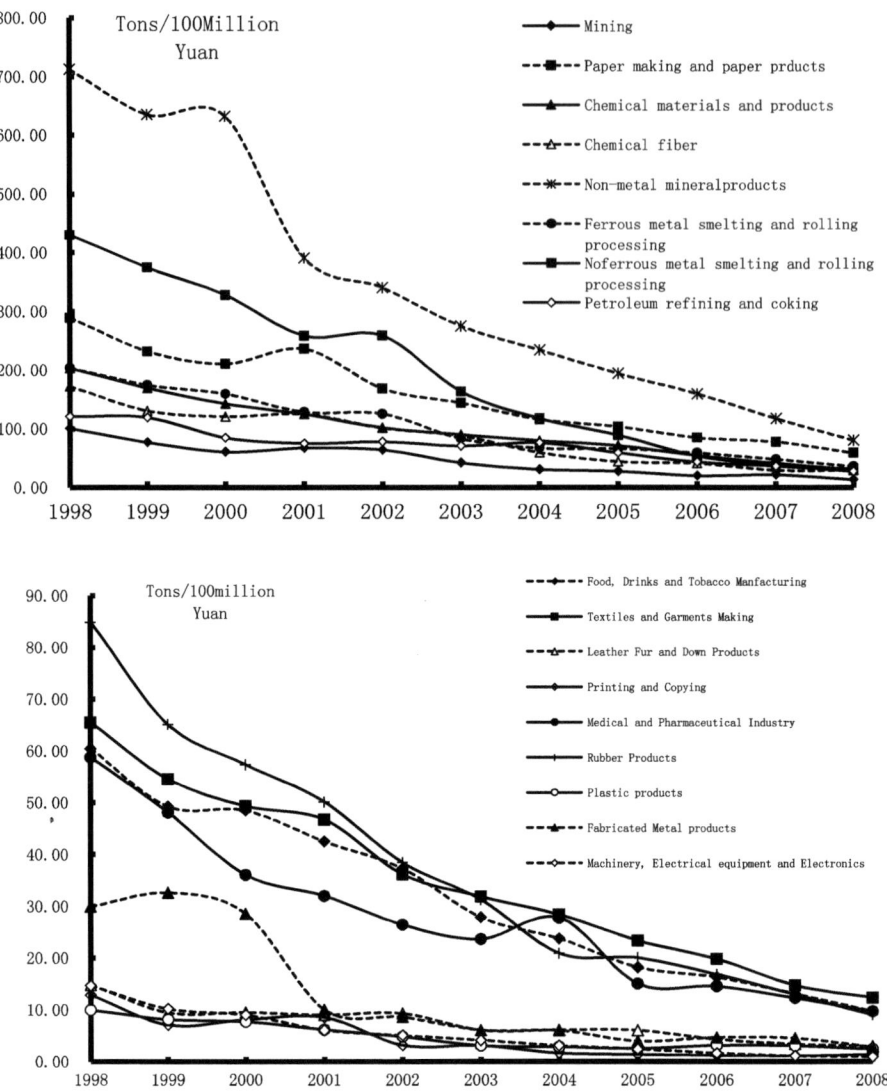

Fig. 6.2 Industrial SO_2 intensity change in high (top) and low (bottom) pollution-intensive industries during 1998–2008

Overall, SO_2 emissions are distributed in clusters. The global Moran's I is 0.1394 and 0.1514 in 2004 and 2007, respectively, indicating a significant spatial auto-correlation pattern. The Yangtze River Delta, the Shandong Peninsula, the Capital region (Beijing–Tianjin–Tangshan), Central-Northern China (including Southern Hebei, Henan, and Shanxi), the Sichuan Basin, and the Pearl River Delta are the hotspots. In 2007, northern China became more polluted and formed a new industrial pollution belt from Chifeng (Eastern Inner Mongolia) to Zhangjiakou (Northern

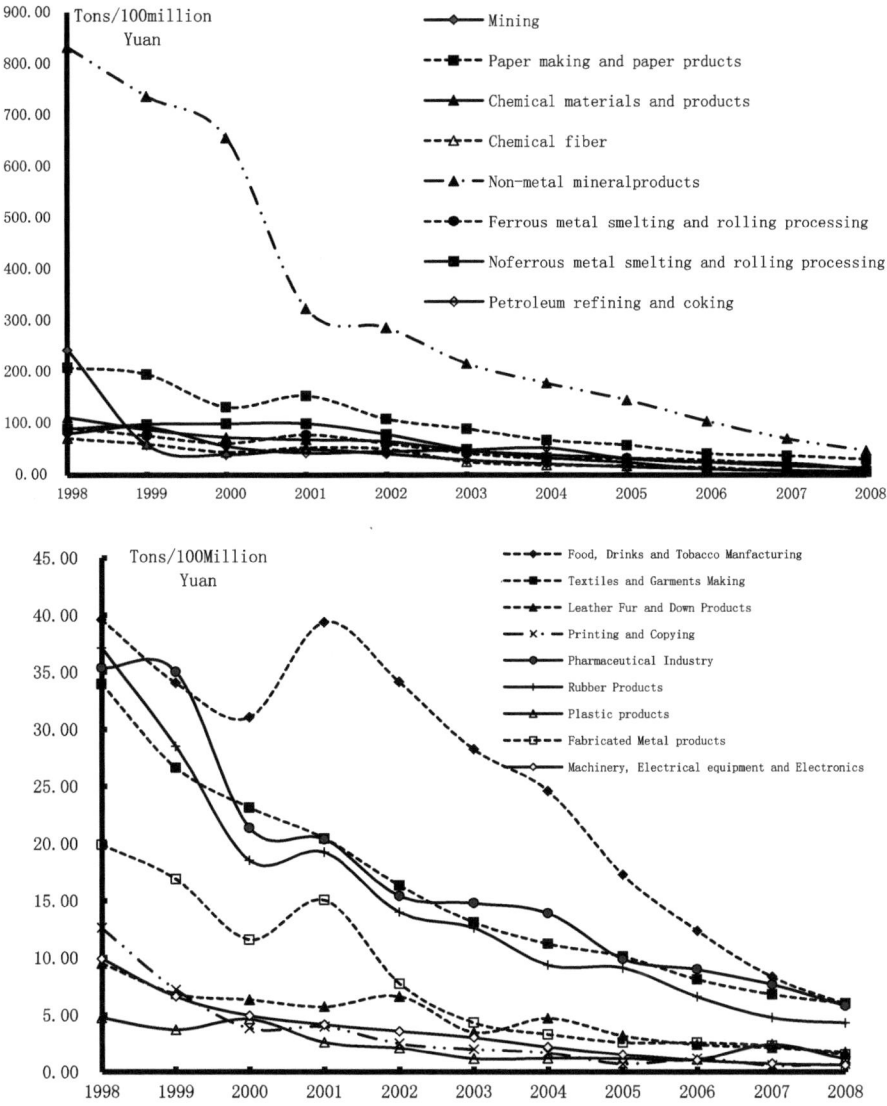

Fig. 6.3 Industrial dust intensity change in high (top) and low (bottom) pollution-intensive industries during 1998–2008

Hebei) to Baotou and from Shizuishan in Ningxia to Baiyin in Gansu. Meanwhile, Northeast China showed the trend of increasing industrial SO_2 emissions. Moreover, there has been a clear pollution dispersion process during 2004–2007.

The good news is that SO_2 emissions intensity has significantly decreased (5.5). During 2004–2007, many cities, especially the most polluted cities, significantly reduced SO_2 emissions per industrial output. The coastal cities bear the lowest SO_2

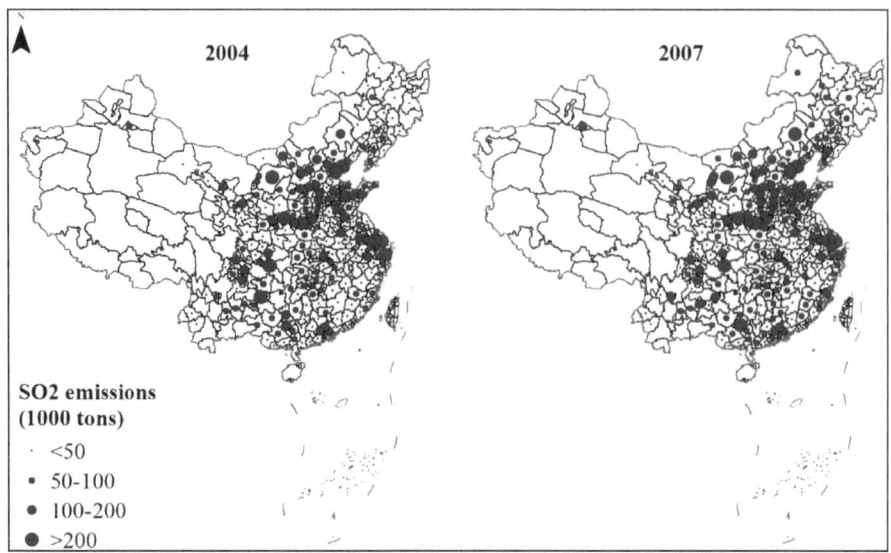

Fig. 6.4 Spatial distribution of SO$_2$ emissions in Chinese cities in 2004 (left) and 2007 (right)

emissions per industrial output. The central part of China, however, has experienced the most environmental degradation. The Yangtze River Delta and the Pearl River Delta are particularly more SO$_2$ effective, with much lower SO$_2$ intensity (Fig. 6.5).

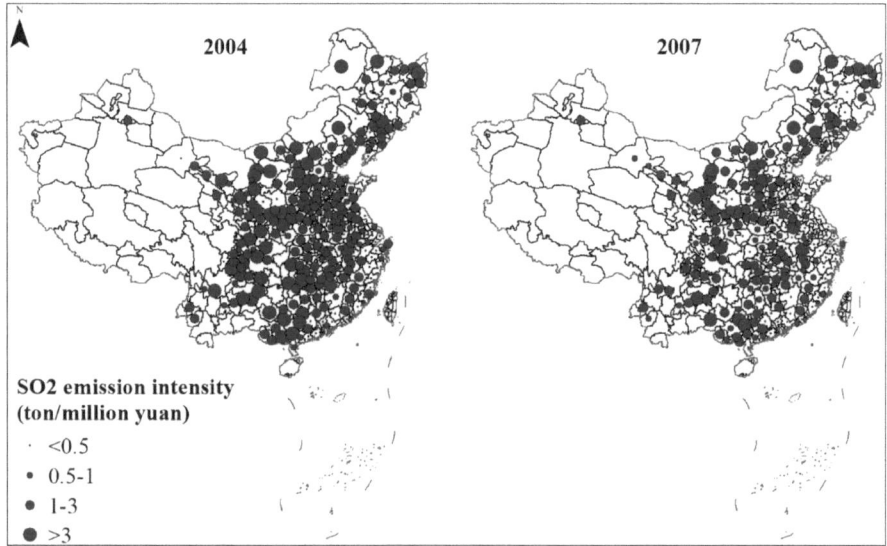

Fig. 6.5 Spatial distribution of SO$_2$ emission intensity in Chinese cities in 2004 (left) and 2007 (right)

There are some differences in the spatial distribution of industrial dust from SO_2 emissions in Chinese cities (5.6). In 2004, Chengdu (Sichuan), Karamay (Xinjiang), and Ordos (Inner Mongolia) ranked as the top three industrial dust emitters, followed by Leshan (Sichuan), Tangshan (Hebei), Luoyang (Henan), Linfen(Shanxi), Chongqing, Datong(Shanxi), and Weinan (Shaanxi), all of which emit more than 100 thousand tons of industrial dusts. In 2007, only Tangshan, Chongqing, and Linfen emitted more than 100 thousand tons of industrial dust. The major dust polluters remained at the top. The least polluted cities by industrial dust include Sanya (Hainan), Haikou (Hainan), Lincang (Yunnan), Lijiang (Yunnan), and Guyuan (Ningxia) in both years.

Like SO_2 emissions, industrial dust is also agglomerated in certain city clusters. The global Moran's I is 0.1994 and 0.2786 in 2004 and 2007, respectively, implying a stronger spatial auto-correlation pattern than SO_2 emissions. The largest concentration of industrial dust is central-northern China covering cities in Tianjin, Shanxi, Henan, and Hebei provinces. Unlike industrial SO_2 emissions, industrial dust is widely distributed in Northeast China. Harbin, Jilin, Tieling, Qiqihar, Jiamusi, Jingzhou, and Daqing are heavily polluted by industrial dust. Other dispersed clusters include the Yangtze River Delta, the Guangdong–Guangxi (along Guangzhou, Shenzhen, Nanning, Liuzhou, Guigang, and Yulin), the eastern Sichuan Basin, the Wuhan city region in Hubei, and the Chang–Zhu–Tan city region in Hunan province (Fig. 6.6).

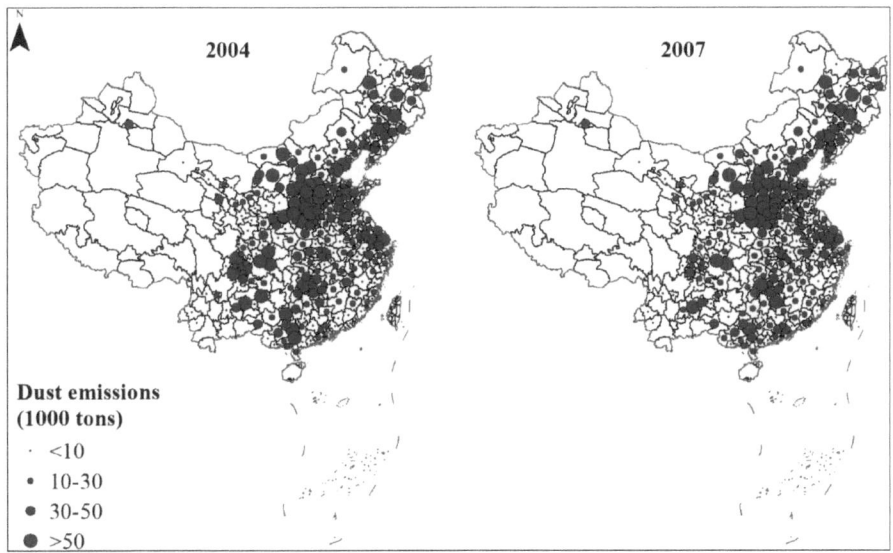

Fig. 6.6 Spatial distribution of industrial dust in Chinese cities in 2004 (left) and 2007 (right)

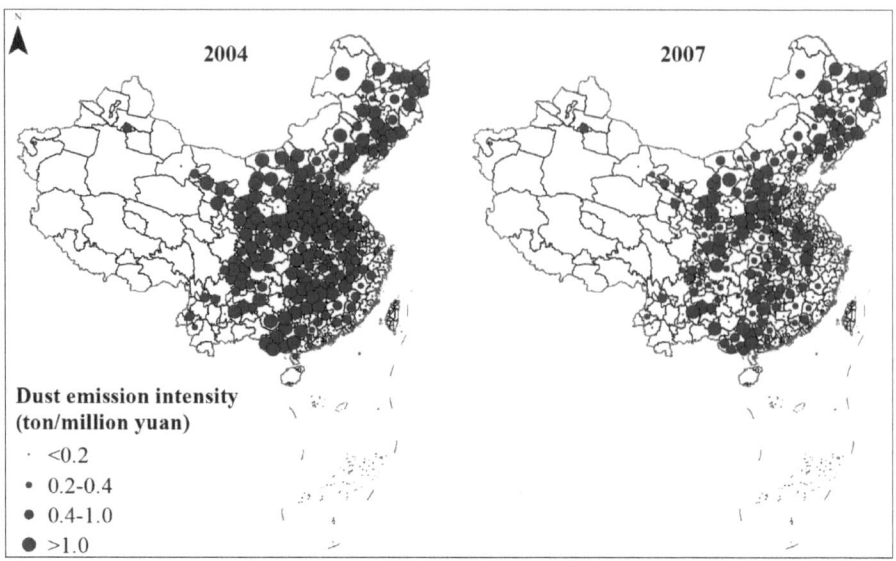

Fig. 6.7 Spatial distribution of industrial dust intensity in Chinese cities in 2004 (left) and 2007 (right)

The spatial pattern of industrial dust intensity is similar to that of SO_2 emission intensity. During 2004–2007, most cities in central China significantly reduced industrial dust per industrial output (Fig. 6.7). However, industrial production in Northeast China remained largely dust intensive, especially in cities with heavy industrial structure, such as Jiamusi, Yichun, Jixi, Fuxin, Mudanjiang, baishan, Qitaihe, Heihe, Shuangyashan, Hegan, Baicheng, and Qiqihar. Also, Guangxi, Shanxi, the north-west Shan–Gan–Ning, and the eastern Sichuan Basin are also still dust intensive. Comparatively, the coastal cities are much cleaner, generating much less industrial dust per industrial output. The Yangtze River Delta and the Pearl River Delta are particularly more dust-effective in their industrial production, with much lower industrial dust per industrial output. Overall, due to the scale, structural, technological, institutional, and locational differences, Chinese cities suffer differently from industrial pollution.

6.3 Empirics and Results

6.3.1 Variables and Models

To understand the significant inter-city variation in industrial air pollution intensity, this chapter conducts a systematic investigation into determinants of industrial air pollution, applying for a panel data regression model during 2004–2007. We particularly focus on the effects of the triple process of economic transition and the explanatory variables include proxies for marketization, globalization, and fiscal decentralization controlling for industrial composition. The model is as follows:

$$LnTSO2_{it}(orLnTDUST_{it}) = \beta_1 LnPGDP_{it} + \beta_2(LnPGDP_{it})^2 + \beta_3 LnPRIV_{it} + \beta_4 LnCOES_{it}$$
$$+\beta_5 LnSOES_{it} + \beta_6 LnFIES_{it} + \beta_7 LnEXPORT_{it} + \beta_8 LnIMPORT_{it} + \beta_9 LnLEXRE_{it}$$
$$+\beta_{10} LnVALTAX_{it} + \sum_{k=1}^{9} \beta_k LnINDU_{itk} + \lambda_t + v_{it}, i = 1, ..., N, t = 1, 2, 3, 4.$$

That is, the pollution intensity (TSO2 or TDUST) in the city i in year t is a function of theoretically discussed variables. i and t denote city and time, λ_t the unobservable time effect, v_{it} the remainder stochastic disturbance term. Note that λ_t is city-invariant and it accounts for any time-specific effect that is not included in the regression. The inclusion of per capita GDP and its square is to test the existence of the environmental Kuznets Curve (EKC) at the city level. It is consistent with the environmental Kuznets curve literature to use per capita GDP and squared per capita GDP to capture scale and technique effects. The inverted U-shaped relationship between per capita GDP and pollution is explained largely in terms of the dominance of scale effects at low levels of income and the dominance of technique effects at high-income levels.

This chapter is particularly interested in the environmental effects of economic transition. As discussed, marketization has played a significant role in changing industrial pollution intensity in Chinese cities although the net effect is unclear. The consequence of marketization is the diversified ownership structure of the Chinese city, with growing shares of non-SOEs. We entertain the percent of state-owned enterprises (SOES), collectively owned enterprises (COES), and privately owned enterprises (PRIV) in gross industrial output to test the environmental impacts of economic liberalization. To investigate the environmental impacts of globalization in Chinese cities, we introduce three variables to quantify trade and investment liberalization, including the percent of industrial output by foreign-invested industrial enterprises (FIES) in gross industrial output, the percent of exports in GDP (EXPORT), and the percent of imports in GDP (IMPORT) in the models. If the pollution haven hypothesis holds, the coefficients on FIES and EXPORT would be positive while IMPORT will have a negative coefficient. Regional decentralization is likely to be harmful to the local environment. This chapter includes the percent of local expenditure in local revenue (LEXRE) and percent of value-added tax in local revenue (VALTAX) to test the impacts of regional decentralization.

Finally, this chapter intends to control industrial composition in Chinese cities by including the percent of the gross industrial output of the pollution-intensive industries (INDU). They include mining, papermaking and paper products, petroleum refining and coking, chemical materials and products, chemical fibre, non-metal mineral products, ferrous metal smelting and pressing, non-ferrous metal smelting and pressing, and power, natural and water production. All variables are summarized in Table 6.3.

Table 6.3 Definitions of dependent and independent variables

Variable	Definition
TSO2	Total industrial SO_2 emissions/gross industrial output
TDUST	Total industrial dust/gross industrial output
PGDP	Per capita GDP
PRIV	Percent of privately owned enterprises in gross industrial output
COES	Percent of collectively owned enterprises in gross industrial output
SOES	Percent of state-owned enterprises in gross industrial output
FIES	Percent of foreign-invested enterprises in gross industrial output
EXPORT	Percent of exports in GDP
IMPORT	Percent of imports in GDP
LEXRE	Percent of local expenditure in local revenue
VALTAX	Percent of value-added tax in local revenue
INDU	Percent of pollution-intensive industries in gross industrial output

6.3.2 Empirical Results

Considering the significant declining trend of industrial air pollution intensity during 2004–2007, this chapter applied the time fixed effect model to estimate the coefficients of explanatory variables. The panel data regression results for both industrial SO_2 and industrial dust intensities are presented in Table 6.4. The Breusch-Pagan tests indicate the existence of heteroscedasticity and all estimates are corrected for heteroscedasticity. This chapter also separately estimates the impacts of marketization, globalization, and decentralization controlling for the EKC effect and industrial structural effects.

Statistical results provide strong evidence to support the EKC effect in Chinese cities. There is a statistically significant inverted U-shaped relationship between LnPGDP and LnTSO2 (Or LnTDUST). Pollution intensity is lower in underdeveloped cities. As cities grow economically, their economies become increasingly pollution-intensive. When per capita GDP reaches a certain level, pollution intensity gradually reduces. The EKC effect suggests that economic development and economic restructuring passing a certain level would do good to improve the environmental quality in Chinese cities.

There is insufficient evidence to show that economic liberalization influences China's environment controlling for industrial structural effects and the EKC effect. The coefficients on LnPRIV and LnCOES are negative while for LnSOES, it is positive in the SO_2 models. Results based on model 1 indicate that developing privately owned enterprises would help to reduce SO_2 emissions per output. The dominance of state-owned enterprises would increase SO_2 emissions per output and hurt the environment. The DUST models confirm a significant positive coefficient on LnSOES. This is consistent with Wang and Wheeler (2000), which found that SOEs are more likely to pollute than privately owned enterprises. As Wang et al. (2003) argued that

Table 6.4 Panel regression results for pollution intensity using the full sample

Variable	Industrial SO$_2$ intensity (LnTSO2)				Industrial dust intensity (LnTDUST)			
	Model 1	Model 2	Model 3	Model 4	Model 1	Model 2	Model 3	Model 4
LnPGDP	2.9942***	3.1294***	3.1147***	3.0950***	3.3798***	2.7686***	4.7135***	3.6245***
LnPGDP*LnPGDP	−0.1848***	−0.1846***	−0.1867*	−0.1812***	−0.2211***	−0.1739***	−0.2777***	−0.2052***
LnPRIV	−0.0849*			−0.0422	0.0304			0.3733
LnCOES	−0.0105			−0.0127	−0.0107			−0.1852
LnSOES	0.0734**			0.0305	0.1342***			0.4881
LnFIES		−0.1075***		−0.0849**		−0.0873**		−0.0451
LnEXPORT		−0.2862***		−0.2772***		−0.4365***		−0.4321***
LnIMPORT		0.1496*		0.1084		−0.0935		−0.1194
LnLEXRE			0.1661*	0.0789			0.5125***	0.4324***
LnVALTAX			0.2476***	0.2571***			0.1202*	0.1847**
LnINDU1	0.2399***	0.1971***	0.2290***	0.1792***	0.2929***	0.2085***	0.2823***	0.1977***
LnINDU2	−0.1375***	−0.1443***	−0.1510***	−0.1262***	−0.1664***	−0.1564***	−0.1678***	−0.1430***
LnINDU3	−0.0333	−0.0348	−0.0284	−0.0352	0.0104	0.0043	0.0098	0.0086
LnINDU4	0.1258***	0.1089***	0.1193***	0.1178***	0.0854*	0.0657	0.0915**	0.0767*
LnINDU5	−0.1220**	−0.1087**	−0.1324***	−0.1065**	−0.0770	−0.0571	−0.0755	−0.0535
LnINDU6	−0.0971*	−0.1115**	−0.1021*	−0.0788*	0.0032	−0.0015	−0.1364	−0.0015

(continued)

Table 6.4 (continued)

Variable	Industrial SO$_2$ intensity (LnTSO2)				Industrial dust intensity (LnTDUST)			
	Model 1	Model 2	Model 3	Model 4	Model 1	Model 2	Model 3	Model 4
LnINDU7	0.0826***	0.0729***	0.0865***	0.0752***	0.0747***	0.0636**	0.0823***	0.0688**
LnINDU8	0.1007***	0.0905***	0.1057***	0.0857***	0.0399	0.0320	0.0524*	0.0441
LnINDU9	0.0944**	0.1320***	0.0992**	0.1174***	0.0229	0.0723*	0.0086	0.0401
Time dummy	Included	Included	Included	Included	Included	Included	Included	Included
# Observations	1144	1144	1144	1144	1144	1144	1144	1144
Adjusted R2	0.3894	0.4056	0.3935	0.4129	0.4680	0.5068	0.4747	0.5153
F-value	43.87	46.88	47.36	37.54	60.15	70.09	65.55	56.22
Breusch-Pagan χ2	187.19	119.16	120.22	200.15	134.71	105.21	95.81	131.63

Notes $*p < 0.10$; $**p < 0.05$; $***p < 0.01$. Results are corrected with heteroscedasticity

SOEs have stronger bargaining power with local governments regarding environmental regulations in contrast to the privately owned enterprises and SOEs are able to elicit a lower pollution payment or punishment and have less incentive to decrease their pollution. The effect of economic liberalization remains the expected sign but turns insignificant when controlling for globalization and decentralization proxies in both SO_2 and DUST models.

There is no statistical support for the pollution haven hypothesis in Chinese cities. Both coefficients on LnFIES and LnEXPORT are negative and significant in Model 2 and Model 4 in the SO_2 and DUST models. More outputs by foreign-invested enterprises and higher shares of exports in GDP are associated with lower levels of pollution intensity, implying that the dependence on international investment and international markets are more pollution effective. Thus, economic globalization has indeed been beneficial to China's environment. Foreign investments and exports are mainly in the labour-intensive and light industries. Foreign investors bring advanced technology to utilize resources more efficiently. Xu et al. (2006) reported that foreign enterprises perform better than domestic enterprises. The results occur inconsistently with He (2009, 2010), which found some evidence of the pollution haven hypothesis. It, however, confirms the argument by Wang and Jin (2007) that foreign-owned enterprises have better environmental performance. LnIMPORT has a negative and insignificant coefficient in the DUST models but has a positive and significant coefficient in the SO_2 models. China largely imports resources and advanced equipment from abroad. More imports possibly reduce the industrial dust emissions since more advanced technologies are embedded in the imported equipment and machinery. Imports mainly go to cities with relative heavy industrial structures or resource-processing economies, generating SO_2 emissions.

As expected, regional decentralization has indeed played a significant role in deteriorating China's environment. Both LnLEXRE and LnVALTAX have positive coefficients controlling for pollution-intensive industries. LEXRE, the percent of local expenditures in local revenues, quantifies the difficulty of local budgets. VALTAX, the percent of value-added tax in local revenues, measures the dependence of local revenues on industrial development. Fiscal decentralization requires the localities to self-finance their budgets and their development (Zhao and Zhang 1999). Fiscal decentralization and designated local tax structures engender strong pressure and incentives to develop industries and enhance local revenues. Fiscal decentralization has triggered the race to the bottom competition to lower environmental regulations and standards to attract highly taxable and value-added industries, which are typically pollution-intensive. Cities suffering from had budgets are also laxer with the enforcement of environmental regulations.

Pollution intensity in Chinese cities is apparently associated with their industrial composition. Resource-based industries, including mining, chemical materials and products, ferrous and non-ferrous metal smelting and rolling processing industries, power, natural gas, and water production, significantly increase pollution intensity in Chinese cities. Interestingly, papermaking and paper products and chemical fibres have deviated from heavily polluted cities.

There are remarkable regional differences in economic development, like the geographical location, industrial structure, technology, institutional environment, and government policies in China. To check the robustness of the estimations, this chapter further divides Chinese cities into three groups: cities located in the coastal, central, and western provinces. Both SO_2 and DUST models are estimated for the three sub-samples. The statistical results are reported in Tables 6.5, 6.6 and 6.7. All models are highly significant. The models perform best in the coastal cities and worst in the western cities. The R-squares in the coastal full models are 0.4694 and 0.5666 for SO_2 and dust, respectively. The values in the western full models are 0.2412 and 0.3124. Panel data regressions reveal some significant differences in the impacts of economic transition on industrial air pollution.

There is a U-shaped rather than inverted U-shaped relationship between LnPGDP and LnTSO2 and LnTDUST in the coastal cities. The EKC effects are more likely to occur in western cities. Controlling for the effects of economic transition and industrial composition, there is no significant relationship between LnPGDP and pollution intensity in the central cities. The inverted EKC effect is possibly related to the recent heavy industrialization in some coastal cities by developing chemical materials and products, steel making, and machinery and equipment. The EKC effect is likely to hold in the western cities, providing that western cities could improve environmental quality by facilitating economic growth.

Three proxies for economic marketization have different effects on urban environment quality. LnPRIV has negative but insignificant coefficients in all three sets of SO_2 models, indicating that economic liberalization is likely to make industrial production less SO_2 intensive and cleaner. LnPRIV has positive coefficients in the DUST models in the coastal and central cities, and negative coefficients in the western cities. The coefficient in the coastal model is significant in model 1, suggesting that privately owned enterprises are associated with more industrial dust. Collectively owned enterprises have different environmental behaviours in central cities than in other cities. Coefficients on LnCOES are positive in the coastal models and but only significant in the SO_2 model, implying that collectively owned enterprises in the coastal cities are environmentally harmful. Some collectively owned enterprises in the coastal cities are previous township enterprises, relatively small and poorly equipped, compared with SOEs and foreign enterprises. They are not environmentally effective. LnCOES is negative and significant in both SO_2 and DUST models in the central cities but is insignificant in the western cities. In the less developed central cities, the state still controls the heavy and high-value added pollution-intensive industries. Collectively owned enterprises are largely in the market-driven light industries and thereby more pollution effective. Statistical results clearly suggest that SOEs hurt the environment in the coastal cities but improve the environment in the western cities controlling for industrial composition. In the coastal cities, SOEs are largely in heavy industries such as chemical materials and chemical products, and ferrous metal mineral smelting and processing, equipment and machinery. They are more likely to pollute. In the underdeveloped western cities, SOEs are relatively more technologically advanced and largely in high-value added industries compared with non-SOEs, generating less pollution.

Table 6.5 Panel regression results for pollution intensity in the coastal cities

Variable	Industrial SO₂ intensity (LnTSO2)				Industrial dust intensity (LnTDUST)			
	Model 1	Model 2	Model 3	Model 4	Model 1	Model 2	Model 3	Model 4
LnPGDP	-5.3657^{***}	-6.4792^{***}	-5.4778^{***}	-6.2419^{***}	-5.1718^{***}	-6.7823^{***}	-4.0652^{**}	-6.2375^{***}
LnPGDP*LnPGDP	0.2292^{***}	0.2910^{***}	0.2325^{***}	0.2782^{***}	0.2032^{**}	0.2992^{***}	0.1461	0.2751^{***}
LnPRIV	-0.0063			-0.0198	0.1939^{**}			0.0995
LnCOES	0.1812^{***}			0.1991^{***}	0.1007			0.0962
LnSOES	0.1810^{***}			0.1267^{**}	0.2884^{***}			0.1622^{***}
LnFIES		-0.1232^{*}		-0.1451^{**}		-0.0618		-0.0661
LnEXPORT		-0.3037^{**}		-0.2608^{**}		-0.4766^{***}		-0.4253^{***}
LnIMPORT		0.0933		0.1452		-0.1028		-0.0124
LnLEXRE			-0.0041	0.0064			0.1896	0.2483
LnVALTAX			0.1988	0.2703^{*}			-0.1073	0.0026
LnINDU1	0.2014^{***}	0.1721^{***}	0.2250^{***}	0.1497^{***}	0.1947^{***}	0.1239^{**}	0.2142^{***}	0.1177^{**}
LnINDU2	0.0082	0.0054	-0.0167	0.0102	-0.2015^{**}	-0.2128^{***}	-0.2142^{***}	-0.1774^{**}
LnINDU3	-0.0654^{*}	-0.0577^{*}	-0.0345	-0.0650^{**}	0.0257	0.0242	0.0352	0.0123
LnINDU4	0.2061^{***}	0.1863^{***}	0.2223^{***}	0.1871^{***}	0.1961^{***}	0.1754^{**}	0.2258^{***}	0.1753^{**}
LnINDU5	-0.0443	-0.0294	-0.0659	-0.0369	-0.0009	0.0223	-0.0053	0.0271
LnINDU6	-0.2607^{***}	-0.2291^{***}	-0.2306^{***}	-0.2292^{***}	-0.0892	-0.0716	-0.0681	-0.0807

(continued)

Table 6.5 (continued)

Variable	Industrial SO$_2$ intensity (LnTSO$_2$)				Industrial dust intensity (LnTDUST)			
	Model 1	Model 2	Model 3	Model 4	Model 1	Model 2	Model 3	Model 4
LnINDU7	0.1603***	0.1729***	0.1900***	0.1414***	0.1894***	0.1916***	0.2157***	0.1709***
LnINDU8	0.0894*	0.0731*	0.0832*	0.0650	0.0819	0.0832	0.1062*	0.0698
LnINDU9	−0.0588	−0.0629	−0.0612	−0.0135	−0.1688**	−0.1285*	−0.1848**	−0.1429*
Time dummy	Included	Included	Included	Included	Included	Included	Included	Included
# Observations	460	460	460	460	460	460	460	460
Adjusted R2	0.4535	0.4541	0.4377	0.4694	0.5464	0.5626	0.5193	0.5666
F−Value	23.41	23.46	23.33	19.46	33.52	35.73	31.99	28.27

Notes *$p < 0.10$; **$p < 0.05$; ***$p < 0.01$. Results are corrected with heteroscedasticity

Table 6.6 Panel data regression results for pollution intensity in the central cities

Variable	Industrial SO$_2$ Intensity (LnTSO2)				Industrial Dust Intensity (LnTDUST)			
	Model 1	Model 2	Model 3	Model 4	Model 1	Model 2	Model 3	Model 4
LnPGDP	3.9237**	3.1633	1.2810	2.8223	4.3359*	3.9539*	2.9214	3.7155*
LnPGDP*LnPGDP	−0.2118**	−0.1700	−0.0624	−0.1445	−0.2424**	−0.2197*	−0.1514	−0.1894*
LnPRIV	−0.0056			−0.0005	0.0866			0.0577
LnCOES	−0.2039***			−0.2066***	−0.1470***			−0.1597***
LnSOES	0.0342			0.0380	−0.0875*			−0.0709
LnFIES		−0.1074**		−0.0878**		−0.881*		−0.0689
LnEXPORT		0.0941		0.9086		0.2750**		0.2420**
LnIMPORT		0.0028		−0.1819*		−0.3473***		−0.4411***
LnLEXRE			0.3509***	0.2674**			0.6538***	0.5817***
LnVALTAX			0.3445***	0.3312***			0.1846*	0.2269**
LnINDU1	0.2575***	0.2241***	0.2220***	0.2136***	0.2816***	0.2520***	0.2688***	0.2426***
LnINDU2	−0.0217	−0.0536	−0.0372	−0.0022	0.0970	0.1219*	0.1185*	0.1320**
LnINDU3	−0.0200	−0.0296	−0.0360	−0.0509*	0.0528*	0.0439	0.0507	0.0304
LnINDU4	0.0006	0.0033*	0.0286	0.0057	−0.0529	−0.0653	−0.0223	−0.0589
LnINDU5	−0.0697	−0.0815	−0.0520	−0.0468	0.0315	0.0021	0.0418	0.0571
LnINDU6	−0.0377	−0.0451	−0.0008	0.0046	−0.0987	−0.0864	−0.5779	−0.0617

(continued)

Table 6.6 (continued)

Variable	Industrial SO$_2$ Intensity (LnTSO2)				Industrial Dust Intensity (LnTDUST)			
	Model 1	Model 2	Model 3	Model 4	Model 1	Model 2	Model 3	Model 4
LnINDU7	0.0715***	0.0557**	0.0709***	0.0769***	0.0709**	0.0623**	0.0774***	0.0865***
LnINDU8	0.0725**	0.0554*	0.0771**	0.0908***	−0.0321	−0.0292	−0.0031	0.0188
LnINDU9	0.1809***	0.2333***	0.1635***	0.1459**	0.1729***	0.1808***	0.0904*	0.0971*
Time dummy	Included	Included	Included	Included	Included	Included	Included	Included
# Observations	440	440	440	440	440	440	440	440
Adjusted R2	0.3681	0.3508	0.3802	0.4013	0.4017	0.4018	0.4311	0.4562
F-value	16.05	14.96	17.83	14.38	18.34	18.35	21.79	17.74

Notes $*p < 0.10$; $**p < 0.05$; $***p < 0.01$. Results are corrected with heteroscedasticity

Table 6.7 Panel data regression results for pollution intensity in the western cities

Variable	Industrial SO₂ Intensity (LnTSO2)				Industrial Dust Intensity (LnTDUST)			
	Model 1	Model 2	Model 3	Model 4	Model 1	Model 2	Model 3	Model 4
LnPGDP	4.6346***	2.9535**	3.5189**	1.9064	0.9754	0.0161	1.2480	0.4419
LnPGDP*LnPGDP	−0.2658***	−0.1844**	−0.2084***	−0.1480	−0.0809	−0.0340	−0.0891	−0.0610
LnPRIV	−0.2107			−0.1395	−0.0901			−0.0436
LnCOES	0.0387			0.0995	−0.0422			−0.0532
LnSOES	−0.2013***			−0.2365***	−0.1611*			−0.1557*
LnFIES		0.0336		0.0309		0.0395		0.0370
LnEXPORT		0.2181		0.3427		0.0304		0.1180
LnIMPORT		0.7640***		0.5891***		0.5929***		0.5119**
LnLEXRE			−0.2829	−0.3423			0.0562	0.0505
LnVALTAX			0.1986	0.2528			0.1730	0.2473
LnINDU1	0.03870	0.1051**	0.0256	0.0698	0.0749	0.1180**	0.0676	0.0967*
LnINDU2	−0.2737**	−0.1917*	−0.2556**	−0.1618	−0.1368	−0.0611	−0.1254	−0.0890
LnINDU3	−0.0024	0.0378	0.0288	0.0308	0.0145	0.0441	0.0309	0.0371
LnINDU4	0.0731	0.0633	0.0440	0.0120	0.0220	0.0115	0.0124	−0.0047
LnINDU5	−0.1760	−0.1410	−0.1334	−0.0920	−0.0556	−0.0303	−0.0396	−0.0197
LnINDU6	0.1864	0.1780	0.2378*	0.2467**	0.3076**	0.3347**	0.3491**	0.3381***
LnINDU7	0.0442	0.0355	0.0295	0.0488	0.0721	0.0592	0.0589	0.0734
LnINDU8	0.0269	−0.0080	0.0522	−0.0431	0.0379	−0.0262	0.0197	−0.0448
LnINDU9	0.1592**	0.1126	0.1427*	0.1221*	−0.0665	−0.0341	−0.0126	−0.0288
Time dummy	Included	Included	Included	Included	Included	Included	Included	Included

(continued)

Table 6.7 (continued)

Variable	Industrial SO$_2$ Intensity (LnTSO2)				Industrial Dust Intensity (LnTDUST)			
	Model 1	Model 2	Model 3	Model 4	Model 1	Model 2	Model 3	Model 4
# Observations	244	244	244	244	244	244	244	244
Adjusted R2	0.1754	0.2100	0.1583	0.2412	0.2958	0.3076	0.2854	0.3124
F-value	4.04	4.80	3.86	4.51	7.00	7.35	7.07	6.02

Notes $*p < 0.10$; $**p < 0.05$; $***p < 0.01$. Results are corrected with heteroscedasticity

Foreign investment and exports help to improve environmental quality in the coastal cities. Both LnFIES and LnEXPORT have significant and negative coefficients in both SO_2 and DUST models. LnIMPORT has an insignificant positive coefficient in SO_2 models but insignificant negative coefficients in DUST models. Overall, dependence on international investment and markets are beneficial for the environment in the coast region. Nevertheless, foreign investment, exports, and imports show mixed impacts on the environment in the central cities. The utilization of foreign investment and imports may reduce pollution intensity but exports are likely to deteriorate the environment in the central cities. Foreign investments in the central cities are either in labour-intensive light industries or in high-tech industries such as electronics, lowering pollution emissions. Imports in the central cities are mainly advanced equipment, which would improve efficiency and productivity and reduce pollution intensity. Exports in the central cities are largely in resource-intensive products, which are more pollution-intensive. LnFIES, LnEXPORT, and LnIMPORT have positive coefficients in both SO_2 and DUST models in the western cities and LnIMPORT has significant coefficients. The results indicate that the pollution haven hypothesis may hold in China's inland regions, especially the landlocked western region. The pressure and incentives in economic development in China's inland region have imposed serious challenges on environment protection.

Fiscal decentralization has significant impacts on the urban environment only in central cities. Coefficients on LnLEXRE and LnVALTAX are barely significant in the coastal and western models. Both variables are highly significant and hold expected signs in the central models. Incentives in economic development and pressure in local revenues would induce the development of pollution-intensive industries in central cities. In the past couple of years, the central region has gained favourable policy support from the central government and thereby realized rapid economic growth. The central cities have competed to attract the relocated industries from the coastal cities, which are often forced to move because of environmental pollution. The regression results suggest that the interregional competition associated with decentralization may deteriorate the environment in the central cities.

Regarding the impacts of industries, mining, chemical materials, and products, ferrous and non-ferrous metal smelting and pressing have indeed made industrial production dirtier in the coastal cities. Petroleum refining and coking, non-metal mineral products and power, natural gas, and water production are not located in heavily polluted cities. Mining, ferrous and non-ferrous metal smelting and pressing, and power, natural gas, and water production have deteriorated the urban environment in the central cities. Industrial structural effects are not very significant in western cities.

6.4   Conclusion and Discussion

The institutional innovations in the last three decades have liberalized the Chinese economy, resulting in eye-catching economic performance. The high-growth, resource-intensive and export-oriented development strategy, coupled with the norms and institutions designed to support this development strategy have certainly played a critical role in deteriorating environmental quality of Chinese cities and imposed a serious challenge on the environment protection in China. This chapter made a special effort to investigate the different environmental effects of marketization, globalization, and decentralization using data on industrial SO_2 emission and dust at the city level.

Empirically, China's industrial production has been cleaner, with decreasing pollution intensity during the last decade. Industrial air pollution has been concentrated in several pollution-intensive industries, including power, natural and water production, chemical materials and products, ferrous and non-ferrous metal smelting and pressing, non-metal mineral products, petroleum refining, and coking. Pollution-intensive industries particularly have made significant achievements in reducing pollution intensity. Industrial pollution is clustered in some Chinese cities. The Yangtze River Delta, the Shandong Peninsula, the Capital region, Central-Northern China, Northeastern China, the Sichuan Basin, and the Pearl River Delta are the hotspots. The coastal region is much less pollution-intensive compared to the inland region.

Statistical results confirm the EKC effect in Chinese cities, implying that economic development is possible to mitigate industrial pollution. The triple process of economic transition shows different environmental effects in Chinese cities. There is only weak evidence that economic liberalization has improved environmental quality. The dominance of SOEs, however, has associated with poorer environmental quality in Chinese cities. No evidence is found to support the pollution haven hypothesis in China. Participation in economic globalization, in turn, is beneficial to the environment. More foreign-invested enterprises and exports are found to be associated with lower pollution intensity. It is likely that decentralization has induced the race to the bottom interregional competition by lowering environmental regulations to attract taxable and high-value added pollution-intensive industries. There, however, remain regional differences in the environmental effects of economic transition. The EKC effect is more likely to occur in the western region. Economic liberalization has made industrial production less SO_2 intensive and cleaner. Collectively owned enterprises are environmentally harmful in coastal cities but environmentally beneficial in the central cities. SOEs hurt the environment in the coastal cities but improve the environment in the western cities. Globalization clearly helps to improve environmental quality in the coastal cities. Pollution haven hypothesis holds in China's inland regions, especially the landlocked western region. Decentralization has significantly contributed to environmental degradation only in the central cities.

As known, China's economic development is accompanying with serious environmental degradation. Existing studies find the effects of scale, technology, and structure on industrial pollution. This chapter identifies institutional effects. The significant EKC effect indicates that economic growth may help to improve environmental quality in Chinese cities. Further economic liberalization has also associated with less environmental degradation. The environmental effects of SOEs and fiscal decentralization are associated with lower environmental standards and lax environmental enforcement and implementation. The direct solution is to strengthen the authority and independence of local environmental protection agencies. It is urgent to enhance the importance of environmental performance in the evaluation system of local officials and grant locales fiscal and tax incentives to pursue environment-friendly economic development models.

References

Antweiler, W., Copeland, B. R., & Taylor, M. S. (2001). Is free trade good for the environment? *American Economic Review, 91*(4), 877–908.

Birdsall, N., & Wheeler, D. (1993). Trade policy and industrial pollution in Latin America: Where are the pollution havens? *Journal of Environment and Development, 2*(1), 137–149.

Cherniwchan, J. (2017). Trade liberalisation and the environment: Evidence from NAFTA and U.S. manufacturing. *Journal of International Economics, 105,* 130–149.

Christmann, P., & Taylor, G. (2001). Globalization and the environment: Determinants of firm self-regulation in China. *Journal of International Business Studies, 32*(3), 439–458.

Cole, M. A. (2003). Development, trade and the environment: How robust is the Environmental Kuznets Curve? *Environment and Development Economics, 8*(4), 557–580.

Cole, M. A. (2004a). Trade, the pollution haven hypothesis and the Environmental Kuznets Curve: Examining the linkages. *Ecological Economics, 48*(1), 71–81.

Cole, M. A. (2004b). US environmental load displacement: Examining consumption, regulations and the role of NAFTA. *Ecological Economics, 48*(4), 439–450.

Dasgupta, S., Laplante, B., Mamingi, N., et al. (2001). Inspections, pollution prices, and environmental performance: Evidence from China. *Ecological Economics, 36*(3), 487–498.

Dean, J. M., Lovely, M. E. (2008) Trade growth, production fragmentation and China's environment. NBER Working Paper # 13860.

Dean, J. M. (2002). Does trade liberalization harm the environment? A new test. *Canadian Journal of Economics, 35*(10), 819–842.

Eiser, J. R., Reicher, S. D., & Podpadec, T. J. (1996). Attitudes to privatization of UK public utilities: Anticipating industrial practice and environmental effects. *Journal of Consumer Policy, 19*(2), 193–208.

Frankel, J., & Rose, A. K. (2005). Is trade good or bad for the environment? Sorting out the causality. *Review of Economics and Statistics, 87*(1), 85–91.

He, J. (2010). What is the role of openness for China's aggregate industrial SO_2 emission? A structural analysis based on the Divisia Decomposition Method. *Ecological Economics, 69*(4), 868–886.

He, J. (2009) China's industrial SO_2 emissions and its economic determinants: EKC's reduced vs. structural model and the role of international trade. Environment and Development Economics, 14(2): 227–262.

He, J. (2006). Pollution haven hypothesis and environmental impacts of foreign direct investment: The case of industrial emission of sulfur dioxide (SO_2) in Chinese provinces. *Ecological Economics, 60*(1), 228–245.

He, C., & Zhu, S. (2007). Economic transition and regional industrial restructuring in China: Structural convergence or divergence? *Post Communist Economies, 19*(3), 321–346.

He, C., Wei, Y. H. D., & Xie, X. (2008). Globalization, institutional change and industrial location: Economic transition and industrial concentration in China. *Regional Studies, 42*(7), 923–945.

Hettige, H., Lucas, R. E. B., & Wheeler, D. (1992). The Toxic intensity of industrial production: Global patterns, trends, and trade policy. *American Economic Review, 82*(2), 478–481.

Jaffe, A., Peterson, S., Portney, P., et al. (1995) Environmental regulation and the competitiveness of U.S. manufacturing: What does the evidence tell us? Journal of Economic Literature, 33(1): 132–163

Jahiel, A. (1997). The contradictory impact of reform on environmental protection in China. *China Quarterly, 149*(149), 81–103.

Jahiel, A. (1998). The organization of environmental protection in China. *China Quarterly, 156,* 757–787.

Jorgenson, A. K. (2007). Does foreign investment harm the air we breathe and the water we drink? A cross-national study of carbon dioxide emissions and organic water pollution in less developed countries, 1975 to 2000. *Organization and Environment, 20*(2), 137–156.

Kellenberg, D. (2008). A reexamination of the role of income for the trade and environment debate. *Ecological Economics, 68*(1–2), 106–115.

Liang, D., & Liu, T. (2017). Does environmental management capability of Chinese industrial firms improve the contribution of corporate environmental performance to economic performance? Evidence from 2010 to 2015. *Journal of Cleaner Production, 142,* 2985–2998.

Lieberthal, K. (1995). *Governing China: From revolution through reform.* New York: W.W. Norton and Co.

Liu, Y., Wang, S., Qiao, Z., et al. (2019). Estimating the dynamic effects of socioeconomic development on industrial SO_2 emissions in Chinese cities using a DPSIR causal framework. *Resources, Conservation and Recycling, 150,* 104450.

Long, X., Chen, Y., Du, J., et al. (2017). Environmental innovation and its impact on economic and environmental performance: Evidence from Korean-owned firms in China. *Energy Policy, 107,* 131–137.

Ma, X., & Ortolano, L. (2000). *Environmental regulation in China.* Lanham: Rowman & Littlefield Publishing Group.

Morin, J.-F., Dur, A., & Lechner, L. (2018). Mapping the trade and environment nexus: Insights from a new data set. *Global Environmental Politics, 18*(1), 122–139.

Oi, J. C. (1995). The role of the local state in China's transitional economy. *the China Quarterly, 144*(144), 1132–1149.

Panayotou, T. (1997). Demystifying the environmental Kuznets curve: Turning a black box into a policy tool. *Environment and Development Economics, 2,* 465–484.

Poon, J., Casas, I., & He, C. (2006). The impact of energy, transport and trade on air pollution in China. *Eurasian Geography and Economics, 47*(5), 568–584.

Qian, Y., & Weingast, B. (1997). Federalism as a commitment to market incentives. *Journal of Economic Perspectives, 11*(4), 83–92.

Sinkule, B. J., & Ortolano, L. (1995). *Implementing environmental policy in China.* Westport: Praeger.

Shin, S. (2004). Economic globalization and the environment in China: A comparative case study of Shenyang and Dalian. *Journal of Environment and Development, 13*(3), 263–294.

Suri, V., & Chapman, D. (1998). Economic growth, trade and energy: Implications for the environmental Kuznets curve. *Ecological Economics, 25,* 195–208.

Swanson, K. E., Kuhn, R. G., & Xu, W. (2001). Environmental policy implementation in rural China: A case study of Yuhang. *Zhejiang. Environmental Management, 27*(4), 481–491.

Talukdar, D., & Meisner, M. C. (2001). Does the private sector help or hurt the environment? Evidence from carbon dioxide pollution in developing countries. *World Development, 29*(5), 827–840.

Tang, S., Lo, C. W. H., & Fryxell, G. E. (2003). Enforcement styles, organizational commitment, and enforcement effectiveness: An empirical study of local environmental protection officials in urban China. *Environment and Planning a, 35*(1), 75–94.

Tang, S., Lo, C. W. H., Cheung, K. C., et al. (1997). Institutional constraints on environmental management in urban China: Environmental impact assessment in Guangzhou and Shanghai. *China Quarterly, 152*(152), 863–874.

Van Rooij, B., & Lo, C. W. (2010). Fragile convergence: Understanding variation in the enforcement of China's industrial pollution law. *Law and Policy, 32*(1), 14–37.

Van Rooij, B. (2006). *Regulating land and pollution in China, Lawmaking, compliance, and enforcement; Theory and cases.* Leiden: Leiden Univ. Press.

Wang, H., & Jin, Y. (2007). Industrial ownership and environmental performance: Evidence from China. *Environmental and Resource Economics, 36*(3), 255–273.

Wang, Y., Bi, F., Zhang, Z., et al. (2017a). Spatial production fragmentation and $PM_{2.5}$ related emissions transfer through three different trade patterns within China. *195*, 703–720.

Wang, Y., Zhang, C., Lu, A., et al. (2017b). A disaggregated analysis of the environmental Kuznets curve for industrial CO_2 emissions in China. *Applied Energy, 190*, 172–180.

Wheeler, D. (2001) Racing to the bottom: Foreign investment and air pollution in developing countries. Policy Research Working Paper Series, 10(3): 225–245.

Xie, Q., Xu, X., & Liu, X. (2019). Is there an EKC between economic growth and smog pollution in China? New evidence from semiparametric spatial autoregressive models. *Journal of Cleaner Production, 220*, 873–883.

Xing, Y., & Kolstad, C. D. (2002). Do lax environmental regulations attract foreign investment. *Environmental Resource Economics, 21*(1), 1–22.

Xu, T. (2018). Investigating environmental Kuznets curve in China-aggregation bias and policy implications. *Energy Policy, 114*, 315–322.

Xu, D., Pan, Y., Wu, C., et al. (2006). Performance of domestic and foreign-invested enterprises in China. *Journal of World Business, 41*(3), 261–274.

Xu, S., Li, Y., Miao., Y., , et al. (2019). Regional differences in nonlinear impacts of economic growth, export and FDI on air pollutants in China based on provincial panel data. *Journal of Cleaner Production, 228*, 455–466.

Young, A. (2000). The Razor's Edge: Distortions and incremental reform in the People's Republic of China. *the Quarterly Journal of Economics, 115*(4), 1091–1135.

Zhang, Z., Zhu, K., & Hewings, G. J. D. (2017). A multi-regional input-output analysis of the pollution haven hypothesis from the perspective of global production fragmentation. *Energy Economics, 64*, 13–23.

Zhao, X. B., & Zhang, L. (1999). Decentralization reforms and regionalism in China: A review. *International Regional Science Review, 22*(3), 251–281.

Chapter 7
How Does Industrial Dynamics Affect Environmental Pollution?

Economic transition is essentially a dynamic term, representing structural changes of multiple facets. Chapter 6 explores the environmental outcomes of economic transition by investigating three key manifestations of transition and unravelling the environmental performance of various driving forces of transition. On this basis, this chapter seeks to demonstrate how the process of transition, in addition to the driving forces, will affect the environment. In other words, this chapter seeks to shift the industrial structure from an exogenous variable to an endogenous one. Conventionally, it is believed that industrial structure seems doom to diversify and upgrade associated with economic growth (Usui 2012). Does the diversification and upgrading of industries necessarily indicate an improvement in environmental performance? This is probable but not always the case. If considering industrial structure as exogenously given, a diversified and sophisticated mix of industries apparently implies better environmental performance. Nevertheless, if industrial structure is considered as endogenous, the trajectories of its diversification and upgrading can be various and then affect the environment in different ways (Dumais et al. 2002; Krafft 2004; Dong et al. 2020).

Looking into the trajectories of industrial diversification of one region is revitalized in recent advances in economic geography (Kogler 2015). There is increasing theoretical thinking regarding how one region's new industries are related to its already-existing industrial bases (Boschma and Frenken 2011; Neffke et al. 2011). Various relationships are identified to categorize the trajectories of industrial development, such as the path extension, path upgrading, path importation, path branching, path diversification, and path creation (Grillitsch et al. 2018). These development "paths" can be roughly summarized by a duality between path dependence and path creation.

Path dependence depicts a dynamic process that new industries rest their development on the already-existing bases, particularly in technological terms (Martin and Sunley 2006). It implies that new industries share similar requirements for knowledge, information, and other inputs so that firms are able to extend or upgrade their incumbent productions to the new ones (Boschma and Fornahl 2011; Kogler

© Springer Nature Singapore Pte Ltd. 2020
C. He and X. Mao, *Environmental Economic Geography in China*,
Economic Geography, https://doi.org/10.1007/978-981-15-8991-1_7

2015). Such an incremental development may improve the environmental performance progressively. On the other hand, it also exposes one region to a risk of being locked-in particular types of production, as the knowledge increment diminishes gradually (Boschma 2017). Once the locked-in situation occurs, the path dependence of industrial dynamics may hinder the improvement of environmental performance.

In contrast, path creation captures the likelihood of drastic innovation that new industries tend to be unrelated to incumbent ones. By the exchange of knowledge and information between sectors and the recombination of dissimilar knowledge, one region is still likely to embrace new industries that seem unrelated to the incumbent ones (Grillitsch et al. 2018). As one place get involved in global economy or regional integration deeply, the establishment of nonlocal linkages and the utilization of nonlocal resources provide another opportunity to develop unrelated industries (Henning et al. 2013). Path breaking allows one place to deviate from its current development paths and prevent one place from development lock-in (Hassink et al. 2019). However, the orientation of path breaking seems less predictable so that its environmental performance is also uncertain.

Overall, the reason why industrial dynamic has long been identified as a panacea for green development is that industrial dynamic is commonly considered as a shift from one particular mix of industries to another one. It has also been taken for granted that economic growth will make the new one more productive and efficient. From an evolutionary view, we will recognize that these industrial mixes at different development stage are interrelated to different extents, then exerting different influences. Their differences are contingent on the heterogeneity of industrial sectors and geographical places (Zhou et al. 2016). As such, the incorporation of economic geography theory allows us to examining the environmental performance of industrial dynamics, which considers industrial dynamics as a genuine "endogenous" factor.

This chapter will first introduce the features of industrial dynamic from the view of path development and then discuss the likely environmental effects of various development paths. Next, we conduct a descriptive analysis to demonstrate the environmental performance of China's industrial dynamics. In the third section, we further construct an empirical model to examine the mechanism between industrial dynamic and environmental performance. The final section concludes.

7.1 Paths of Industrial Dynamic and Environmental Performance

Previous studies regarding growth-environment relationship have never underestimated the role of industrial dynamic (Wagner and Timmins 2009; Cole et al. 2010). Even in most cases, industrial dynamic is a frequently used term in most studies' "policy implications". This is because these studies consider the industrial dynamic from a static way. Industrial dynamic is assumed to be upgraded definitely. However, industrial dynamic is a dynamic term, taking various forms as well as following different trajectories. Various forms and trajectories can generate utterly different effects, including the environmental effects. Thus, industrial dynamic should be endogenous rather than exogenous to the growth-environment linkages. In this regard, the concept of path development (or path creation) in Economic Geography can make a contribution.

As already discussed in Chap. 2, path development has been a key concept to analyse regional development for economic geographer. Path development concerns how economic novelty (e.g. new industries, new firms, and new products) occurs in one place. Generally, economic novelty is usually an incremental process led by firms and based on the accumulation of knowledge capital (Boschma 2017; Trippl et al. 2018). By recombining the existing local knowledge and adjusting to external conditions, firms are expected to develop new productions (Isaksen 2014; MacKinnon et al. 2019a, b). In such a case, economic novelty represents path dependence feature largely that the emergence of new production is related to pre-existing bases.

Although the incremental development is more common, recent findings have further highlighted the possibility of more radical processes that economic novelty is manifested by the new but unrelated productions (Essletzbichler 2012; Simmie 2012; Castaldi et al. 2015; He et al. 2016). The knowledge for innovation can be substantially different, such as the analytical knowledge, synthetic knowledge, and symbolic knowledge (Asheim et al. 2017). Synthetic knowledge is created by modifying the pre-existing ones, but the production of analytical knowledge and symbolic requires creativity. As a result, the recombination of knowledge is not confined to the related knowledge. The recombination of unrelated knowledge by repeated experiment can lead to drastic innovation, then allowing more radical development (Grillitsch et al. 2017). In such a case, economic novelty in one region is largely represented as path creation rather than path dependence.

Industrial dynamic is essentially a process of structural changes, manifested by the development of new production as well as the likely depression of old ones. Therefore, the environmental implications of path dependence and path creation illustrate the environmental performance of industrial dynamic endogenously.

What are the environmental implications of path dependence and path creation? Regarding the path dependence, the development of related industries may produce some positive effects. "Relatedness" between industries implies either similarities in internal inputs and external demands or complementaries in input–output relations (Neffke et al. 2011; Tanner 2014; Essletzbichler 2015). Such "relatedness" increases the probability for joint actions and knowledge spill-overs and then results in collective efficiencies. Firms are expected to benefit from their "relatedness" to already-existing bases and increase their productivity (Boschma and Iammarino 2009; Boschma et al. 2013). In other words, the "relatedness" may guarantee or even amplify the agglomeration economies.

However, path dependence may also expose one region to a risk of being locked-in particular types of production. Incremental development rests its innovation on the recombination of pre-existing knowledge so that the increment of new knowledge is largely diminishing (Boschma et al. 2017). As the increment of knowledge becomes trivial, the development will hit a bottleneck. A "product space" approach may help to visualize such a possibility (Hausmann et al. 2007; Hidalgo et al. 2007). Product space consists of various types of incumbent products. Products are unevenly distributed within the space due to their different levels of bilateral relatedness. Path dependence indicates that new products tend to occur in product clusters more densely. Development lock-in will be manifested by a crowded product cluster, even leaving no room for new ones. As such, a positive effect of path dependence requires a proper degree of "relatedness" to pre-existing development bases, neither too close nor too far (Nooteboom et al. 2007). In this chapter, we may assume that this principle also applies to environmental effects, and then develop the first hypothesis.

Hypothesis 1. A moderate degree of path dependence due to incremental knowledge accumulation can promote environmental performance.

From an evolutionary view, knowledge production is not the only source of path dependence. Another key source is the institution. As shown in Chaps. 5 and 6, looking into the role of institution is especially relevant in China's case. The rise and fall of SOEs capture the path dependence of institutions exactly. However, as revealed in Chap. 3, SOEs in China are found to be less environment-friendly than their private counterparts. Similar points are also reported by Meyer and Pac (2013) and Audretsch et al. (2016). As the result, the institutional path dependence in China is likely to impact the improvement of environmental performance. To put it another way: if industrial dynamics refers to a process of marketization, environmental quality improvement is more expected to be seen. Thus, we develop the second hypothesis as follow:

Hypothesis 2. Path dependence of institutions can hold back the improvement of environmental performance.

Regarding path creation, the recombination of unrelated knowledge determines that path creation faces uncertainties. To look into its likely effects, we may trace the primary sources of path creation. Recent studies highlight the role of nonlocal linkages in path creation (Miguelez and Moreno 2018; MacKinnon et al. 2019a, b). The establishment of nonlocal linkage or the utilisation of nonlocal inputs will import new knowledge into the region (Bathelt et al. 2004; Fitjar and Rodriguez-Pose 2013). The inflows of technological contents are expected to improve the productivity and thereby benefit the environmental performance. Correspondingly, the third hypothesis is formulated as follow:

Hypothesis 3. Path creation by nonlocal linkages can input technological contents into the region and then improve the environmental performance.

Another key source of path creation is the regional institutions, particularly for lagging regions (Essletzbichler 2012; Dawley 2014; Eraydin 2016). Particularly, state-led, policy-driven models of regional development prevail in catching-up regions. This is especially the case during the central planning phases of China, when the development trajectories are largely determined by the central government (Wei 1999). Nowadays, policy supports are still considered as strategic resources for regional development. The local government is also influential in industrial selection

(Hu and Lin 2013). On the other hand, regarding the environmental issues, environmental regulation is indispensable. As already discussed in Chap. 5, environmental regulation in China has a mixed effect on environmental performance. Regarding the positive side, environmental regulation will filter out polluting productions by either command-and-control regulation or market-oriented regulation. It may further encourage the green innovation. As for the negative side, there are also risks of firms to increase competitiveness by manipulating environmental regulation. Taken together, the institution may help to alter the development trajectory and even prevent regions from being locked-in. However, its environmental effects are uncertain, depending on the specific context of development. The last hypothesis can be developed as follow:

Hypothesis 4. Although policy instruments contribute to path creation by selection effect, they do not necessarily improve environmental performance.

7.2 Industrial Dynamics and Environmental Performance in China

7.2.1 Data and Methods

Like previous chapters, the industry-related data are primarily aggregated from the firm-level database, China's Annual Survey of Industrial Firms (ASIF), which is administered by the National Bureau of Statistics of China. It also enables the estimation of total factor productivity (TFP) following the approach by Levinsohn and Petrin (2003).

Region-specific data, including gross domestic product, foreign investment, imports and exports, and fiscal expenditure are taken from *China Statistical Yearbook for Regional Economy*. Industry-specific emission of Sulphur Dioxide at the country level is taken from *China Statistical Yearbook on Environment*. Industrial emissions of Sulphur dioxide at the prefecture-level are collected from *China City Statistical Yearbook*.

With regard to environmental regulation, we use the ratio of domestic water centralized sewage treatment. Why can this performance-based indicator serve as a proxy for environmental regulation? The reason is related to the context of China. Environmental regulation in China largely relies on the command-and-control modes, representing the scope and efforts of both the central and local government. The centralized sewage treatment is a public service, directly reflecting the desire and capacity of local government to improve environmental performance. As such, a high ratio of centralized sewage treatment mirrors a tendency to highlight the role of environmental regulation. On the other hand, this proxy is based on the performance of wastewater discharges. It can avoid the likely endogeneity issues since the dependent variable is based on the performance of SO_2 emissions.

Industrial dynamics in this chapter primarily focus on the entry and exit of industries in cities. We define the entry and exit of the industry with the following rules. For industry i and city c, industry i is identified as a new industry to city c if it does not occur in the base year but exists in the target year. Likewise, if industry i can be found in the base year but disappear in the target year, it is defined as the exit. Considering the data accessibility and quality, the base year and target year are 2003 and 2007, respectively.

In order to capture the changes in pollution intensity in the wake of industrial dynamics, we apply the weighted average to evaluate the average pollution intensity of new industries, pre-existing industries, and exiting industries, respectively. Industry-specific pollution intensity of the whole nation is used as the weight. This is built on the premise that the level of technology utilization within one sector is relatively similar across the nation. However, due to the scarcity of more specific data, the weight has to be at the two-digit level. In such a case, this approach only captures the between-industry difference, not the within-industry one, since the entry and exit of industries are defined at the four-digit level.

We introduce an indicator, *density*, proposed by Hidalgo et al. (2007), to the measure proximity between new industries and the pre-existing industrial structure. It is formulated as Eq. (7.1).

$$\text{density}_{i,c} = \sum_j x_{j,c}\phi_{i,j} / \sum_j \phi_{i,j} \tag{7.1}$$

where for industry i and j in city c, x indicates whether one city has revealed comparative advantages (RCA) in a certain industry; if so, it takes a value of 1. ϕ measure inter-industry technological relatedness based on the RCA, as can be seen in Eq. (7.2).

$$\phi_{i,j} = \min\{P(\text{RCA}_{c,i} > 1|\text{RCA}_{c,j} > 1), P(\text{RCA}_{c,j} > 1|\text{RCA}_{c,i} > 1)\} \tag{7.2}$$

where RCA is measured based on employment (E) as follows:

$$\text{RCA}_{i,c} = E_{i,c} / \sum_i E_{i,c} / (\sum_c E_{i,c} / \sum_{c,i} E_{i,c}) \tag{7.3}$$

7.2.2 Changes of Pollution Intensity in the Wake of Industrial Dynamics

Theoretically, industrial dynamics (in terms of the entry and exit of industries) is expected to improve the environment if it is able to drive out the dirtier industries or to replace the pre-existing industries with cleaner ones. However, as can be seen in

Fig. 7.1, evidence from Chinese cities shows that this is not always the case. Dots in the two charts of Fig. 7.1 are scattered above and below the line of equality ($X = Y$), suggesting that the average pollution intensity of new industries or exit industries can be either larger or smaller than that of pre-existing industries. That is to say, industrial dynamics in Chinese cities do not guarantee a process of replacing dirty with clean industries. Hence, the question is what kinds of industries and regions will be able to witness such industrial dynamics.

The upper graphs in Fig. 7.2 reveal that a cumulative effect can be seen in the case of China. It shows a negative relationship between technological relatedness and pollution intensity, suggesting that cities with higher levels of technological

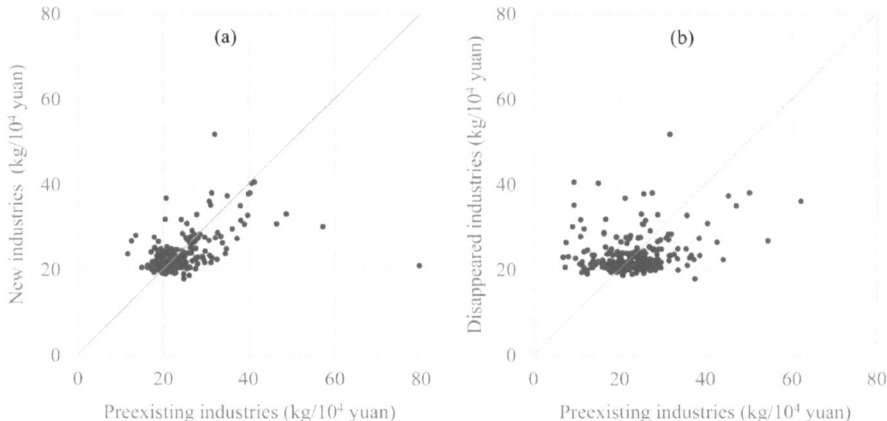

Fig. 7.1 The average pollution intensity of pre-existing industries and new industries (**a**)/exit industries (**b**)

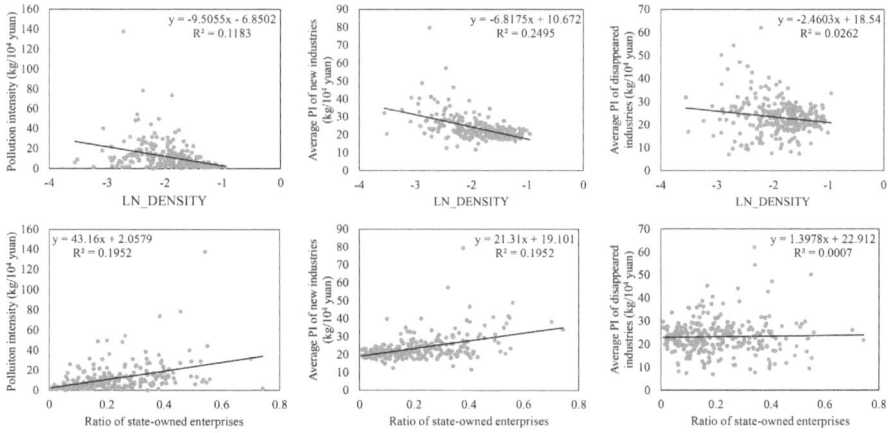

Fig. 7.2 Pollution intensity and path dependence

relatedness tend to have lower average pollution intensity. This effect is more clear in the new industries than the exit ones; that is to say, technological relatedness helps to introduce cleaner industries rather than drive out dirty ones. On the contrary, the lower graphs in Fig. 7.2 support the second hypothesis. The ratio of state-owned enterprises is positively correlated with pollution intensity, indicating a likely lock-in effects for regions. This kind of path dependence will keep new industries from enhancing environmental performance, while its effect on the exit industries is less significant.

The effect of foreign investment is similar to technological relatedness. The upper graphs in Fig. 7.3 show that the level of foreign investment is negatively correlated with the average pollution intensity of cities and new industries, while its relationship with the exit of industries is rather weak. It captures the fact that external linkages like foreign investment are likely to foster environmental improvement through enhancing the environmental performance of new industries. As can be seen in the lower graphs in Fig. 7.3, the relationship between pollution intensity and environmental regulation is negative but weak. It looks more likely that environmental regulation reduces pollution intensity by introducing cleaner industries rather than evicting dirtier industries. Overall, it turns out that path creation is likely to be equipped with capacities to reduce pollution intensity, as predicted by the theoretical discussed above.

It should also be noted that the level of regional inequality in China keeps expanding associated with the drastic integration with the global market (Zhang and Zhang 2003). A wide range of industrial relocation has contributed to regional inequality, which has been well documented by previous studies (Huang and Li 2007; Wen 2004). Subsequently, more efforts have been made to investigate whether pollution production transfers in the wake of industrial mobility (Cole et al. 2011; Zhao and Yin 2011). As a result, there are increasing concerns about the likely occurrence of the pollution haven effect in lagging regions (He et al. 2014), in particular, the inland of China. We argue that the spatial mobility of some pollution-intensive industries does

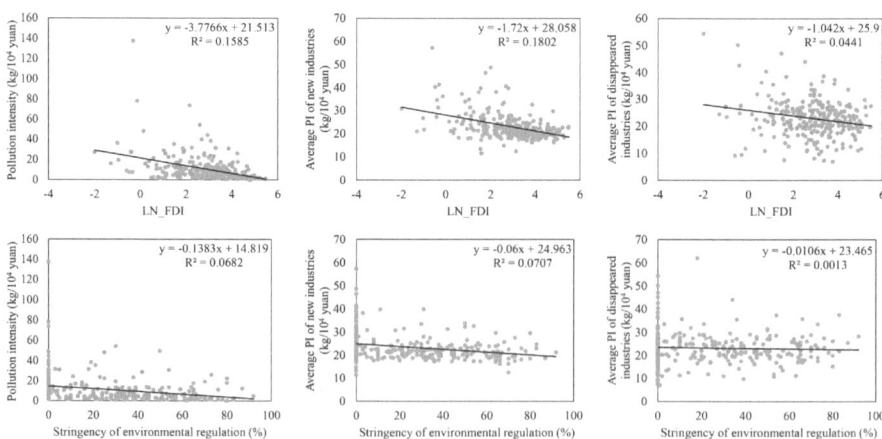

Fig. 7.3 Pollution intensity and path creation

Table 7.1 Distribution of the average pollution intensity of pre-existing, new, and disappeared industries across three Chinese Economic Regions

	Three Chinese economic regions		
	East	Central	West
Average pollution intensity of pre-existing industries (PRAPI)	20.70	22.44	26.43
Average pollution intensity of new industries (ENAPI)	21.86	22.11	26.88
Average pollution intensity of disappeared industries (EXAPI)	22.47	23.34	23.78
Number of cities	87	95	90
No. of cities with ENAPI<PRAPI	41	54	42
No. of cities with EXAPI>PRAPI	57	51	39
No. of cities with ENAPI>EXAPI	33	41	50

not necessarily refer to the pollution haven effect. Firstly, we can learn from Table 7.1 that it is not only the inland cities but also the coastal cities that have welcomed new industries with higher pollution intensity than existing ones. The developed coastal region does not filter out the dirty industries as predicted by the pollution haven effect. Secondly, there are still a considerable number of coastal cities having the exit of industries with lower average pollution intensity than the pre-existing ones. It captures the fact that the coastal cities do not only export pollution-intensive industries to the inland. Thirdly, almost half of the cities in the east, central, and west regions, respectively, introduced new industries with higher pollution intensity than their disappeared counterparts. That is to say, the rising coastal cities are not at the expense of the inland cities' equity of development and environment.

Hence, the developing-developed division, which has been widely adopted in existing hypotheses, may not be a robust perspective to reveal the role of industrial dynamics in the environment. In contrast, the evolutionary perspective is comprised of path dependence and path creation, which may exert opposing influences on the environment. In such a case, pollution intensity can be seen as the equilibrium of path dependence and path creation. Since path dependence and path creation are determined by a wide range of factors, endogenously or exogenously, the evolutionary perspective enables the incorporation of more factors than the developing-developed divisions. As a result, industrial dynamics from the evolutionary perspective would be authentically "dynamic".

7.3 Empirical Analysis

7.3.1 Model Specification

The empirical tests are to achieve two objectives. One is to examine how industrial dynamics affect industrial pollution emissions at the prefectural level, and the other is to investigate its mechanism via path dependence and path creation. More specifically, we seek to investigate two pairs of differential effects. The first is the difference between and within the effects of path dependence and path creation on pollution intensity. The second is between the coastal (relatively developed) and inland (relatively developing) regions, regarding the effect of industrial dynamics on pollution intensity.

Equation (7.4) is used to achieve the first objective. It is based on the idea of the classical decomposition of scale–composition–technique, proposed in the early 1990s (Grossman and Krueger 1993; Copeland and Taylor 1994).

$$\text{pollution}_{c,t2} = \alpha + \beta f\left(\text{scale}_{c,t2}\right) + \gamma f\left(\text{composition}_{c,t2}\right) + \delta f\left(\text{technique}_{c,t2}\right) + e_{c,t2} \tag{7.4}$$

where c refers to the prefecture, t_2 is the year 2007. To be specific, the industrial SO_2 emission per capita is employed to denote the level of industrial pollution. Regarding the scale effect, gross domestic product per capita (GDPPC) and its quadratic term are incorporated (GDPPCSQ). Regarding the technique effect, previous studies prefer two kinds of variables. The first refers to endogenous development, applying the productivity-related ones (Grether et al. 2009). In contrast, the second turns to the external inflows of foreign investment (Bao et al. 2011). Thus, the average total factor productivity (TFP) and the ratio of FDI to GDP (FDI) are applied to represent the technique effect. Last, the entry and exit of industries exactly indicate the compositional changes. On the one hand, the average pollution intensity of new industries (ENAPI) and exit industries (EXAPI) measure the quality of industrial dynamics. On the other hand, the ratios of new industries (ENRATE) and exit industries (EXRATE) in one prefecture represent between industrial dynamics. Meanwhile, to capture the within-industrial dynamics, the ratios of new firms and exit firms within a particular industry in one prefecture are incorporated. According to the distribution of industrial pollution intensity at the 2-digit level, those manufacturing sectors are subdivided into four categories based on the first quartile, the median, and the third quartile. Thus, the within-industrial entry rate and exit rate are calculated based on these four categories, respectively (Fig. 7.4).

Equation (7.5) is constructed for investigating the hybrid mechanism of path dependence and path creation.

$$\begin{aligned} \text{API}_{c,t2} = {} & \alpha + \beta f\left(\text{technological relateness}_{c,t1}, \text{institutional legacy}_{c,t1}\right) \\ & + \gamma f\left(\text{external linkages}_{c,t1}, \text{regional institutions}_{c,t1}\right) + X_{c,t2}\delta + e_c \end{aligned} \tag{7.5}$$

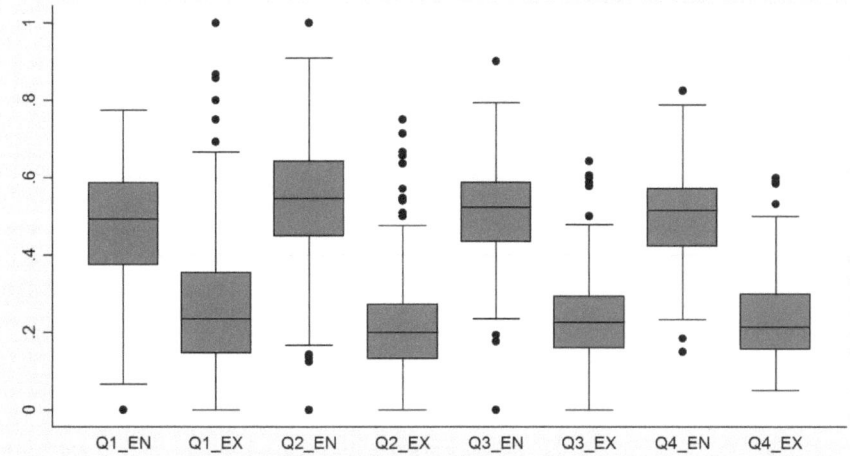

Fig. 7.4 The distribution of within-industrial entry rate and exit rate based on the industrial subcategories by pollution intensity at the 2-digit level. *Note* Q1, Q2, Q3, and Q4 refers to four groups of manufacturing sectors divided by the quartiles of their industrial pollution intensity. EN and EX refer to the proportion of new firms and exit firms in the sectors, respectively. Thus, Q1_EN indicates the entry rate of firms within industries that pollution intensity lies below the first quartile. All else follows

where API denotes the average pollution intensity, t_1 and t_2 are the year 2003 and 2007, respectively. It can be either the API of new industries or the exit industries. It can be seen that this model is operated at the prefectural level. This is because what makes industrial dynamics vary from one region to another can be highly region-specific. Although industrial dynamics in most regions are featured as path creation amidst path dependence, industrial dynamics in some regions can be much more path dependent than in the others.

Following studies on evolutionary economic geography, we identify technological relatedness as one of the primary sources of path dependence, measured by the DENSITY at the city level. A higher level of density indicates a larger path-dependent effect. Another source of path dependence is identified based on economic transition in China, namely, the state-dominated ownership structure of certain industries. The strong state-owned enterprises are distributed in a wide range of industrial sectors, in particular, most of the capital-intensive sectors. This phenomenon is seen as the legacy of the central planning system in Chinese early development. Since SOEs tend to be less profit-oriented but with more bargaining power than their private counterparts (Wang et al. 2003), SOEs have become more sensitive to the government than the market. In this regard, SOEs tend to witness a belated entry or exit to become path-dependent. At the city level, we use the ratio of SOEs in terms of gross output (STATE) to represent this effect. The higher the ratio, the more probable the city will be trapped by pre-existing production.

With regard to the path creation effect, we have also identified two sources, namely, external linkages and regional institutions. The role of external linkages in path creation has been supported by a series of empirical studies (He et al. 2016). It can be achieved through the utilization of foreign investment, the integration of the global market, and the migration of the labour force. China's economic take-off has long been identified as an outcome of export-oriented development strategy (Gereffi 2009). Through transforming China into a so-called "world factory", FDI and international trade make an essential contribution to its industrial restructuring and relocation (He et al. 2008). Under this condition, we use the ratio of FDI to GDP (FDI) and the trade openness (OPEN) to indicate the path creation effect from external linkages. Better performance in the utilization of FDI and participation in international trade may refer to a larger probability of path creation.

Moreover, the transitional development in China highlights the role of institutions, especially the public sectors. Although the process of marketization in China has gone deeper, the government is still equipped with great motivations and capacities in regulating industrial dynamics, directly or indirectly (Barbieri et al. 2012; Zhou et al. 2016). On the one hand, the local government is still likely to introduce or protect preferred industries through policy supports, fee waivers, and tax refund, under the pressure of interregional competition (Thun 2004). On the other hand, the government may also use regulatory toolkit to filter out the underqualified industries (Sengupta 2010). Thus, we incorporate fiscal expenditure per capita (PUB) and firm subsidies from the government (SUB) to indicate the industrial selection via economic policy tools. The environmental regulation, which is measured by the ratio of wastewater centralized sewage treatment (ER), is also applied to denote industrial selection via regulatory policy tools. Overall, a higher level of these policies may increase the probability of path creation.

7.3.2   Empirical Results

The summary statistics of the variables are shown in Table 7.2. All of the models are estimated by OLS based on several preliminary tests. Since the Breusch-Pagan test for heteroscedasticity rejects the null hypothesis and concludes heteroscedasticity, the robust regression is applied to estimate the models. Moreover, the correlation between variables is overall lower than 0.5, and variance inflation factors (VIFs) for the independent variables are below 5, suggesting the estimations are free from severe multicollinearity. Results of estimation are listed in Table 7.3 and Table 7.4, respectively.

Table 7.3 reports the effect of industrial dynamics on pollution emissions at the prefectural level. The first and second models examine the effect due to the magnitude and quality of industrial dynamics, whereas the third and fourth models regard the effect of industry entry and exit, respectively. The results clearly show that active industrial dynamics in one city has led to the increase of industrial pollution emissions, as predicted by environmental studies and claimed by official documents of

Table 7.2 Summary statistics of the variables

Variables	Description	Mean	Std. dev	Min	Max
SEPC	Industrial Sulphur dioxide emissions per capita in 2007	0	1	−0.76	7.12
GDPPC	Gross domestic production per capita in 2007	0	1	−1.10	4.79
GDPPCSQ	Quadratic term of GDPPC	1	2.38	0	22.97
ENRATE	The ratio of new industries in prefectures	0	1	−2.05	3.67
EXRATE	The ratio of disappeared industries in prefectures	0	1	−1.41	5.82
ENAPI	The average pollution intensity of new industries in prefectures	0	1	−1.78	8.32
EXAPI	The average pollution intensity of disappeared industries in prefectures	0	1	−2.17	5.18
Q1_EN	The entry rate of firms within industries that pollution intensity lies below the first quartile	0	1	−2.69	1.79
Q1_EX	The exit rate of firms within industries that pollution intensity lies below the first quartile	0	1	−1.51	3.99
Q2_EN	The entry rate of firms within industries that pollution intensity lies between the first quartile and median	0	1	−3.41	2.93
Q2_EX	The exit rate of firms within industries that pollution intensity lies between the first quartile and median	0	1	−1.69	3.96
Q3_EN	The entry rate of firms within industries that pollution intensity lies between the median and third quartile	0	1	−4.23	3.21
Q3_EX	The exit rate of firms within industries that pollution intensity lies between the median and third quartile	0	1	−2.25	3.81
Q4_EN	The entry rate of firms within industries that pollution intensity lies between the third quartile and maximum	0	1	−3.00	2.72
Q4_EX	The exit rate of firms within industries that pollution intensity lies between the third quartile and maximum	0	1	−1.80	3.60
PRAPI	The average pollution intensity of pre-existing industries in prefectures	0	1	−1.21	6.51
FDI_t2	The ratio of foreign investment over GDP in 2007	0	1	−0.97	4.12
TFP	The total factor productivity estimated by LP approach in 2007	0	1	−3.28	2.70
DENSITY	Indicator proposed by Hidalgo et al. (2007)	0	1	−1.86	2.95
STATE	Ratio of the output of state-owned firms over gross output in prefectures in 2003	0	1	−1.46	3.81

(continued)

Table 7.2 (continued)

Variables	Description	Mean	Std. dev	Min	Max
FDI_t1	The ratio of foreign investment over GDP in 2003	0	1	−0.72	8.21
OPEN	The ratio of imports and exports over GDP in prefectures in 2003	0	1	−0.47	8.36
PUB	Prefectural public expenditure per capita	0	1	−0.51	13.22
SUB	Average firm subsidies from government	0	1	−0.56	8.53
ER	Ratio of wastewater centralized sewage treatment in 2003	0	1	−1.01	2.55

Table 7.3 Industrial dynamics as the composition effect

	Industrial SO_2 emissions per capita in prefectures (SEPC)					
	1	2	3	4	5	6
GDPPC	0.729^{***}	0.619^{***}	0.715^{***}	0.648^{***}	0.592^{***}	0.638^{***}
GDPPCSQ	−0.072	−0.058	−0.071	−0.063	−0.040	−0.058
ENRATE	0.187^{***}		0.208^{***}			
EXRATE	0.286^{**}			0.191^{**}		
ENAPI		0.173^{**}	0.040^{**}			
EXAPI		0.098		0.114		
Q1_EN					-0.415^{***}	
Q2_EN					0.150	
Q3_EN					0.002	
Q4_EN					0.165^{*}	
Q1_EX						0.371^{**}
Q2_EX						−0.123
Q 3_EX						−0.064
Q4_EX						−0.044
FDI	-0.270^{***}	-0.275^{***}	-0.272^{***}	-0.265^{***}	-0.250^{***}	-0.279^{***}
TFP	-0.113^{**}	−0.079	-0.104^{*}	-0.079^{*}	−0.072	-0.101^{**}
Observations	272	272	272	272	272	272
R-squired	0.30	0.27	0.30	0.28	0.33	0.30
F	5.09	4.48	4.75	4.94	3.58	4.49

Note $*p < 0.10$; $**p < 0.05$; $***p < 0.01$

the government (He et al. 2012; Vennemo et al. 2009). The coefficients of ENRATE and EXRATE are statistically significant and positive. As expected, the inflow of new industries will bring in more pollution emissions simultaneously. Nevertheless, the local environment does not benefit from the exit of industries. On the contrary, the more industries exit, the higher the pollution emissions would be. One likely

Table 7.4 Effect of path dependence and path creation on the pollution intensity of new industries

	Average pollution intensity of new industries (ENAPI)				
	Nation	Developed	Developing	Coastal	Inland
Path dependence					
DENSITY	-0.366^{***}	-0.365^{***}	-0.295^{***}	-0.380^{***}	-0.324^{***}
STATE	0.137^{**}	0.165	0.132^{**}	0.268^{**}	0.115^{**}
Path creation					
FDI	0.416^{*}	0.588^{**}	-0.001	0.523^{**}	-0.042
OPEN	-0.210	-0.316^{*}	0.018	-0.278	-0.213
PUB	0.059	0.120	-0.232	0.061	0.153
SUB	-0.069^{**}	-0.058^{*}	-0.085	-0.023	-0.124^{**}
ER	-0.093^{**}	-0.165^{*}	-0.090	-0.127	-0.105^{**}
Control variables					
PRAPI	0.438^{***}	0.359^{***}	0.428^{***}	0.099	0.405^{***}
ENRATE	-0.369^{***}	-0.334^{**}	-0.328^{***}	-0.386^{***}	-0.320^{***}
Observations	272	90	182	87	185
R-squired	0.52	0.62	0.52	0.60	0.53
F	14.39^{***}	5.71^{***}	11.95^{***}	2.32^{**}	13.83^{***}

Note $*p < 0.10$; $**p < 0.05$; $***p < 0.01$

reason is that the motivation of industrial relocation is diversified, including industrial upgrading and regional rebalance in spite of the sustainable environment (Liao and Chan 2011), indicating that the exit industries do not have to be dirtier. Even so, geographical relocation is not the only way for dirty industries to respond to the city, as innovation, upgrading, and outsourcing are feasible alternatives (Zhu et al. 2014).

Through delving into the within-industrial dynamics, the results of the fifth and sixth models show that cities in China primarily benefit from the entry of the cleaner industries. Meanwhile, the entry of the dirtier industries has also increased pollution emissions to some extent. Its coefficient is smaller and less significant than Q1_EN. With regard to the exit of firms within one industry, only the coefficient of Q1_EX is statistically significant and positive. The results clearly reveal that the exit of firms within industries does not benefit the urban environment in China. The exit of firms within the cleanest industries increases the environmental pressure, while the exit of firms within other dirty industries is unable to efficiently benefit the environment.

This trend also can be seen through the results of the average pollution intensity. The results of the second model showed that higher average pollution intensity will naturally lead to a higher level of pollution emissions, while higher average pollution intensity of the exit industries is not necessarily able to reduce the pollution emissions. Overall, industrial dynamics have contributed to the increase in industrial pollution emissions across the whole nation. The entry of new industries has pushed up the level of pollution emissions, while introducing new firms within those cleanest industries has alleviated the environmental pressure to a large extent. In contrast, Chinese cities do not benefit from the exit of existing industries environmentally. Obviously, the industrial dynamics of introducing the clean and evicting the dirty is not always the case in China.

Since the effect of industrial dynamics primarily stems from the entry of new industries, we further examine how path dependence and path creation contribute to the average pollution intensity of new industries. The results are reported in Table 7.4. Firstly, the observations across the whole nation are incorporated into the model. Path dependence due to technological relatedness is able to reduce the pollution intensity, while the one due to institutional legacy will increase the pollution intensity. A similar trend was also found previously within industrialized provinces (Shin 2004). In contrast to the path dependence, the effect of path creation is relatively weak. It mainly comes from regional institutions. The coefficients of the SUB and ER are significant and negative, suggesting a positive role of new industries in enhancing the environmental performance. This is because the role of external linkages is somehow conditional by institutional factors. Evidence at the firm level (Child and Tsai 2005) and the provincial level (Lan et al. 2012) has been documented.

Secondly, we operate the model based on the developed-developing division of cities. It turns out that the positive effect of technological relatedness on environmental improvement holds in developing and developed cities, while the negative effect of institutional legacy can only be found in developing cities. It also discloses the fact that the environment of transition economies may benefit from their privatization process. Likewise, the role of path creation can only be seen in developed cities, too. Regional institutions may exert positive but still weak influences in environmental improvement in developed cities. However, the effect of external linkages is mixed. Integration with the global market may help to reduce the pollution intensity of new industries to a limited extent, while the utilization of foreign investment does increase pollution intensity to a considerable extent. This is partially related to the measure of average pollution intensity. Due to the constraints of data availability, the ENAPI is weighted by the pollution intensity at the 2-digit level, thus, it only represents the between industry effect. Since the utilization of FDI has been identified as the essential force to drive the industrial upgrading (Bao et al. 2011; Gereffi 2009), the positive correlation between ENAPI and FDI indicates a significant upgrading process in the developed cities, shifting from a labour-intensive dominated to a capital-intensive dominated industrial mix. However, the difference along the value chain within one industry also matters, which is unable to be captured by the weighted average pollution intensity.

Thirdly, we further estimate the model with observations based on the coastal-inland division of cities. Regarding the effect of path dependence, the results are consistent with other models; that is, technological relatedness reduces the pollution intensity, while institutional legacy has witnessed the opposite. The coastal-inland division is clearly manifested in the effect of path creation. In the coastal cities, the effect of path creation is generated by the utilization of foreign investment, which is inconsistent with the advantage of coastal cities in external linkages. In stark contrast, the inland cities may take advantage of path creation through regional institutions, including firm subsidies and environmental regulations. Firm subsidies allow the government to select the preferred industries so that to introduce clean industries, while stringent environmental regulation may serve as a filter to keep the dirty industries away.

7.4 Conclusion and Discussion

In most growth-environment studies, industrial dynamics in the wake of economic growth tend to be hypothesized as a process of introducing the clean and evicting the dirty. It entails a developing-developed impasse, confronting the lagging regions a trade-off between the equity of development and the environment. In this chapter, we argue that the paths of industrial dynamics are diversified and highly region-specific, therefore, the impasse is not always the case. Industrial dynamics in most Chinese cities is featured as path creation amidst path dependence, while some cities are prone to be more path dependent than the others. In this regard, not only the industrial mix is able to alter the environmental performance, but its paths also matter.

By using data covering 272 prefecture-level cities during 2003–2007 in China, this chapter has evidenced that relatively developed cities may welcome dirty industries, while their developing counterparts may filter them out. The pollution haven effect does not hold across the whole country. Introducing new industries across the whole nation has overall increased industrial pollution emissions, while at the same time, the pollution emissions have also been effectively alleviated by introducing new firms within the clean industries. Moreover, the exit industries, especially the exit firms within those clean industries, have increased pollution emissions to a large extent. It can be seen that China does not witness a process of filtering out the dirty industries.

The effect of path dependence can be seen as a trade-off between the technological relatedness and institutional legacy. In Chinese cities, technological relatedness can generate a cumulative effect to lead to stepwise improvement of the environmental performance. On the contrary, the institutional legacy has inevitably curbed environmental quality improvement due to a lack of productivity. In this regard, path dependence is one of the channels where the marketization process may affect the environment. In contrast, the effect of path creation is much more region-specific than path dependence, leaving more room for the process of globalization and decentralization in China.

The developed-developing division and the coastal-inland division of empirical results may offer insights into urban sustainable development. Firstly, taking advantage of the path creation effect is much difficult in developing cities than their developed counterparts. Evidence from Chinese cities has even shown that path creation only can be seen in developed cities. If this is the case, this finding may highlight the role of path dependence in the environmental improvement of lagging regions. Taking advantage of the stepwise improvement due to technological relatedness and fostering the process of marketization may be a more efficient way for lagging regions.

Secondly, the coastal and inland cities have their own way to achieve path creation. As coastal cities have better access to the global market as well as a better domestic industrial base (Coşar and Fajgelbaum 2016), the coastal cities may have comparative advantages in the utilization of external linkages. In contrast, policy supports and regulations from the government may help the inland cities to create new paths, then improving the environmental performance of industries.

Thirdly, regarding the role of environmental regulation, evidence from China supports the positive role of stringent environmental regulation in enhancing environmental performance. However, its effect is conditional by developing status and geographical location. As a source of the path creation effect, it will achieve better results in developed cities. Meanwhile, as a component of regional institutions, it will serve as an effective tool for the inland cities.

References

Asheim, B., Grillitsch, M., & Trippl, M. (2017). Introduction: Combinatorial knowledge bases, regional innovation, and development dynamics. *Economic Geography, 93,* 429–435.

Audretsch, D., Guo, X. D., Hepfer, A., et al. (2016). Ownership, productivity and firm survival in China. *Economia e Politica Industriale, 43*(1), 67–83.

Bao, Q., Chen, Y. Y., & Song, L. G. (2011). Foreign direct investment and environmental pollution in China: A simultaneous equations estimation. *Environment and Development Economics, 16*(1), 71–92.

Barbieri, E., Di Tommaso, M. R., & Bonnini, S. (2012). Industrial development policies and performances in Southern China: Beyond the specialized industrial cluster program. *China Economic Review, 23*(3), 613–625.

Bathelt, H., Malmberg, A., & Maskell, P. (2004). Clusters and knowledge: Local buzz, global pipelines and the process of knowledge creation. *Progress in Human Geography, 28*(1), 31–56.

Boschma, R. (2017). Relatedness as driver of regional diversification: A research agenda. *Regional Studies, 51*(3), 351–364.

Boschma, R., Coenen, L., Frenken, K., et al. (2017). Towards a theory of regional diversification: Combining insights from evolutionary economic geography and transition studies. *Regional Studies, 51*(1), 31–45.

Boschma, R., & Fornahl, D. (2011). Cluster evolution and a roadmap for future research. *Regional Studies, 45*(10), 1295–1298.

Boschma, R., & Frenken, K. (2011). The emerging empirics of evolutionary economic geography. *Journal of Economic Geography, 11*(2), 295–307.

Boschma, R., & Iammarino, S. (2009). Related variety, trade linkages, and regional growth in Italy. *Economic Geography, 85*(3), 289–311.

Boschma, R., Minondo, A., & Navarro, M. (2013). The emergence of new industries at the regional level in Spain: A proximity approach based on product relatedness. *Economic Geography, 89*(1), 29–51.

Castaldi, C., Frenken, K., & Los, B. (2015). Related variety, unrelated variety and technological breakthroughs: An analysis of US state-level patenting. *Regional Studies, 49*(5), 767–781.

Child, J., & Tsai, T. (2005). The dynamic between firms' environmental strategies and institutional constraints in emerging economies: Evidence from China and Taiwan. *Journal of Management Studies, 42*(1), 95–125.

Cole, M. A., Elliott, R. J. R., & Okubo, T. (2010). Trade, environmental regulations and industrial mobility: An industry-level study of Japan. *Ecological Economics, 69*(10), 1995–2002.

Cole, M. A., Elliott, R. J. R., & Zhang, J. (2011). Growth, foreign direct investment, and the environment: Evidence from Chinese cities. *Journal of Regional Science, 51*(1), 121–138.

Copeland, B. R., & Taylor, M. S. (1994). North-south trade and the environment. *The Quarterly Journal of Economics, 109*(3), 755–787.

Coşar, A. K., & Fajgelbaum, P. D. (2016). Internal geography, international trade, and regional specialization. *American Economic Journal: Microeconomics, 8*(1), 24–56.

Dawley, S. (2014). Creating new paths? Offshore wind, policy activism, and peripheral region development. *Economic Geography, 90*(1), 91–112.

Dong, Z., Chen, W., & Wang, S. (2020). Emission reduction target, complexity and industrial performance. *Journal of Environmental Management, 260,* 110148.

Dumais, G., Ellison, G., & Glaeser, E. L. (2002). Geographic concentration as a dynamic process. *The Review of Economics and Statistics, 84*(2), 193–204.

Eraydin, A. (2016). The role of regional policies along with the external and endogenous factors in the resilience of regions. *Cambridge Journal of Regions, Economic, and Society, 9*(1), 217–234.

Essletzbichler, J. (2012). Renewable energy technology and path creation: A multi-scalar approach to energy transition in the UK. *European Planning Studies, 20*(5), 791–816.

Essletzbichler, J. (2015). Relatedness, industrial branching and technological cohesion in US metropolitan areas. *Regional Studies, 49*(5), 752–766.

Fitjar, R., & Rodriguez-Pose, A. (2013). When local interaction does not suffice: Sources of firm innovation in urban Norway. *Environment and Planning A, 43,* 1248–1267.

Gereffi, G. (2009). Development models and industrial upgrading in China and Mexico. *European Sociological Review, 25*(1), 37–51.

Grether, J.-M., Mathys, N. A., & Melo, J. (2009). Scale, technique and composition effects in manufacturing SO_2 emissions. *Environmental and Resource Economics, 43*(2), 257–274.

Grillitsch, M., Asheim, B., & Trippl, M. (2018). Unrelated knowledge combinations: The unexplored potential for regional industrial path development. *Cambridge Journal of Regions, Economy and Society, 11,* 257–274.

Grillitsch, M., Martin, R., & Srholec, M. (2017). Knowledge base combinations and innovation performance in Swedish regions. *Economic Geography, 93,* 458–479.

Grossman, G. M., & Krueger, A. B. (1993). Environmental impacts of a North American free trade agreement. In P. M. Garber (Ed.), *The U.S.-Mexico free trade agreement* (pp. 13–56). Cambridge MA: MIT Press.

Hassink, R., Isaksen, A., & Trippl., M. (2019). Towards a comprehensive understanding of new regional industrial path development. *Regional Studies.* https://doi.org/10.1080/00343404.2019. 1566704.

Hausmann, R., Hwang, J., & Rodrik, D. (2007). What you export matters. *Journal of Economic Growth, 12*(1), 1–25.

He, C. F., Huang, Z. J., & Ye, X. Y. (2014). Spatial heterogeneity of economic development and industrial pollution in urban China. *Stochastic Environmental Research and Risk Assessment, 28*(4), 767–781.

He, C. F., Pan, F. H., & Yan, Y. (2012). Is economic transition harmful to China's Urban environment? Evidence from industrial air pollution in Chinese cities. *Urban Studies, 49*(8), 1767–1790.

He, C. F., Wei, Y. H., & Xie, X. Z. (2008). Globalization, institutional change and industrial location: Economic transition and industrial concentration in China. *Regional Studies, 42*(7), 923–945.

He, C. F., Yan, Y., & Rigby, D. (2016). Regional industrial evolution in China. *Papers in Regional Science.* https://doi.org/10.1111/pirs.12246.

Henning, M., Stam, E., & Wenting, R. (2013). Path dependence research in regional economic development: Cacophony or knowledge accumulation? *Regional Studies, 47*(8), 1348–1362.

Hidalgo, C. A., Klinger, B., Barabási, A. L., et al. (2007). The product space conditions the development of nations. *Science, 317*(5837), 482–487.

Hu, F. Z. Y., & Lin, G. C. S. (2013). Placing the transformation of state-owned enterprises in North-East China: The state, region and firm in a transitional economy. *Regional Studies, 47*(4), 563–579.

Huang, J. L., & Li, K. W. (2007). Foreign trade, local protectionism and industrial location in China. *Frontiers in Economics in China, 2*(1), 24–56.

Isaksen, A. (2014). Industrial development in thin regions: Trapped in path extension? *Journal of Economic Geography, 15,* 585–600.

Kogler, D. F. (2015). Editorial: Evolutionary economic geography–theoretical and empirical progress. *Regional Studies, 49*(5), 705–711.

Krafft, J. (2004). Entry, exit and knowledge: Evidence from a cluster in the info-communications industry. *Research Policy, 33*(10), 1687–1706.

Lan, J., Kakinaka, M., & Huang, X. G. (2012). Foreign direct investment, human capital and environmental pollution in China. *Environmental and Resource Economics, 51*(2), 255–275.

Levinsohn, J., & Petrin, A. (2003). Estimating production functions using inputs to control for unobservables. *Review of Economic Studies, 70*(2), 317–341.

Liao, H. F., & Chan, R. C. K. (2011). Industrial relocation of Hong Kong manufacturing firms: Towards an expanding industrial space beyond the Pearl river delta. *GeoJournal, 76*(6), 623–639.

MacKinnon, D., Dawley, S., Pike, A., et al. (2019a). Rethinking path creation: A geographical political economy approach. *Economic Geography, 95*(2), 113–135.

MacKinnon, D., Dawley, S., Steen, M., et al. (2019b). Path creation, global production networks and regional development: A comparative international analysis of the offshore wind sector. *Progress in Planning, 130,* 1–32.

Martin, R., & Sunley, P. (2006). Path dependence and regional economic evolution. *Journal of Economic Geography, 6*(4), 395–437.

Meyer, A., & Pac, G. (2013). Environmental performance of state-owned and privatized eastern European energy utilities. *Energy Economics, 36,* 205–214.

Miguelez, E., & Moreno, R. (2018). Relatedness, external linkages and regional innovation in Europe. *Regional Studies, 52*(5), 688–701.

Neffke, F., Henning, M., & Boschma, R. (2011). How do regions diversify over time? Industry relatedness and the development of new growth paths in regions. *Economic Geography, 87*(3), 237–265.

Nooteboom, B., Haverbeke, W. V., Duysters, G., et al. (2007). Optimal cognitive distance and absorptive capacity. *Research Policy, 36*(7), 1016–1034.

Sengupta, A. (2010). Environmental regulation and industry dynamics. *The B.E. Journal of Economic Analysis and Policy, 10*(1). Article 52.

Shin, S. B. (2004). Economic globalization and the environment in China: A comparative case study of Shenyang and Dalian. *Journal of Environment and Development, 13*(3), 263–294.

Simmie, J. (2012). Path dependence and new technological path creation in the Danish wind power industry. *European Planning Studies, 20*(5), 753–772.

Tanner, A. N. (2014). Regional branching reconsidered: Emergence of the fuel cell industry in European regions. *Economic Geography, 90*(4), 403–427.

Thun, E. (2004). Keeping up with the Jones': Decentralization, policy imitation, and industrial development in China. *World Development, 32*(8), 1289–1308.

Trippl, M., Grillitsch, M., & Isaksen, A. (2018). Exogenous sources of regional industrial change: Attraction and absorption of non-local knowledge for new path development. *Progress in Human Geography, 42*(5), 687–705.

Usui, N. (2012). *Taking the right road to inclusive growth: Industrial upgrading and diversification in the philippines.* Asian Development Bank.

Vennemo, H., Aunan, K., Lindhjem, H., et al. (2009). Environmental pollution in China: Status and trends. *Review of Environmental Economics and Policy, 3*(2), 209–230.

Wagner, U. J., & Timmins, C. D. (2009). Agglomeration effects in foreign direct investment and the pollution haven hypothesis. *Environmental and Resource Economics, 43*(2), 231–256.

Wang, H., Mamingi, N., Laplante, B., et al. (2003). Incomplete enforcement of pollution regulation: Bargaining power of Chinese factories. *Environmental and Resource Economics, 24*(3), 245–262.

Wei, Y. H. D. (1999). Regional inequality in China. *Progress in Human Geography, 23*(1), 48–58.

Wen, M. (2004). Relocation and agglomeration of Chinese industry. *Journal of Development Economics, 73*(1), 329–347.

Zhang, X. B., & Zhang, K. H. (2003). How does globalization affect regional inequality within a developing country? Evidence from China. *Journal of Development Studies, 39*(4), 47–67.

Zhao, X. L., & Yin, H. T. (2011). Industrial relocation and energy consumption: Evidence from China. *Energy Policy, 39*(5), 2944–2956.

Zhou, Y., He, C. F., & Zhu, S. J. (2016) Does creative destruction work for Chinese regions. *Growth and Change*. https://doi.org/10.1111/grow.12168.

Zhu, S. J., He, C. F., & Liu, Y. (2014). Going green or going way: Environmental regulation, economic geography and firms' strategies in China's pollution-intensive industries. *Geoforum, 55*, 53–65.

**Part II
Global Shift of Environmental Burdens
and the Geography of Industrial
Environmental Performance in China**

Chapter 8
Is There a Trade-Related Pollution Trap for China?

As discussed in Chap. 2, the geography of industrial pollution in China is determined by not only how places differ from each other, but also how they are interrelated. Both eco-environmental and socio-economic interrelations between places create various channels for sharing or shifting environmental burdens. In this regard, the second part of this book turns to examine how China's integration with the global economy affects its environmental performance. As the first chapter in Part II, this chapter starts with the interrelations between nations at the global scale. By revisiting those classical but controversial hypotheses on trade-environmental effects, this chapter intends to enrich our understanding of the global shift of environmental burdens, going beyond a view centered on environmental regulations.

International trade has long been identified as a potential channel for shifting environmental burdens across trading partners. A bunch of theoretical hypotheses has been proposed to infer the environmental performance of trade liberaliza-tion. Typical hypotheses include the environmental Kuznets curve hypothesis, the pollution haven/halo hypothesis, the factor endowment hypothesis, the race to the bottom/race to the top hypothesis, the Porter hypothesis, and the eco-dumping hypothesis. Although these hypotheses are controversial or even conflicting, they overall care whether environmental burdens would be shifted between trading economies or be addressed by trade-induced economic growth.

A positive view like the Environmental Kuznets Curve Hypothesis insists that the trade-induced economic growth will eventually contribute to the reduction of envi-ronmental impacts in the long run, although there are "growing pains" at the early stage of development (Stern 2004; Sarkodie and Strezov 2019). Economic growth accelerates the diffusion and application of green technology and promotes the effi-ciencies and productivities of economic activities (Wang et al. 2013). On the other hand, economic growth potentially alters consumers' attitudes and their consump-tion behaviours, then forcing the improvement of environmental performance (Huang et al. 2016). Taken together, economic growth equips one nation with both motivation and capabilities to be environmentally benign.

© Springer Nature Singapore Pte Ltd. 2020
C. He and X. Mao, *Environmental Economic Geography in China*,
Economic Geography, https://doi.org/10.1007/978-981-15-8991-1_8

On the contrary, sceptics seriously doubt that economic growth will absorb the environmental burdens instead of simply shifting them. In their view, according to the international division of labour, the developed economies are able to outsource traditional manufacturing, which tends to be resource-based, energy-intensive, and labour-intensive. As these activities move offshore, the home economies are expected to be more environmentally benign than before. However, one the other end, the host economies bear the rising environmental costs while they witness the economic take-off (Li and Wang 2020). During these processes, international trade provides a channel to shift environmental burdens from home economies to host ones. Both theoretical prediction (the location of FDI and the creation of international trade) and empirical evidence echo the concerns that such environmental burdens will be shifted from developed economies to developing ones, not the other way around (Dietzenbacher et al. 2012; Bakirtas and Cetin 2017; Solarin et al. 2017).

As the largest developing economies around the world, China has been integrated with the global economy through its advantages on mass production since its reform and opening up in 1978. This process has accelerated after China's accession to the World Trade Organisation in 2001, growing to become the "world factory". Even under the shock of global economic crisis in 2008, China's exports recovered within a short time and then became the top exporter in 2012 (Wang and Li 2017). Nevertheless, environmental challenges and controversies simultaneously arise behind the prosperity of trade. On the one hand, "world factory" implies the fact that one economy has to serve a geographically extended market with more intensified utilization of indigenous resources and endure the rising environmental impacts on its own. On the other hand, one should also note the exchange of knowledge and information embodied in trade, which has great potential to enhance the capabilities in addressing environmental issues. Empirical evidence in Chap. 6 has already shown that the net effects of trade liberalization on China's environmental performance can be rather mixed. Therefore, as China's economic growth steps from early take-offs into the stage of transition, industrial activities in this huge country have stood at the cross-roads of "going green" or "moving away". Looking into the controversial theoretical hypotheses in Chap. 2 and the context-contingent empirical results in Chap. 6, there is no simple answer for decision-making.

This chapter revisits the question of how China's foreign trade in the global market affects its environmental performance by challenging the duality between developed and developing economies, which has overlooked the various contexts of nations and the complicated interrelations between them. The classical hypotheses seek to examine whether and how developed economies shift their environmental burdens to the rest of the world, then emphasizing their gaps on the stringency and enforcement of environmental regulation. Their results create an illusion that environmental regulation is not A factor but THE factor driving the global shift of environmental burdens (Mao and He 2018). If this is the case, this chapter seeks to ask the same question in a different way: As one developing economy grows to become relatively developed will it start to shift environmental burdens to other developing or even less developed economies?

To answer this question, this chapter starts by proposing a hypothesis named pollution trap hypothesis, based on the classical pollution haven hypothesis. This hypothesis incorporates the effects of both spatial differences (income level and regulation stringency) and spatial interrelations (vertical specialization and horizontal specialization), proposing that when one developing economy becomes relatively developed, it is not necessarily able to further shift their environmental burdens to other developing economies. Next, we introduce the term "pollution embodied in trade", which provides an approach to measure the pollution content in trade. The third section introduces research design to examine the pollution trap hypothesis. The fourth section reports the evidence from the Chinese case. The last section concludes.

8.1 Developing Economies: Pollution Haven or Pollution Trap?

The division of developed and developing economies is a common way for simplifying the geography of world development. In terms of the international division of labour, the dynamic of comparative advantages triggers a global industrial shift from developed economies to developing ones. During the global industrial shift, the relocation of industries allows firms to utilize the cheapest input factors and improve their production efficiency. Moreover, the transfer of industries creates an opportunity for developing economies to diversify and upgrade their production, then starting their catching-up processes (Akamatsu 1962). Development of economies in East Asia are particularly the cases (Kojima 2000; Ginzburg and Simonazzi 2005).

The spatial shift of industries includes the relocation of polluting industries, then creating opportunites to shift environmental burdens simultaneously (Shen et al. 2019). Why do polluting industries also tend to relocate from developed economies to developing economies during the global industrial shift? A cost analysis may help to reveal the answers.

Theoretical thinking in environmental economics believes that environmental regulation alters the comparative advantages by imposing compliance costs on polluting productions. If the compliance costs are difficult to overcome by internalized efforts (e.g.), the relocation will become an option for polluting production in response to environmental regulation (Zhu et al. 2014; Wu et al. 2019). However, when it comes to moving abroad, firms face higher fixed costs for facilities as well as higher transaction costs due to the unfamiliar environment (Davies and Kristjáns-dóttir 2010; Boateng et al. 2015). Polluting production will transfer internationally only when the new location can save costs large enough to overcome the relocation costs to host places as well as the compliance costs with environmental regulation in home places. In such a case, the gap between host and home places becomes important, since it suggests the potential of cost-saving for polluting firms. Developing economies thereby become ideal host economies for relocated polluting firms, since the gap between developed and developing economies is large enough.

Existing studies are used to describe the gap between developed and developing economies to their different comparative advantages in production and distinct priorities of environmental regulation (Cole and Elliott 2005; Marconi 2012). Regarding the comparative advantages, inputs for mass production are much cheaper in developing economies than developed ones, such as natural resources, fossil energy, and unskilled labour. Therefore, mass production is able to save production costs to a larger extent in developing economies, leaving room for the rising costs of relocation and regulation. Such a cost-saving is less likely to be achieved in other developed economies. Likewise, developing economies are more willing to prioritize economic promotion rather than environmental conservation during their economic take-offs (Hong 2016). Besides, the environmental-relevant technological bases in developing economies are relatively weak. As a result, environmental standards in developing economies are overall lower than developed economies. The enforcement of environmental regulation usually gives way to investment invitation and growth promotion. The compliance costs with environmental regulation are therefore lower in developing economies, which provides an attractive location for polluting production.

Taken together, differences in comparative advantages and gaps of environmental regulation inevitably expose developing economies to a risk of becoming the pollution haven of developed economies. Such a logic seems to be quite solid. If this is the case, it is natural to infer that the dynamic of comparative advantages and the improvement of environmental regulation allow one economy to shift its environmental burdens to other developing or underdeveloped economies (Cai et al. 2018), putting an end to the state of pollution haven. However, it is not always the case.

From a geographical perspective, this chapter argues that the duality between developed and developing economies may oversimplify the map of global shift of environmental burdens, overlooking some valuable details as well as some crucial determinants. First, developing economies do not catch up in the same way. According to Lee and Lim (2001), the latecomers can either follow the paths of forerunners or explore their own paths. Even by following forerunners' path, the latecomers can still skip some stages within the path. Path selections vary across developing economies due to the internal accumulation of development bases, the external exchange of tangible or intangible assets (Yeung 2015), and even the chance events (Boschma and Frenken 2006). Even for one particular economy, the path for its catching-up is not static, it may change as the economy steps into a new development stage (Yeung 2015). Also, one economy can get stuck in a path for a long time, resulting in the development lock-in (Hassink et al. 2019). Overall, this point reminds us that the catching-up of developing economies is not necessarily at the cost of becoming a pollution haven for developed economies. On the other side, the state of pollution haven can be altered but not necessarily be reversed.

Second, conventional thinking focuses on the difference between developed and developing economies but somehow overlooks the different ways that economies interconnect. In the case of international trade, trade patterns matters. An important facet to investigate is the rise of intra-industry trade. Intra-industry trade refers to the exchange of products belong to the same sectors between two economies. Relative to inter-industry trade, intra-industry trade exchanges similar products. However, for intra-industry trade, these products are different since they do not produce under identical technical conditions. Therefore, the exchange of these products also creates a channel for the shift in pollution emissions (Benarroch and Weder 2006; Jakob and Marschinski 2013).

More importantly, products exchanged by intra-industry trade can be differentiated horizontally or vertically (Greenaway et al. 1995). Therefore, intra-industry trade occurs between the Global North-North, the Global North-South, and even the Global South-South economies. In such a case, the global shift of environmental burdens by international trade are no longer confined to the interactions between developing and developed economies. Particularly in the context that the South-South trade rises rapidly, the shift of environmental burdens between developing economies becomes increasingly significant (Hochstetler 2013; Meng et al. 2018).

Combining the above two points, it is insufficient to identify developing economies as pollution haven simply based on the developing-developed duality. The transition of developing economies and their rising connections with other developing economies will further make a difference. It thereby raises a question of how these factors will affect the pollution haven hypothesis. This chapter proposes a hypothesis that developing economies fact the risk of becoming not only a pollution haven but also a pollution trap.

In this chapter, we assume that one developing economy does not necessarily shift its environmental burdens to other developing economies during its catching-up process. On the contrary, it faces a risk of converging the flows of embodied pollution from both developed economies and developing economies. If this is the case, the catching-up economy will be caught by a pollution trap.

Why can the pollution trap phenomenon occur? The answer is the shifting trade pattern. At the early stage of the catching-up process, economic growth of the developing economies relies on the utilization of foreign investment and the expansion of exporting sectors. On the one hand, comparative advantages determine the inter-industry trade between developing and developed economies (Cole and Elliott 2003). Since the developing economies overall specialize in primary goods with low added value, the pollution intensity of exports from developing economies would be higher than developed economies.

On the other hand, in line with the international fragmentation of production, the processing trade burgeons between developing and developed economies (Fung and Maechler 2005; Zhao et al. 2014; Zhang et al. 2017). In other words, there is rising vertical intra-industry trade between them. Developing economies import intermediated inputs from developed economies and then re-export the final goods. Such a trade pattern enlarges the emissions in developing economies (Swart 2013). Taken together, both the inter-industry trade and intra-industry trade between developing and developed economies contribute to the rise of pollution haven phenomenon in developing economies.

As the catching-up process continues and the income level increases quickly, the trade pattern of developing economies will change accordingly. Firstly, the proportion of intra-industry trade increases significantly. Moreover, the products exchanged by intra-industry trade also upgrade to different extents, exhibiting a higher level of quality and productivity. This process will effectively reduce the pollution intensity of its exports to developed economies (Roy 2017; Zhang et al. 2017).

Secondly, their trade with other developing economies set out to expand rapidly. The catching-up economies and other developing economies still stay in the relatively lower value-added position in comparison with the developed economies. Both of them tend to be vertically integrated with developed economies. However, the catching-up economies start to have comparative advantages in capital-intensive industries, compared with other developing economies. Thus, they may further be horizontally integrated with other developing economies. Given the manufacturing processes, the catching-up economies are more capable of specializing in capital-intensive parts, whereas other developing economies or underdeveloped economies tend to specialize in resource-based and labour-intensive parts. Since the capital-intensive industries tend to be more energy-intensive and emission-intensive (Mukhopadhyay and Chakraborty 2014), the pollution intensity of exports from catching-up economies remains to be high. The emissions of production emissions do not shift to other developing economies.

This trade pattern helps to explain why environmental regulation does not necessarily prevent catching-up economies from becoming pollution havens. Admittedly, increasing the standards and improving the enforcement of environmental regulation allow catching-up economies to shrink their gap between developed economies. This is beneficial for catching-up economies to avoid the shift of environmental burdens from developed economies (Abdo et al. 2020). However, even the catching-up economies have already had more stringent environmental regulation than other developing economies or underdeveloped economies. The shift of environmental burdens from these economies still occurs, because increasing compliance costs is unable to alter their comparative advantages (Lu 2009).

Taken together, for catching-up economies, although the pollution intensity of their exports to developed economies is expected to reduce to some extent, the pollution intensity of exports to other developing economies may remain high or even increase. Since the trade patterns and environmental regulations work together to determine the pollution intensity of exports and imports, their trade-off may expose catching-up economies to a risk of the pollution trap phenomenon. The following question is how to identify whether the pollution trap phenomenon occurs or not. The literature on embodied pollution may help. We will address this question in next section.

8.2   How to Compute Pollution Contents in Trade?

International trade has separated the production from its final consumption geographically. This is especially the case with the rise of international fragmentation of production (Johnson and Noguera 2012). The movement of trading goods from one place to another essentially represents a spatial displacement of resources or emissions between two places, which are used to produce the goods or emitted by the production. Such a spatial displacement of resources/emissions can be seen as the resource contents/emission contents in trading goods that relocate between places associated with the exchanges of goods. A developing strand of the literature investigates this issue by measuring the emission contents in trade and mapping the flows of emission contents (Subak 1995; Peters 2008; Wu et al. 2016). These studies include many facets, such as the embodied pollution/emission, carbon leakage, and virtual water (see Chap. 2). The approaches of embodied pollution can support the examination of pollution trap hypothesis in this chapter.

Pollution Embodied in Trade (PET) assesses the pollution contents in trading goods based on consumption (Wyckoff and Roop 1994; Peters and Hertwich 2006). For a particular economy, the domestic pollution emissions due to overseas consumption can be seen as its Pollution Embodied in Exports (PEE). The overseas pollution emissions due to its domestic consumption can be seen as its Pollution Embodied in Imports (PEI). Building on this premise, the flows of embodied pollution can be analysed by an input-output table. The input-output table documents the material flows or value changes along the value chain (Timmer et al. 2015). Given the pollution intensity of each intermediated inputs, the input-output analysis can further assess the pollution emissions distributed along the production process (Xie et al. 2018). Regarding the international division of labour, productions are fragmented and scattered in the cheapest locations. Few regions can cover the whole process of manufacturing a final goods. In such a case, the Multi-Regional Input-Output (MRIO) shows its capabilities in tracing the flows across industries and countries (Acquaye et al. 2017). The measurements of PEE and PEI thereby become feasible.

Equation (8.1) shows the identical equation of a basic Input-Output (I-O) model, pointing to the fact that total output equals the sum of internal consumption and external demand.

$$X = AX + B \tag{8.1}$$

where X denotes the output vector, including the total output of each industry. A denotes the input-output matrix, representing the input level per unit of output. B denotes the external demand vector. Equation (8.1) can be transformed to Equation (8.2).

$$X = (I - A)^{(-1)} B = L B \tag{8.2}$$

where L denotes Leontief's inverse matrix, measuring all the direct and indirect inputs from other sectors for one unit of output in one particular sector. It represents the ripple effects between industries.

Equations (8.1) and (8.2) provide the basic setting of an open I-O model. By incorporating the pollution intensity (i.e. pollution emissions per unit output), PEE and PEI can be measured. Equation (8.3) calculates the PEE.

$$\text{PEE}^D = P^D \big(I - A^d\big)^{-1} E^D \tag{8.3}$$

where D denotes the economy. PEE denotes the pollution emission vector of the economy D. P denotes the pollution intensity matrix, consisting of each industry's pollution intensity in economy D. P is a diagonal matrix. E denotes the export vector, including the added value of exports in each industry. A^d denotes the domestic input-output matrix, representing the domestic input level per unit of output. Likewise, there is also A^i indicating the imported input level per unit of output. A equals the sum of A^d and A^i. If the inputs for manufacturing exports are purchased overseas, they do not increase the domestic emissions. The reason for using A^d other than A is to avoid an overestimation of PEE.

Equation (8.4) calculates the PEI.

$$\text{PEI}^D = P^F \left(I - A^{Fd} \right)^{-1} I^D \tag{8.4}$$

where F denotes other economies in the world except for economy D. P is a block diagonal matrix, consisting of each industry's pollution intensity in each foreign economy. A^{Fd} denotes the foreign economy's domestic input-output matrix. I denotes the import vector of economy D.

The imbalance between PEE and PEI indicates whether one economy gains or loses in environmental terms during the international division of labour. A direct way to measure such imbalance is to calculate the difference between PEE and PEI. Related studies report that there are significant flows of embodied pollution from developed economies to developing ones (Peters et al. 2011; Lin et al. 2014; Jiborn et al. 2018).

However, although measuring the difference between PEE and PEI is simple and practicable, the defect is quite obvious that the net flows of embodied pollution are highly sensitive to the trade volume imbalance (Antweiler 1996; Grether and Mathys 2013). In such a case, Antweiler (1996) proposes a PTT index to indicate the imbalance between PEE and PEI, named Pollution Terms of Trade (PTT). The PTT index seeks to eliminate the effects of trade imbalance by dividing PEE and PEI by export volume and import volume, respectively. In other words, PTT is essentially a ratio of pollution intensity of exports to the one of imports. Equation (8.5) shows the formula of PTT.

$$\text{PTT} = \text{PIE}/\text{PII} = \frac{\text{PEE}}{E} \Big/ \frac{\text{PEI}}{I} \tag{8.5}$$

where PIE denotes the pollution intensity of exports, PII denotes the pollution intensity of imports, E denotes the value of exports, I denotes the value of imports. Grether and Mathys (2013) suggest to replace the value of exports/imports with the added value of exports/imports in response to the rise of intermediate products trade. This revision is especially meaningful in the case of China since the processing trade accounts for an enlarging portion of China's foreign trade. Besides, the added value may avoid double-counting the trade flows between two economies.

Since the PTT is a relative pollution intensity of exports to imports, we can assume the value of 1 as a threshold. If one economy has a PTT above 1, its integration with the world economy has relatively large environmental costs. In other words, this economy is exposed to a risk of becoming the pollution haven.

8.3 Examine the Pollution Trap Hypothesis

8.3.1 Identifying the Pollution Trap Phenomenon Using PTT Index

The approach of embodied pollution provides an analytical tool to examine the pollution trap hypothesis, which can be conducted in two steps. The first step is to examine the pollution trap phenomenon. As mentioned above, the pollution trap phenomenon indicates that a catching-up economy converges the flows of pollution contents in trade. The PTT index provides a simple condition to tell whether one economy becomes a pollution trap. If one economy's trade with both developed and developing economies has a PTT above 1, this economy would have witnessed the pollution trap phenomenon. In other words, this catching-up economy becomes a pollution haven of developed economies but is simultaneously unable to shift environmental burdens to other developing economies.

8.3.2 Examining the Determinants of Pollution Trap Phenomenon

The second step is to examine the determinants of pollution trap phenomenon. Based on the theoretical discussion above, the determinants of pollution trap phenomenon should at least cover three dimensions, namely, the environmental regulation, the intra-industry trade, and the inter-industry trade. Equation (8.6) shows the basic model for empirical studies.

$$\mathrm{PI}_{ict} = a_{ic} + b_1 \mathrm{ER}_{ct} + \mathrm{INTRA}_{ict} b_2 + \mathrm{INTER}_{ict} b_3 + X_{ict} b_4 + b_5 C_{ict} + u_{ict}$$
$$(8.6)$$

where i, c, t denote industry, economy, and year, respectively. PI denotes the pollution intensity, including the PIE, PII, and PTT above.

ER denotes the relative stringency of environmental regulation. In this chapter, we use the stringency of China's environmental regulation as a benchmark so that ER is the ratio of other economies' environmental regulation stringency to the stringency of China. According to the pollution haven hypothesis, the coefficients of ER for PIE and PTT should be positive, while the coefficients of ER for PII should be negative. This indicates that if one economy's environmental regulation is at an advantage, it is confronted with inflows of pollution contents but difficult to transfer the pollution contents backward. However, if the pollution trap hypothesis holds, such a relationship can be reversed for the trade with other developing economies.

INTRA denotes the pattern of intra-industry trade. In this chapter, we apply the Grubel-Lloyd (G-L) index to measure the vertical and horizontal intra-industry trade. Equation (8.7) demonstrates the G-L index (Grubel and Lloyd 1975).

$$GL_{ic} = 1 - |E_{ic} - I_{ic}|/(E_{ic} + I_{ic}) \tag{8.7}$$

where i and c denotes the industry and economy, respectively. E denotes the value of exports. I denotes the value of imports. GL denotes the $G\text{-}L$ index, which ranges from zero to one. The higher the GL is, the more the intra-industry trade accounts for in the total value of trade. If GL equals zero (one), foreign trade is thoroughly inter-industry trade (intra-industry trade).

By considering the value differences between exports and imports, the G-L index can distinguish between vertical and horizontal differentiation of intra-industry trade and measure their levels. Equation (8.8) shows this approach (Greenaway et al. 1995; Rivera 2014).

$$\begin{cases} GL\_H_{ic} = 1 - |E_{ic} - I_{ic}|/(E_{ic} + I_{ic}), 0.85 \leq R \leq 1.15 \\ GL\_VH_{ic} = 1 - |E_{ic} - I_{ic}|/(E_{ic} + I_{ic}), R > 1.15 \\ GL\_VL_{ic} = 1 - |E_{ic} - I_{ic}|/(E_{ic} + I_{ic}), R < 0.85 \end{cases} \tag{8.8}$$

where GL_H denotes the level of horizontal intra-industry trade, indicating those exchanging products that are similar in types but different in varieties. GL_VH and GL_VL denote the level of vertical intra-industry trade with high-value products and low-value products, respectively. R is the ratio of exports to imports in terms of the unit value of products.

INTER denotes the pattern of inter-industry trade, representing the role of comparative advantages. In this chapter, we consider three facets that manifest comparative advantages, including labour, capital, and technology. Like the indicator of environmental regulation, we apply relative terms to indicate the labour, capital, and technology. The level of China is set as the benchmark. Then the variables LAB, CAP, and TECH represent the ratio of other economies' level of labour, capital, and technology to China, respectively.

X denotes the matrix consisting of various interaction terms. Considering the ER, LAB, CAP, and TECH are relative terms, it is necessary to consider the likelihood that their effects are piecewise. In this regard, we further incorporate four interaction terms, marked as ER×er, LAB×lab, CAP×cap, TECH×tech. The lowercase variables are dummy variables. They equal one, if the value of their corresponding uppercase variables is above one.

C denotes a vector of the control variables. Based on the theoretical discussion above, the empirical model primarily controls the difference of two economies in their economic scales. Besides, the fixed effect model is used to control the inherent differences among industries.

8.3.3   Data

To support the multi-regional I-O analysis in this chapter, we use the input-output table provided by the World Input-Output Database (WIOD). The WIOD has released two versions of I-O tables so far. The early version, released in 2013, includes 35 sectors and 40 economies, spanning from 1995 to 2011. Besides, the early version also released socio-economic accounts and environmental accounts. The socio-economic accounts provide data about labour, capital, technology in each economy. The environmental accounts provide sector-specific emissions data in each economy. Also of note, data in the environmental accounts span from 1995 to 2009. The updated version is released in 2016, spanning from 2000 to 2014. The number of sectors and economies has increased to 56 and 43, respectively. Even so, the updated version has not yet released the environmental accounts. Thus, this chapter applies the early version. The 40 economies in this version of the I-O table accounts for at least 85% of the global gross domestic production (Timmer et al. 2015). Empirical results based on these economies are still representative.

Regarding the environmental regulation, the data used in this chapter is different from other chapters in this book, as this chapter focuses on the gap between trading economies. The World Economic Forum (WEF) has conducted a survey about the international competitiveness, including two questions related to environmental regulation. These questions inquire about firms' assessment of the stringency and enforcement of environmental regulations in economies where they are located. This survey covers over 100 economies around the world since 2004. Thus, this chapter applies the data from this survey to measure the stringency and enforcement of environmental regulation.

8.4   Evidence from China's Foreign Trade

8.4.1   The Pollution Trap Phenomenon in China

Figure 8.1 shows the pollution embodied in China's foreign trade from 2003 to 2009, which is aggregated by the results of multi-regional I-O analysis. It is evident that China faces a severe trade imbalance in terms of embodied pollution, manifested by the fact that PEE is observably higher than PEI. The average of PEE even octuple the average PEI during 2003–2009. The figures are 9.19×10^5 t and 1.23×10^5 t, respectively. The value of PTT index remains to be above one, even taking the added value into consideration, suggesting a disadvantageous position in environmental terms. Besides, one should further note that PEE has represented over a quarter of domestic emissions during this period. All these results point to the fact that the pollution haven phenomenon has occurred during China's catching-up by export-oriented growth model. The role of "world factory" enlarges its domestic pollution emissions, making it bear more environmental burdens shifted from overseas.

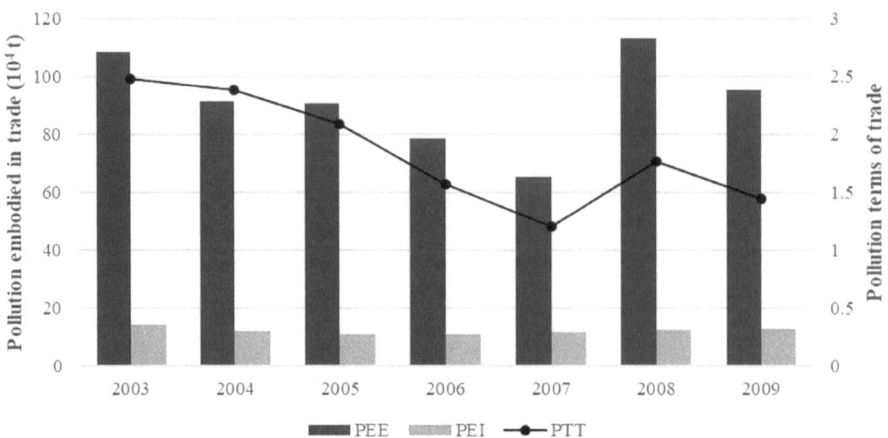

Fig. 8.1 Pollution embodied in China's foreign trade and its pollution terms of trade, 2003–2009

However, from a dynamic view, the pollution haven condition has improved gradually. The value of PTT index demonstrates a clear downward trend in spite of an upswing in 2008 temporarily. Although the value of PTT is still above one, the imbalance of embodied pollution in China's foreign trade has been ameliorated. The change of PEE contributes more to such amelioration. The annual average changing rate of PEE is 11.9%, which is higher than the 4.6% of PEI.

Figure 8.2 shows each sector's PEE, PEI, and PTT. According to Figure 8.2, PEE and PEI are distributed across sectors unevenly. The two largest sectors in terms of PEE and PEI are the Basic Metals and Fabricated Metal (27t28) and the Chemical and Chemical Products (24). Their combined proportions of PEE and PEI reach 50% and 60%, respectively. Regarding PEE, there are two more sectors accounting for a percentage higher than 10%, namely, the Textiles and Textile Products (17t18) and the Other Non-metallic Mineral (26). As for the PEI, the Coke, Refined Petroleum, and Nuclear Fuel (23) ranks the third with a percentage of 17.5%.

The value of PTT further reveals the different conditions for pollution haven among sectors. According to the results, the pollution haven phenomenon is most manifested by three sectors, namely, the Transport Equipment (34t35), the Machinery (29), and the Recycling and Other Manufacturing (36t37). By considering the value of exchanging products, the results show that trade with the most pollution contents does not equal to the trade with the worst pollution terms of trade. Neither the deficit/surplus of embodied pollution nor the relative pollution intensity between exports and imports is sufficient to show the global shift of environmental burdens.

Figure 8.3 distinguishes between China's trade with developed economies and other developing economies. The results demonstrate that pollution emissions embodied in trade with developed economies and developing economies are very different. PEE of the trade with developed economies occupies the largest shares but exhibits a downward trend. On the contrary, PEI of the trade with developed

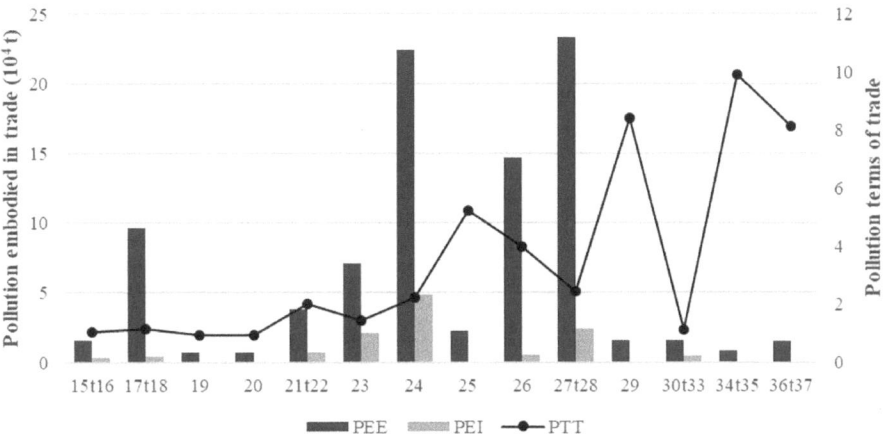

Fig. 8.2 Average pollution embodied in China's foreign trade by sectors, 2003–2009. *Note* The horizontal axis labels denote industrial sectors. 15t16 denotes Food, Beverages, and Tobacco. 17t18 denotes Textiles and Textile Products. 19 denotes Leather and Footwear. 20 denotes Wood and Products of Wood and Cork. 21t22 denotes Pulp, Paper, Printing, and Publishing. 23 denotes Coke, Refined Petroleum, and Nuclear Fuel. 24 denotes Chemicals and Chemical Products. 25 denotes Rubber and Plastics. 26 denotes Other Non-Metallic Mineral. 27t28 denotes Basic Metals and Fabricated Metal. 29 denotes Machinery. 30t33 denotes Electrical and Optical Equipment. 34t35 denotes Transport Equipment. 36t37 denotes Recycling and Other Manufacturing

economies continues to increase. These results point to the fact that the pollution haven condition between China and developed economies has been improved gradually. The decline of PTT further supports this point.

Regarding trade with other developing economies, although PEE is relatively small and waved, PEI exhibits a clear trend of reduction. Moreover, PTT increases and becomes above one since 2008. These results reveal that China does not shift their environmental burdens to other developing economies. On the contrary, its pollution terms of trade have been getting worse. Overall, the results in Figure 8.3 observe the pollution trap phenomenon. Although the pollution haven condition of China seems to be improved, the pollution terms of trade with other developing economies deteriorate.

8.4.2 Determinants of the Pollution Trap Phenomenon

This section investigates the pollution trap phenomenon from three different perspectives, namely, the pollution intensity of exports (PIE), the pollution intensity of imports (PII), and the pollution terms of trade (PTT). The pollution trap phenomenon suggests that the pollution intensity of exports to developing economies increases, while the pollution intensity of imports from developing economies decreases. Regarding the PTT, the pollution trap phenomenon indicates the increase of PTT,

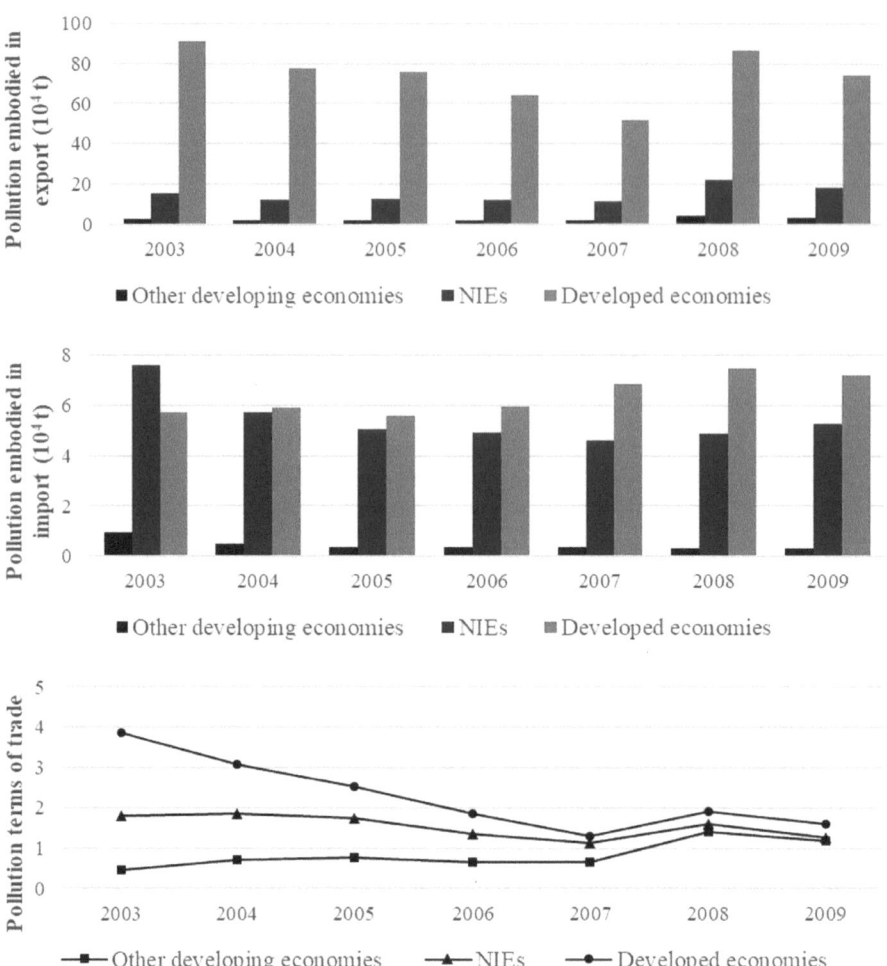

Fig. 8.3 Pollution embodied in China's foreign trade by economies, 2003–2009

particularly the PTT of trade with developing economies. As for the pollution trap hypothesis, it highlights the role of trade patterns that may increase the PTT of catching-up economies.

Table 8.1 reports the estimation results of empirical models regarding PIE. Firstly, the exports of low-value products to developed economies increase China's pollution intensity of exports. This result corresponds to the pollution haven conditions. The exports of similar products and low-value products to developing economies decrease the pollution intensity of exports. However, the intra-industry exports of similar products and low-value products to developing economies account for a relatively proportion in China's exports. Thus, the overall effects of intra-industry trade on the pollution intensity of exports to developing economies are insignificant. Instead, the

Table 8.1 Determinants of pollution intensity of China's exports

	Whole sample		Trade with developed economies		Trade with developing economies	
	[1]	[2]	[3]	[4]	[5]	[6]
ln*GL*	−0.004		0.630***		-0.474	
ln*GL_H*		−0.505		0.264		−2.21***
ln*GL_VH*		−0.052		0.119		−0.001
ln*GL_VL*		0.043		0.721***		−0.534**
ln*ER*	2.585***	2.588***	2.43***	2.43***	2.88***	2.87***
ln*ER*er*	−0.079***	−0.079***	−0.216***	−0.214***	−0.096***	−0.095***
ln*CAP*	−0.012	−0.011	−0.024	−0.021	0.039	0.036
ln*CAP*cap*	−0.171**	−0.213***	−0.135*	−0.170**		
ln*LAB*	0.115	0.116	0.640**	0.647**	−0.101	−0.109
ln*LAB*lab*	0.513***	0.520***	0.485***	0.485***		
ln*TECH*	0.672	0.662	1.693***	1.615***	0.451	0.446
ln*TECH*tech*	−0.659	−0.648	−1.819***	−1.740***	−0.344	−0.330
ln*ADD*	0.093	0.090	−0.269**	−0.280**	0.230	0.243
Obs.	2639	2639	1211	1211	1428	1428
Groups	538	538	247	247	291	291
R-sq	0.433	0.433	0.522	0.522	0.385	0.387

Note $*, p < 0.10, **, p < 0.05, ***, p < 0.01$

effects on exports to developed economies are statistically significant and positive. To sum up, intra-industry trade continues the pollution haven conditions of China by exporting low-value products to developed economies.

Secondly, the gap of environmental regulation plays a crucial role in affecting the pollution intensity of exports. The coefficients are significant and positive, suggesting that the regulation gap between China and other economies increases the pollution intensity of China's exports. This partly supports the occurrence of pollution haven phenomenon. Moreover, the coefficients of interaction terms are negative, suggesting that the shift of environmental burdens due to China's lax environmental regulation has weakened. On the other hand, as the China's environmental regulation become stricter than other economies, their gaps would reduce the pollution intensity of China's exports. This result rejects that environmental regulation will result in the pollution trap phenomenon. Overall, the role of environmental regulation supports the pollution haven phenomenon.

Thirdly, the effects of inter-industry trade are confined to the exports to developed economies. This is in line with the theoretical thinking that inter-industry trade represents the division of labour based on comparative advantages, occurring between developed and developing economies primarily. The gap of labour costs between China and developed economies would lead to an increase in the pollution intensity of China's exports. On the country, the gap of technological levels between them would contribute to the reduction of pollution intensity, exhibiting a positive technique effect.

Overall, the findings above support that the pollution haven conditions between China and developed economies have been improved by environmental regulations and inter-industry trade. However, the intra-industry trade with low-value products still contributes to the pollution haven conditions.

Table 8.2 reports the estimation results of empirical models regarding PII. According to the coefficients of *GL_H*, *GL_VH*, and *GL_VL*, China's imports of similar products and low-value products from developed economies increase the pollution intensity of imports. Combined with the results of exports, it is apparent that the changing pattern of intra-industry trade contributes to the amelioration of China's pollution haven condition. The exchange of low-value products and similar products between China and developed economies prevents the previously unidirectional transfer of environmental burdens. Nevertheless, China's imports from other developing economies through intra-industry trade decrease the pollution intensity of imports, rejecting the overall shift of environmental burdens. Combined with the results of exports, the intra-industry trade between China and other developing economies rejects the pollution haven phenomenon.

The role of environmental regulation in PII keeps the same as its role in PIE. If China's environmental regulation is stricter than its trading partners, the pollution

Table 8.2 Determinants of pollution intensity of China's imports

	Whole sample		Trade with developed economies		Trade with developing economies	
	[7]	[8]	[9]	[10]	[11]	[12]
ln*GL*	-0.266^{***}		0.178^{***}		-0.570^{***}	
ln*GL_H*		0.073		0.494^{***}		-0.597^{***}
ln*GL_VH*		-0.256^{***}		-0.172		-0.162^{**}
ln*GL_VL*		-0.320^{***}		0.122^{***}		-0.681^{***}
ln*ER*	0.516^{***}	0.513^{***}	0.212^{***}	0.208^{***}	0.977^{***}	0.973^{***}
ln*ER*er*	0.002	0.003	-0.011^{***}	-0.012^{***}	-0.042^{***}	-0.041^{***}
ln*CAP*	-0.015	-0.016	0.008^{*}	0.005	-0.009	-0.010
ln*CAP*cap*	-0.007	0.022	0.010	0.040		
ln*LAB*	-0.198^{***}	-0.200^{***}	-0.012	-0.015	-0.248^{***}	-0.253^{***}
ln*LAB*lab*	-0.049^{***}	-0.053^{***}	-0.015^{*}	-0.023^{*}		
ln*TECH*	-0.630	-0.625	0.208	0.145	-0.888	-0.883
ln*TECH*tech*	0.767^{**}	0.760^{**}	-0.160	-0.100	1.10^{**}	1.10^{**}
ln*ADD*	0.076^{**}	0.078^{**}	-0.035^{**}	-0.028^{*}	0.077	0.084
Obs.	2639	2639	1211	1211	1428	1428
Groups	538	538	247	247	291	291
R-sq	0.130	0.127	0.136	0.159	0.193	0.194

*Note * p < 0.10, ** p < 0.05, *** p < 0.01*

intensity of its imports tends to be higher. The coefficients of interaction terms are negative. That is to say, if China's trading partners have stricter environmental regulations, a wider gap of environmental regulation reduces the pollution intensity of China's imports.

Regarding the role of inter-industry trade in PII, the effects are confined to the imports from developing economies and the crucial determinant is the gap of labour costs. The wider the gap of labour costs between China and other developing economies is, the lower the pollution intensity of China's imports from these economies would be.

Overall, the results of PII reveal that the occurrence of pollution trap phenomenon is ascribed to trade patterns rather than environmental regulations. Environmental regulation is a key determinant of the pollution haven phenomenon. However, the rising intra-industry trade prevents the further shift of environmental burdens to other developing economies. The exchanges of products make China and other developing economies share the environmental burdens.

Table 8.3 reports the results of empirical models regarding PTT. PTT represents the integrated effects of PIE and PII, which capture the determinants of pollution trap conditions more specifically. The results echo the findings above that it is trade patterns that result in the pollution trap phenomenon rather than environmental regulations. Intra-industry trade and inter-industry trade contribute to the

Table 8.3 Determinants of China's pollution terms of trade

	Whole sample		Trade with developed economies		Trade with developing economies	
	[13]	[14]	[15]	[16]	[17]	[18]
lnGL	0.328^{**}		0.058		0.432	
lnGL_H		-0.769^{***}		-0.768^{***}		-1.28
lnGL_VH		0.855^{***}		0.575		0.780^{***}
lnGL_VL		0.365^{***}		0.193		0.393
lnER	0.480	0.489	1.12^{**}	1.13^{**}	-0.511^{**}	-0.515^{**}
ln$ER*er$	-0.026	-0.026	-0.202^{**}	-0.200^{**}	0.090^{***}	0.090^{***}
lnCAP	-0.055^{**}	-0.055^{***}	-0.117^{***}	-0.111^{***}	-0.031^{*}	-0.034^{**}
ln$CAP*cap$	-0.174^{**}	-0.255^{***}	-0.194^{***}	-0.262^{***}		
lnLAB	0.045	0.116	0.117	0.128	-0.056	-0.063
ln$LAB*lab$	0.768^{***}	0.788^{***}	0.575^{***}	0.592^{***}		
ln$TECH$	-0.162	-0.111	0.971^{***}	1.07^{***}	-0.227	-0.231
ln$TECH*tech$	-0.286	-0.288	-1.24^{**}	-1.33^{**}	0.105	0.118
lnADD	0.215^{***}	0.214^{***}	0.208^{*}	0.197^{**}	0.327^{***}	0.337^{***}
Obs.	2639	2639	1211	1211	1428	1428
Groups	538	538	247	247	291	291
R-sq	0.044	0.054	0.173	0.176	0.017	0.018

Note $*, p < 0.10, **, p < 0.05, ***, p < 0.01$

occurrence of pollution trap. The exchange of horizontally differentiated products between China and the world economy helps to improve China's pollution terms of trade. This effect is confined within the trade with developed economies. However, the exchange of vertically differentiated products will significantly worsen China's pollution terms of trade, particular the exchange of high-value products with other developing economies.

As for inter-industry trade, comparative advantages in the capital and labour play a significant role in altering the pollution terms of trade. The gap of capital costs between China and other economies will improve the pollution terms of trade, while the gap of labour costs will worsen the pollution terms of trade. This mechanism is largely confined to the trade with developed economies. As a result, inter-industry trade helps to prevent China from the pollution haven conditions due to the change of comparative advantages. However, its effects on the trade with other developing economies are insignificant. Thus, although the inter-industry trade can alleviate the pollution haven issues, it has little impact on the trade with other developing economies so that the dynamic of comparative advantages can not prevent the pollution trap.

Unlike the results of PIE and PII, the effects of environmental regulation on PTT are insignificant, suggesting an offset effect between PIE and PII. Moreover, its effects are exactly the opposite between the trade with developed economies and the trade with developing economies. For the trade with developed economies, the gap of environmental regulation worsens China's pollution terms of trade. However, for the trade with developing economies, the gap of environmental regulation will improve the pollution conditions of China. Again, these results support the finding above that environmental regulation is an important determinant of pollution haven phenomenon.

Generally, empirical evidence from China echoes the classical pollution haven hypothesis and also highlights the conditions for pollution trap hypothesis. The gap of environmental regulation among economies drives the environmental burdens shifting from strict economies to lax ones. However, the change of trade patterns make the global shift of environmental burdens become more complicated than the prediction of pollution have hypothesis. One the one hand, the dynamic of comparative advantages allows one economy to prevent or even reverse the previously unidirectional shift of environmental burdens. On the other hand, the rising intra-industry trade promotes the exchange of products between developing economies. The exchange of vertically differentiated products determines that a catching-up economy is not necessarily able to shift its environmental burdens further to other developing/underdeveloped economies. During the intra-industry trade, the catching-up economies may even stay in a disadvantageous position in environmental terms. Thus, the intra-industry trade is the crucial determinant of pollution trap phenomenon.

8.5   Conclusion

This chapter seeks to change the perception that environmental burdens shift from developed economies to developing ones, which is subject to the gap of environmental regulation between economies. From a geographical perspective, this chapter argues that the global shift of environmental burdens does not have to be unidirectional. It can occur between developing economies, between developed economies, and between developed and developing economies. The geographical perspective also points to the fact that spatial patterns are outcomes of not only spatial differences but also spatial interconnections. Conventional thinking focus on the spatial differences in income level and environmental regulation, but somehow overlook the role spatial interconnections. In the case of international trade, trade patterns represent the spatial interconnection typically. Therefore, this chapter revisits the pollution haven hypothesis by incorporating the role of trade patterns and then proposes the pollution trap hypothesis for catching-up economies.

The pollution trap hypothesis essentially posits a possibility that the environmental burden shift can be reversed between a developing/underdeveloped economy and a catching-up economy, even if the catching-up economy may have higher income level and stricter environmental regulation. The reason behind this is the changing trade patterns, particularly the rise of intra-industry trade. According to the theory of spatial division of labour, the developing economies (including the catching-up economies) are vertically differentiated with developed economies, which is subject to their comparative advantages. The catching-up process triggers the dynamic of comparative advantages so that catching-up economies is more capable of improving their pollution haven conditions than other developing economies. However, the dynamic of comparative advantages does not necessarily mean that they can further shift their environmental burdens to other developing economies, because the exchange of products between catching-up economies and other developing economies is dominated by intra-industry trade. The exchange of products makes these economies share the environmental burdens instead of shifting the burdens from one to the other. If this is the case, there would be a special period during the catching-up process, during which the catching-up economies still serve as the pollution haven for developed economies (although the condition is improving) and simultaneously face the increasing shift of environmental burdens from other developing economies. Empirical evidence from China supports the theoretical prediction above.

Overall, in terms of economic geography, the context of development can lead to various trajectories and growth models. In response to the duality of developed and developing economies, one should notice that the developing economies (or developed economies) are different from each other. The differences can be great enough to trigger internal shifts of environmental burdens that should not be overlooked. From a geographical view, another facet that should not be overlooked is the various ways of spatial interconnections, which determine the spatial configuration of both eco-environmental and socio-economic activities. This chapter uses the global shift of environmental burdens as a case, showing that how the geographical perspective would further enrich (or even challenge) our understanding of the theoretical thinking that tends to be taken for granted.

References

Abdo, A.-B., Li, B., Zhang, X., et al. (2020). Influence of FDI on environmental pollution in selected Arab countries: A spatial econometric analysis perspective. *Environmental Science and Pollution Research International*. https://doi.org/10.1007/s11356-020-08810-4.

Acquaye, A., Feng, K., Oppon, E., et al. (2017). Measuring the environmental sustainability performance of global supply chains: A multi-regional input-output analysis for carbon, sulphur oxide and water footprints. *Journal of Environmental Management, 187*, 571–585.

Akamatsu, K. (1962). A historical pattern of economic growth in developing countries. *The Developing Economies, 1*(S1), 3–25.

Antweiler, W. (1996). The pollution terms of trade. *Economic Systems Research, 8*(4), 361–366.

Bakirtas, I., & Cetin, M. A. (2017). Revisiting the environmental Kuznets curve and pollution haven hypotheses: MIKTA sample. *Environmental Science and Pollution Research, 24*, 18273–18283.

Benarroch, M., & Weder, R. (2006). Intra-industry trade in intermediate products, pollution and internationally increasing returns. *Journal of Environmental Economics and Management, 52*(3), 675–689.

Boateng, A., Hua, X., Nisar, S., et al. (2015). Examining the determinants of inward FDI: Evidence from Norway. *Economic Modelling, 47*, 118–127.

Boschma, R. A., & Frenken, K. (2006). Why is economic geography not an evolutionary science? Towards an evolutionary science? Towards an evolutionary economic geography. *Journal of Economic Geography, 6*(3), 273–302.

Cai, X., Che, X., Zhu, B., et al. (2018). Will developing countries become pollution havens for developed countries? *An empirical investigation in the Belt and Road., 198*, 624–632.

Cole, M. A., & Elliott, R. J. R. (2003). Determining the trade-environment composition effect: The role of capital, labor and environmental regulations. *Journal of Environmental Economics and Management, 46*(3), 363–383.

Cole, M. A., & Elliott, R. J. R. (2005). FDI and the capital intensity of "dirty" sectors: A missing piece of the pollution haven puzzle. *Review of Development Economics, 9*(4), 530–548.

Davies, R. B., & Kristjánsdóttir, H. (2010). Fixed costs, foreign direct investment, and gravity with zeros. *Review of International Economics, 18*(1), 47–62.

Dietzenbacher, E., Pei, J., & Yang, C. (2012). Trade, production fragmentation, and China's carbon dioxide emissions. *Journal of Environmental Economics and Management, 64*(1), 88–101.

Fung, K. C., & Maechler, A. M., (2005). The impact of intra-industry trade on the environment. In G. S. Heiduk, K. Wong (Eds.), *WTO and World Trade*. Physica-Verlag HD: 87–123.

Ginzburg, A., & Simonazzi, A. (2005). Patterns of industrialization and the flying geese model: The case of electronics in East Asia. *Journal of Asian Economics, 15*(6), 1051–1078.

Greenaway, D., Hine, R., & Milner, C. (1995). Vertical and horizontal intra-industry trade: A cross industry analysis for the United Kingdom. *The Economic Journal, 105*(433), 1505–1518.

Grether, J.-M., & Mathys, N. A. (2013). The pollution terms of trade and its five components. *Journal of Development Economics, 100*(1), 19–31.

Grubel, H., & Lloyd, P. (1975) Intra-industry trade: The theory and measurement of international trade in differentiated products. Macmillan.

Hassink, R., Isaksen, A., & Trippl, M. (2019). Towards a comprehensive understanding of new regional industrial path development. *Regional Studies*. https://doi.org/10.1080/00343404.2019.1566704.

Hochstetler, K. (2013). South-South trade and the environment: A Brazilian case study. *Global Environmental Politics, 13*(1), 30–48.

Hong, Y. (2016). Sustaining factors for China's economic growth. *The China Path to Economic Transition and Development* (pp. 95–107). Singapore: Springer.

Huang, X., Hu, Z., Liu, C., et al. (2016). The relationship between regulatory and customer pressure, green organizational responses, and green innovation performance. *Journal of Cleaner Production, 112*(4), 3423–3433.

Jakob, M., & Marschinski, R. (2013). Interpreting trade-related CO_2 emission transfers. *Nature Climate Change, 3,* 19–23.

Jiborn, M., Kander, A., Kulionis, V., et al. (2018). Decoupling or delusion? Measuring emissions displacement in foreign trade. *Global Environmental Change, 49,* 27–34.

Johnson, R. C., & Noguera, G. (2012). Accounting for intermediates: Production sharing and trade in value added. *Journal of International Economics, 86*(2), 224–236.

Kojima, K. (2000). The "flying geese" model of Asian economic development: Origin, theoretical extensions, and regional policy implications. *Journal of Asian Economics, 11*(4), 375–401.

Lee, K., & Lim, C. S. (2001). Technological regimes, catching-up and leapfrogging: Findings from the Korean industries. *Research Policy, 30,* 459–483.

Li, M., & Wang, Q. (2020). Does industrial relocation alleviate environmental pollution? A mathematical economics analysis. *Environment Development and Sustainability., 22*(5), 4673–4698.

Lin, J., Pan, D., Davis, S., et al. (2014). China's international trade and air pollution in the United States. *Proceedings of the National Academy of Sciences of the United States of America, 111*(5), 1736–1741.

Lu, Y. (2009). Do environmental regulations influence the competitiveness of pollution-intensive products? *Economic Research Journal, 44*(4), 28–40.

Mao, X., & He, C. (2018). A trade-related pollution trap for economies in transition? Evidence from China. *Journal of Cleaner Production, 200,* 781–790.

Marconi, D. (2012). Environmental regulation and revealed comparative advantages in Europe: Is China a pollution haven? *Review of International Economics, 20*(3), 616–635.

Meng, J., Mi, Z., Guan, D., et al. (2018). The rise of South-South trade and its effect on global CO_2 emissions. *Nature Communications, 9,* 1871.

Mukhopadhyay, K., & Chakraborty, D. (2014). Is liberalisation of trade good for the environment? Evidence from India. *Asia-Pacific Journal of Rural Development, 12*(1), 109–136.

Peters, G. P. (2008). From production-based to consumption-based national emission inventories. *Ecological Economics, 65*(1), 13–23.

Peters, G. P., & Hertwich, E. G. (2006). Pollution embodied in trade: The Norwegian case. *Global Environmental Change, 16*(4), 379–387.

Peters, G. P., Minx, J. C., Weber, C. L., et al. (2011). Growth in emission transfers via international trade from 1990 to 2008. *Proceedings of the National Academy of Sciences of the United States of America, 108*(21), 8903–8908.

Rivera, M. (2014). Trade patterns in the process of European integration: Evidence for the intra-industrial exchanges of a Mediterranean peripheral region. *Annals of Regional Science, 52*(1), 227–249.

Roy, J. (2017). On the environmental consequences of intra-industry trade. *Journal of Environmental Economics and Management, 83,* 50–67.

Sarkodie, S. A., & Strezov, V. (2019). A review on environmental Kuznets Curve hypothesis using bibliometric and meta-analysis. *Science of the Total Environment, 649,* 128–145.

Shen, J., Wang, S., Liu, W., et al. (2019). Does migration of pollution-intensive industries impact environmental efficiency? Evidence supporting "Pollution Haven Hypothesis". *Journal of Environmental Management, 242,* 142–152.

Solarin, S. A., Al-Mulali, U., Musah, I., et al. (2017). Investigating the pollution haven hypothesis in Ghana: An empirical investigation. *Energy, 124,* 706–719.

Stern, D. I. (2004). The rise and fall of the Environmental Kuznets Curve. *World Development, 32*(8), 1419–1439.

Subak, S. (1995). Methane embodied in the international trade of commodities: Implication for global emissions. *Global Environmental Change, 5*(5), 433–446.

Swart, J. (2013). Intra-industry trade and heterogeneity in pollution emission. *The Journal of International Trade and Economic Development, 22*(1), 129–156.

Timmer, M. P., Dietzenbacher, E., Los, B., et al. (2015). An illustrated user guide to the World Input-Output Database: The case of global automotive production. *Review of International Economics, 23*(3), 575–605.

Wang, B., & Li, X. (2017). From world factory to world investor: The new way of China integrating into the world. *China Economic Journal, 10*(2), 175–193.

Wang, Y., Kang, X., Wu, X., et al. (2013). Estimating the environmental Kuznets curve for ecological footprint at the global level: A spatial econometric approach. *Ecological Indicators, 34,* 15–21.

Wu, J., Wei, Y. H. D., Chen, W., et al. (2019). Environmental regulations and redistribution of polluting industries in transitional China: Understanding regional and industrial differences. *Journal of Cleaner Production, 206,* 142–155.

Wu, R., Geng, Y., Dong, H., et al. (2016). Changes of CO_2 emissions embodied in China–Japan trade: Drivers and implications. *Journal of Cleaner Production, 112,* 4151–4158.

Wyckoff, A. W., & Roop, J. M. (1994). The embodiment of carbon in imports of manufactured products: Implications for international agreements on greenhouse gas emissions. *Energy Policy, 22*(3), 187–194.

Xie, R., Zhao, G., Zhu, B., et al. (2018). Examining the factors affecting air pollution emission growth in China. *Environmental Modeling & Assessment, 23,* 389–400.

Yeung, H. W. C. (2015). Regional development in the global economy: A dynamic perspective of strategic coupling in global production networks. *Regional Science Policy & Practice, 7*(1), 1–24.

Zhang, Z., Zhu, K., & Hewings, G. J. D. (2017). A multi-regional input–output analysis of the pollution haven hypothesis from the perspective of global production fragmentation. *Energy Economics, 64,* 13–23.

Zhao, Y., Zhang, Z., Wang, S., et al. (2014). CO_2 emissions embodied in China's foreign trade: An investigation from the perspective of global vertical specialization. *China and World Economy, 22*(4), 102–120.

Zhu, S., He, C., & Liu, Y. (2014). Going green or going away: Environmental regulation, economic geography and firms' strategies in China's pollution-intensive industries. *Geoforum, 55,* 53–65.

Chapter 9
How Does Spatial Division of Labour Relate to Industrial Pollution?

Foreign trade reallocates resources across sectors and spaces at multiple levels, which is subject to the spatial division of labour. This fact thereby draws more attention to the inequality of development in terms of economic performance, social welfare, and environmental challenges (Zhang and Zhang 2003). Regarding the internal geography of one country, one of the main concerns is that foreign trade is likely to promote the development of some places at the sacrifice of the others' resources and opportunities. If so, foreign trade may expose the internal geography to a higher risk of polarisation that lagging regions may be trapped in underdevelopment or even recession (Rodríguez-Pose 2012).

Regarding environmental performance, the trade-induced polarisation among domestic regions is manifested by the sustained development of polluting industries in those specialized regions. Theoretical thinking based on the pollution haven phenomenon believes that foreign trade will make some regions specialized in primary or traditional sectors, which are lower value-added and more pollution-intensive (Ren et al. 2014). On the other hand, foreign trade will also make other regions specialized in emerging industries, which tend to be higher value-added and cleaner. As a result, such a spatial division of labour will lead to the inequality of environmental performance within the nation. As foreign trade continues growing and promoting the regional specialization, regional inequality of environmental performance is expected to increase. If this is the case, the growth of foreign trade will expose regions to a higher risk of polarisation. Put it another way, the trade-induced spatial division of labour will allow some regions to improve their environmental performance at the sacrifice of the others.

Such concern is especially relevant to a country with a vast territory like China. Since the 1980s, China has turned to an outward-oriented growth model, which triggers economic growth through utilizing foreign investment and encouraging foreign trade. This growth model has gradually moved the position of China within the global trade network, manifested by the changing flows and patterns of trading goods and services at the global level (Yang 2005). In line with this process, foreign trade exposes China's internal regions to a higher risk of becoming a pollution haven

© Springer Nature Singapore Pte Ltd. 2020
C. He and X. Mao, *Environmental Economic Geography in China*,
Economic Geography, https://doi.org/10.1007/978-981-15-8991-1_9

(López et al. 2018). It is also noteworthy that foreign trade further causes factor relocations from one region to another region, altering the internal spatial division of labour. As a result, the internal economic geography has witnessed a rising level of inequality in terms of economic growth and social welfare (Huang 2011), which also reinforces the coastal-inland division to different extents.

However, if we look into the spatial dynamics of regional environmental performance, there is no clear spatial divergence among Chinese regions. Instead, either the pollution emissions or the pollution intensities among regions are converging from 2003 to 2011. Among the prefectural-level cities in China, the Gini coefficient of their industrial sulphur dioxides (SO_2) emissions decreases from 0.492 in 2003 to 0.445 in 2011. The lowest value appeared in 2009, which is 0.399. Similarly, the Gini coefficient of pollution intensities (measured by SO_2) also continuously reduces from 0.519 in 2003 to 0.464 in 2011.

This thereby raises a question of why trade-induced spatial division of labour does not result in the regional polarisation of environmental performance. Previous studies tend to investigate this question from the perspective of the pollution haven effect, highlighting the role of external forces (especially the policies and regulations) in hindering spatial division of labour (Zhang and Zhang 2003). For example, studies on the spatial pattern of polluting industries argue that the development of polluting industries is influenced by various determinants, such as factor endowment, local institutions, and nonlocal linkages (Marconi 2012; Zhou et al. 2017). As such, during the spatial relocation of polluting industries, both the pollution haven effect and the factor endowment effect work together. The relocation of polluting industries is also subject to their trade-offs (Shen 2008).

Empirical evidence from the Yangtze River Delta and the Pearl River Delta in China also reveals that leading regions still have the motivations to maintain the development of polluting industries during their transition development, even though they have comparative advantages in developing cleaner industries (Cui and Zhao 2015; Shen et al. 2012). As for the lagging regions that are not attractive enough to those emerging industries, they have the motivation to participate in the interregional competition to attract those traditional relocating industries by preferential policies (Woods 2006). This is also known as environmental race to the bottom. Besides, since lagging regions tend to be disadvantaged in benefiting from a trade-induced division of labour, they may also turn to regional protectionism and then hinder the spatial division of labour (Bai et al. 2004). Both the race to the bottom and the regional protectionism may result in the homogeneity of industrial development among regions (Lu et al. 2004).

However, not just the external forces matter. There are also internal forces that will restrict the rigidification of incumbent spatial division of labour. In other words, the formation of regional development paths is not only subject to external forces, but also the internal forces. The development of one region entails new demands as well as new capacities, then shifting towards a new path internally.

Building on this premise, this chapter revisits the environmental effect of the trade-induced spatial division of labour from the evolutionary perspective. This chapter categorizes the spatial division of labour into path extension and path branching of regions. Regarding the environmental performance, path extension is the rigid-ification of the current development path, which suggests the sustained develop-ment of polluting/non-polluting industries in those already specialized regions. Path branching is the development of new specialized industries, given its already specialized industries. In this chapter, path branching means one region becomes co-specialized in both polluting and non-polluting industries.

On this basis, this chapter rephrases the environmental effect of the trade-induced spatial division of labour by three questions. First, how will foreign trade affect the path extension/path branching? Answers to this question will help to clarify how foreign trade reshapes the spatial division of labour. Second, how does the path extension/path branching affect environmental performance? Answers to this question may reveal the environmental outcome due to the change of spatial division of labour. Third, what determines the path extension/path branching during regional development? Answers to this question capture the conditions for path creation and predict the change of spatial division of labour. These questions and answers interact so that this chapter applies the simultaneous equation models for empirical analysis.

Overall, the evolutionary perspective allows this chapter to contribute to the liter-ature in twofold. First of all, this chapter considers the trade-induced spatial division of labour as diversified paths of regional development in response to trade openness, so that we essentially investigate the environmental effects of regional development paths. The rationale behind this is that foreign trade allows regions to rigidify its current development path, upgrade their current path, and branch into new paths, or even replace the current path by a new one. In this regard, the spatial division of labour is no longer a static pattern but a dynamic process. Its environmental effect does not only depend on which region specialised in polluting industries to what extent. More importantly, its environmental effect is a matter of how those specialised regions will change in the future.

Secondly, this chapter investigates the internal forces by combining the theory of agglomeration externalities and evolutionary economic geography. The develop-ment path is closely related to the dynamic of industrial agglomeration. This chapter highlights that the scale diseconomies serve as internal forces for path creation, which will restrict the rigidification of incumbent development path and simultane-ously leave more space for possible paths. It is also notable that the environmental degradation due to the sustained concentration of polluting industries will produce a significant crowding-out effect (Cheng 2016; Drut and Mahieux 2017). Hence, the regional specialization in polluting industries is essentially an endogenous force to prevent the sustained specialization process. Moreover, it will also reduce the proba-bility of developing new paths. Empirical evidence supports that scale diseconomies and increasing pollution intensity are crucial internal forces that restrict the regional polarisation of environmental performance.

9.1   Foreign Trade, Spatial Division of Labour, and the Environment

9.1.1   Foreign Trade and the Internal Geography of Trading Countries

Foreign trade can alter the division of labour between nations by its capacity in reallocating resources and factors. Recently, there is also increasing awareness of the effect of foreign trade in reshaping the internal geography of trading countries (Farole 2013; Yao and Sun 2016). Particularly, more efforts have been made to examine whether trade growth contributes to the rising of inequality (Rodríguez-Pose 2013). This is especially the case in developing economies that benefit from their export-oriented growth model since the 1980s. Empirical evidence has already supported that foreign trade attracts more resources and factors to concentrate in border regions, and then largely promotes the development of border regions. Typical cases include the US-Mexico border regions and the coastal region of China (Coşar and Fajgelbaum 2016). Besides, empirical studies on Southeast Asia and South America also find that foreign trade significantly reshapes the industrial geography within trading countries (Mukim 2013; Rodríguez-Pose et al. 2013; Sanguinetti and Martincus 2009).

How does foreign trade reshape the internal geography of trading countries? Conventionally, theoretical thinking refers to the comparative advantage theory. When trade openness exposes internal regions to overseas markets, their comparative advantages will change accordingly. A prototype is the changing locational advantage of border regions, which shifts from the periphery to the core due to their proximities to overseas markets (Villar 1999). The changing comparative advantages allow internal regions to develop new activities, and then reshape the domestic economic geography. The new trade theory further explores the endogenous mechanisms besides the exogenous endowment and location (Markusen and Venables 1998), and highlights the role of scale economies (Amiti 1998). Because of the scale economies, foreign trade will promote the resources and firms to concentrate in some regions to reduce the production costs and support their continuous development, then leading to the core-periphery pattern of industries (Krugman 1998, 2011).

Regarding the internal geography, empirical studies have explored the effects of foreign trade on industrial agglomeration, regional division of labour, and regional inequality. Firstly, as predicted by the new trade theory and new economic geography model, foreign trade is able to shift the spatial pattern of industries from one to another. Such a shift is manifested by the relocation of resources and activities from place to place (Coşar and Fajgelbaum 2016). China has witnessed a shift towards the coastal regions since the 1980s, which attracts the concentration of capital-intensive and labour-intensive industries (Ge 2009). Empirical studies ascribe this shift to the locational advantages of coastal regions, and the coast-biased policy supports from the government. Since the 2000s, there is another shift that labour-intensive industries move out of the coastal regions and relocate to the inland regions, which are caused by the rising costs of land resources and the labour force (Zhu and Wang 2018). Thus, theoretical thinking posits that the effects of foreign trade on industrial agglomeration depend on the trade-offs between factors of economic geography, increasing returns to scale, and economic policy (Jin et al. 2006).

Next, previous studies further explore the changes in the regional division of labour and regional integration in response to trade openness. Empirical findings reveal that foreign trade highlights the importance of overseas markets and then intensifies the interregional competition for overseas markets. As a result, it is likely to strengthen the segmentation of domestic markets and lead to homogeneity of regional industries. Poncet (2005) reveals that China's provinces embrace the regional protectionism during their foreign trade growth, resulting in the domestic market segmentation. Although market segmentation allows provinces to benefit from foreign trade, it also leads to the scattered distribution of industries as well as the redundancy of industrial development. Huang (2011) find that foreign trade is one of the key determinants of the regional division of labour in China. However, there is also a U-shaped curve relationship between trade openness and domestic market integration.

Finally, based on the findings of industrial agglomeration and regional integration, studies revisit the inequality issue of regional development in exposure to foreign trade. Particularly, they focus on the role of foreign trade in triggering the catch up of lagging regions. In the *World Development Report 2009: Reshaping Economic Geography* proposed by the World Bank, the trio of density-distance-division is developed to explain the "unbalanced growth" (World Bank 2009). Based on this framework, foreign trade will impact the development of lagging regions, which tend to be distant with low density. It confronts lagging regions with a high level of trade costs and a low level of economic efficiencies. According to the scale economies and the home market effect in the new economic geography model, foreign trade may reinforce the core-periphery pattern and rigidify the unfavourable position of lagging regions. Fujita and Hu (2001) report the increasing level of inequality between the coastal and the inland regions in China. Associated with the continuous flows of capital and labour from the inland to the coastal regions, foreign trade widens the gap of industrial agglomeration between them (Wei and Wang 2011).

However, things have changed since the 2000s. As the coastal regions step into a new stage of development, factor prices increase drastically. At the same time, the demand from overseas markets keeps shrinking. In such a context, empirical findings disclose that some exporting industries start to relocate to the inland regions of China or even leave China. As a result, the gap between the coastal and the inland regions narrows to some extent (Wei and Bai 2009). On this basis, theoretical thinking highlights the importance of free mobility of factors, regional specialization, and domestic market integration in reducing regional inequality (Jin et al. 2006; Farole 2013).

9.1.2 An Evolutionary Perspective for Spatial Division of Labour and Environmental Effects

Overall, based on the theoretical thinking and empirical findings above, foreign trade provides more possibilities for regional development. From a dynamic perspective, it alters the already-existing development paths of regions. The enlarging market allows more regions to get involved and reap the benefits of trade openness. In this way, some regions are able to expand their incumbent production and become specialized, while other regions are able to create new development paths and participate in the competition for overseas markets (Mao and He 2019). On the other hand, the new demands, technologies, and standards embedded in the trade flows will promote economic novelty (Lin and Lin 2010), which would then allow regions to either upgrade their incumbent activities towards higher value-added ones or branch into the new but related fields (Boschma and Iammarino 2009).

Advances in Evolutionary Economic Geography provide various models of developing new paths, including the path extension, path upgrading, path importation, path branching, path diversification, and path creation (Tanner 2014; Isaksen 2015; Grillitsch et al. 2018). "Path" in the literature essentially depicts the relationship between the incumbent industries and new ones. For example, regarding the three pairs of path development models, both path extension and path upgrading indicate the growth of incumbent industries. Path extension denotes the similarity between them and emphasize that their co-development strengthens the regional base in this field. Path upgrading accentuates the higher added value of new industries than the incumbent ones.

Both path branching and path diversification acknowledge the role of knowledge for incumbent industries in supporting new industries, such as the recombination of existing knowledge. However, path branching indicates that new industries are still related to incumbent industries, while path diversification suggests that new industries are unrelated.

Both path creation and path importation describe the emergence of unrelated industries, which is also irrelevant to the incumbent industries. However, path creation is endogenous, indicating the role of radical innovation in promoting the emergence of new industries. In contrast, path importation is exogenous and suggests that the emergence of new industries stems from nonlocal forces.

The development of one region never follows a particular path solely. It keeps shifting from path to path, associated with the changes in its development states (Hassink et al. 2019). Hence, although conventional studies notice that foreign trade is likely to promote regional specialization and regional division of labour, they overlook that regional specialization and regional division of labour are dynamic essentially. A high level of specialization will suggest either the potential for growth or the signal of recession. Regional division of labour will be an outcome of either market segmentation or regional integration. In this regard, it is less likely to assess the performance of regional specialization or regional division of labour simply by their levels. Thus, it is preliminary to identify their likely changes. Otherwise, the results of the performance assessment will be misleading.

Under this condition, a spatial division of labour is essentially an outcome of regional specialization that regions become specialized in various but interconnected industries (He et al. 2018). The change of spatial division of labour is manifested by the dynamic of regional specialization patterns and suggests primarily the changes in existing industries. In this regard, path extension, path upgrading, and path branching are related to the spatial division of labour. In this chapter, we focus on the role of path extension, and path branching, while the next chapter will regard the role of upgrading.

Regarding the environmental performance of development paths, this chapter primarily looks into the structural changes in regional industries between polluting and non-polluting industries. Path extension describes the continuous growth of polluting industries or non-polluting industries in already specialised regions. The effects of path extension of polluting industries on environmental performance should be quite negative, since enlarging scales of polluting industries suggest an increase in production capacity and energy consumption (Cheng 2016). By contrast, the path extension of non-polluting industries is supposed to improve environmental performance. Regarding the path branching, the ones based on polluting industries towards non-polluting will significantly improve environmental performance.

Building on this premise, we can answer the question of why the trade-induced spatial division of labour does not lead to regional polarisation. On the one hand, although the path extension of polluting industries will worsen the environmental performance, path extension has its limit. According to the agglomeration externalities, the increasing scale of industrial agglomeration will result in more fierce competition for both input factors and markets (Lin et al. 2011). The scale diseconomies thereby emerge and hinder the development of similar firms and industries, which is also known as the crowding-out effects. The recession of polluting industries leaves more spaces for the development of non-polluting industries so that promotes path branching.

On the other hand, there are also external forces hindering path extension of polluting industries and simultaneously promoting path branching into non-polluting industries. Typical external forces include environmental regulations, industrial policies, foreign investment, and trade linkages (Zheng and Shi 2017). On this basis, we develop two research hypotheses regarding the spatial division of labour and environmental performance.

Hypothesis 1: Internal forces stemming from scale diseconomies of specialisation hinder path extension and promote path branching.

Hypothesis 2: External forces, including environmental regulation and nonlocal linkages, restrict the path extension of polluting industries, and promote path branching into non-polluting industries.

9.2   Research Design

9.2.1   Measurement of Regional Specialization

This chapter uses the locational quotient (LQ) to indicate the regional specialization, which measures how concentrated one industry is in a region compared to the whole nation. LQ is formulated as the share of a particular industry in one region divided by its share in the whole nation (Eq. 9.1). Using the national level as a benchmark, an LQ of 1.00 indicates that the shares of a particular industry are the same between the region and the nation. If one region has an LQ above 1.00, the industry takes a larger share in the region than the national benchmark, indicating a regional specialization in this industry. In general, an LQ above 1.50 is supposed to indicate a significant level o f specialization.

$$\mathrm{LQ}_{ic} = (p_{ic}/p_c)/(p_i/p_n) \tag{9.1}$$

where i, j, n denotes the industry, city, and the nation. p indicates the gross outputs.

9.2.2   Measurement of Path Extension and Path Branching

This chapter applies the co-occurrence probability to measure path extension and path branching. This chapter considers two paths for regional development, namely, the development of polluting industries and non-polluting industries. Path extension indicates the reinforcement of incumbent industries, whereas path branching suggests the creation of a new development path based on the incumbent ones.

In terms of probability, path extension can be measured by the co-occurrence probability of incumbent industries. A larger value means that incumbent industries will continue to grow and then reinforce the current development path.

Similarly, path branching can be measured by the co-occurrence probability of incumbent industries and new industries, which means polluting industries and non-polluting industries in this chapter. A larger probability suggests that the development of incumbent industries will trigger the development of other industries, leading to the creation of a new development path.

This chapter calculates the conditional probability of one region to specialized in two industries simultaneously (Eq. 9.2).

$$P(A|B) = P(AB)/P(B) \tag{9.2}$$

where A refers to the event that $LQ_{ic} > 1$ and B refers to the event that $LQ_{jc} > 1$. P (.) refers to the probability of events based on the classical models of probability, which is based on the 261 prefectural-level cities in China.

Then, we select the minimum value between $P(A|B)$ and $P(B|A)$ as the probability of path development based on the interaction between industry i and j (Eq. 9.3). Considering the industry i and j belong to the same development path (either polluting or non-polluting ones), this probability indicates the path extension. As for the i and j belong to different paths, this probability indicates the path branching.

$$\begin{cases} PE_{ij} = \min\{P(A|B), P(B|A)\} if\ i,\ j \text{ belongs to the same path} \\ PB_{ij} = \min\{P(A|B), P(B|A)\} if\ i,\ j \text{ belongs to different paths} \end{cases} \quad (9.3)$$

Finally, we aggregate the bilateral probabilities to the city level, which is labelled as D. D indicates the probability of one region to develop its paths by industry i, given its already-existing specialization in industry j (Eq. 9.4).

$$\begin{cases} D_p = \sum_i \sum_j x_{ic} x_{jc} PE_{ij} / \sum_j PE_{ij} \text{ if } i,\ j \text{ are polluting} \\ D_{np} = \sum_i \sum_j x_{ic} x_{jc} PE_{ij} / \sum_j PE_{ij} \text{ if } i,\ j \text{ are non} - \text{polluting} \\ D_{tr} = \sum_i \sum_j x_{ic} x_{jc} PB_{ij} / \sum_j PB_{ij} \text{ if } i,\ j \text{ are different} \end{cases} \quad (9.4)$$

where D_p indicates the path extension of polluting industries; D_{np} represents the path extension of non-polluting industries; D_{tr} denotes the path branching.

9.2.3 Identification of Polluting and Non-polluting Industries

In this chapter, we use the pollution intensity at the sector level to indicate industrial environmental performance. Pollution intensity is measured by the industrial pollution emissions per unit of gross output. We calculate the annual average of pollution intensity of each sector from 2003 to 2009, ranking the sectors by intensities from high to low. We categorize these sectors into four types based on their ranks. Sectors in the upper quartile are marketed as Q1, indicating the most pollution-intensive ones. Sectors in the following quartiles are marked as Q2, Q3, and Q4 sequentially, indicating the relatively polluting, relatively clean, and clean sectors respectively. Moreover, the Q1 and Q2 sectors are identified as polluting sectors, while the Q3 and Q4 sectors are non-polluting ones. We also refer to the First National General Survey of Pollution Sources to calibrate the results of categorization. Finally, we identify the polluting sectors for air pollution and water pollution separately (Table 9.1).

Table 9.1 Identification of polluting industries

Industrial sulphur dioxides emissions			Industrial wastewater discharges		
Code	Sectors	Pollution intensity (t/10^8 yuan)	Code	Sectors	Pollution intensity (t/10^8 yuan)
31	Non-metallic mineral products	151.60	22	Paper and paper products	68.22
22	Paper and paper products	88.77	28	Chemical fibres	15.55
33	Smelting and processing of non-ferrous metals	67.58	26	Raw chemical materials and chemical products	14.87
26	Raw chemical materials and chemical products	50.89	15	Beverages	12.30
32	Smelting and processing of ferrous metals	49.76	17	Textile	12.15
28	Chemical fibres	45.29	14	Foods	8.47
30	Plastics	45.21	13	Processing of food from agricultural products	8.25
25	Petroleum, coking, processing of nuclear fuel	43.16	27	Medicines	8.20
15	Beverages	27.70	32	Smelting and processing of ferrous metals	5.95
14	Foods	20.09	19	Leather, fur, feather, and related products	4.73
29	Rubber	18.44	25	Petroleum, coking, processing of nuclear fuel	4.56
17	Textile	18.32	31	Non-metallic mineral products	3.22

(continued)

Table 9.1 (continued)

Industrial sulphur dioxides emissions			Industrial wastewater discharges		
Code	Sectors	Pollution intensity (t/10^8 yuan)	Code	Sectors	Pollution intensity (t/10^8 yuan)
20	Processing of timber, manufacture of wood, bamboo, rattan, palm, and straw products	17.49	33	Smelting and processing of non-ferrous metals	2.85
27	Medicines	14.30	34	Metal products	2.51
13	Processing of food from agricultural products	14.17	20	Processing of timber, manufacture of wood, bamboo, rattan, palm, and straw products	2.22

9.2.4 Model Specifications

Empirical models in this chapter should achieve three basic targets. First, the models should examine the effects of development paths on environmental performance. Second, the models should test the effect of external and internal forces on development paths. Third, the models have to figure out the effective external forces and internal forces. It is also noteworthy that there are significant feedbacks between industrial pollution emissions and industrial agglomeration. We apply the simultaneous equation models (SEM) to establish the empirical models.

The basic model consists of two equations. One regards the environmental performance and its determinants and captures the environmental effects of path extension or path branching. The other one considers the development paths and their driving forces, identifying the internal forces and external forces.

The first equation is based on the stochastic impacts by regression on population, affluence and technology (STIRPAT) model. This is a revised version of the IPAT model, proposed by Dietz and Rosa (1997). It allows for the non-proportional, non-monotonic effects of population, affluence, and technology, which has been widely used in studies on the determinants of pollution emissions and greenhouse gas emissions (York et al. 2003; Rosa and Dietz 2012). The model is formulated as Eq. (9.5).

$$I_{ct} = \alpha P_{ct}^{\beta} A_{ct}^{\gamma} T_{ct}^{\delta} e_{ct} \tag{9.5}$$

where for city c and year t, I represents the industrial pollution emissions (or pollution intensities); P denotes the population; A indicates affluence; T represents the technology; e is the error term.

This chapter incorporates the proxies and control variables (Eq. 9.6).

$$ln S_{ct} = \alpha_0 + f(D_{ct}) + \beta_1 ln P_{ct} + \beta_2 ln Y_{ct} + \beta_3 ln T_{ct} + \beta_4 G_{ct} + \beta_5 I_{ct} + u_c + e_{ct}$$
$$(9.6)$$

where S represents the industrial pollution intensities of either industrial sulphur dioxides emissions or industrial wastewater discharges. As for the control variables, P denotes the labour in manufacturing sectors, controlling the city-specific scales of pollution sources. Y indicates the GDP per capita, controlling the city-specific level of affluence. T denotes the labour in information and communications technology (ICT) industries. This serves as a proxy to control the differences in technological base among cities. This proxy can be efficient because intelligence manufacturing leads to the promotion of productivities and the upgrading of procedures. ICT industries underlie the growth of intelligent manufacturing. Thus, the growth of ICT industries in one city suggests its knowledge base and innovation capacity to some extent.

Besides the internal factors P, A, and T, we further incorporate two influential external factors. One regards the city-specific stringency of environmental regulation, which is measured by the performance-based indicators (I). Specifically, the treatment ratio of industrial sulphur dioxides emissions is used for the regulation on air pollution, while the compliance rate of industrial wastewater discharge is for water pollution regulation. The other one regards the role of foreign trade. We use the share of new export linkage in all exports (G) as a proxy to control the inter-city differences in export expansion. As for the identification of new trade linkage, we construct a three-dimensional panel of "domestic localities-overseas markets-industries". By comparing the cross-sections in two years, the trade linkage that is absent in the former year but appears in the latter year is identified as a new trade linkage.

D represents a set of explanatory variables regarding development paths, including the D_p for path extension of polluting industries, the D_{np} for path extension of non-polluting industries, and the D_{tr} for path branching. A higher D_p indicates that one region is likely to be locked in the development of polluting industries, worsening environmental performance. Its coefficients are expected to be positive. By contrast, the coefficients of D_{np} should be negative. In theory, the coefficients of D_{tr} can be mixed. Conventionally, path branching implies the escape of development lock-in of polluting industries, which can improve the environmental performance. On the other hand, the results in Fig. 10.2 reveal that the development of polluting and non-polluting industries are discordant. The development of polluting industries prevails, while one of the non-polluting industries is still fledgling. The incumbent polluting industries have an advantage during path branching. If so, path branching will still hinder the improvement of environmental performance.

The second equation regarding the determinants of development paths is formulated as Eq. 9.7.

$$D_{ct} = a_0 + f\left(X_{ct}, X_{ct}^2\right) + f(\ln S_{ct}) + f(G_{ct}, I_{ct}) + u_c + e_{ct} \qquad (9.7)$$

where X denotes a set of internal forces that affects the dynamic of industrial agglomeration, including the changing factor densities and firm competitions. According to the Cobb–Douglas production function, we focus on the effects of the labour force (X_L) and the capital (X_K). X_L indicates the labour density, which is the number of labours divided by the city area. Similarly, X_K is the stock of capital divided by the city area. As for the level of firm competition, following the approach of Zhou and Zhu (2013), we use the density of firms in one city as a proxy (X_F). X_F is the number of firms in one city divided by its area. According to the theory of agglomeration externalities, the scale diseconomies can emerge when the agglomeration scale grows too large. On this basis, we incorporate the quadratic terms of X_L, X_K, X_F to capture the likely scale diseconomies. If the scale diseconomies come into being, the coefficient of X should be positive and the one of X^2 should be negative.

S, G, and I in the model represent the external forces that affect the dynamic of industrial agglomeration. S in Eqs. 9.6 and 9.7 represents the same thing—industrial pollution intensity, which serves as an endogenous variable in the SEM. Since the deterioration of environmental performance will crowd out economic activities and hinder further growth of agglomeration, the coefficient of S should be negative.

G denotes the trade expansion, which is also the same as the one in Eq. 9.6. The coefficients of G tend to be positive. Regarding path branching, trade expansion can introduce new resources, technologies, and markets for the development of new paths. If so, the coefficient of G should be positive for path branching. As for path extension, trade expansion would promote regional specialization by reallocating resources based on their comparative advantages. In this regard, the coefficient of G for path extension is also likely to be positive.

Overall, there are three groups of SEMs for empirical analysis. One regards the interaction between environmental performance and path extension, including the path extension of both polluting and non-polluting industries. The second one looks into how environmental performance interacts with path branching. The former consists of three equations, incorporating three endogenous variables, namely, $\ln S$, D_p, and D_{np}. The latter has two equations, including two endogenous variables, which are $\ln S$ and D_{tr}. Both the order and rank condition of identification support that the SEMs in this chapter are over-identified. Both the two-stage least square (2SLS) and the three-stage least square (3SLS) approaches can be used for coefficient estimations.

Furthermore, we apply the Hausman test to examine the simultaneity of endogenous variables. Specifically, we regress one particular endogenous variable by all predetermined variables. Using ordinary least squares (OLS), we predict the residuals. Then, we regress other endogenous variables by this endogenous variable and the residuals. We can also predict the residuals at the second stage. If the second-stage residuals are statistically significant, there is simultaneity between two endogenous variables (Pindyck and Rubinfeld 1997). The results support the bilateral simultaneity between the endogenous variables in our SEMs. Besides, we also conduct the test for heterogeneity and multicollinearity issues. The variation inflation factors are smaller than 5.0, suggesting that the models are free of severe multicollinearity issues. However, the results of the Breusch-Pagan test conclude heterogeneity. Thus, we use the 2SLS for coefficient estimations.

9.2.5   Data

Trade data for export expansion are obtained from the Statistics of China Customs. This dataset is product-specific utilizing the Harmonized System (HS) and covers the period 2000–2013. Since we primarily focus on the production of exporting goods, the observations of trade agencies were deleted. Then we aggregate the exports by the pairs of localities and overseas markets to the six-digit level of HS code.

Industry data are aggregated from the China Annual Survey of Industrial Firms. This is a firm-specific dataset that covers all Chinese industrial state-owned enterprises and those non-state-owned enterprises with annual sales of five million yuan or above. It utilizes the Standard Industrial Classification Codes of China (GB/T 4754–2002) covering the period 1998–2013.

The matching of these two datasets utilize the product concordance provided by the World Bank and shift the product nomenclature of HS code at the six-digit level to the International Standard Industrial Classification Revision 3 (ISIC Rev. 3) at the four-digit level. We further link the four-digit ISIC code to the two-digit GB/T code.

As for the environmental performance, we select industrial sulphur dioxides emissions to indicate air pollution. This is because the sulphur dioxides are primarily industry-driven rather than emitted by transportation or household activities (Grether and Mathys 2013). According to the National Bulletin of Environmental Statistics of China, the industrial sulphur dioxides emission is 1.79×10^7 t in 2003, accounting for 83% of the total emissions of sulphur dioxides. In 2011, these figures reached 2.02×10^7 t and 91%, respectively. Therefore, industrial sulphur dioxides emissions can reflect the air pollution induced by industrialization. As for water pollution, we select the industrial wastewater discharges as an indicator. Correspondingly, we use the treatment ratio of industrial sulphur dioxides emissions and the compliance ratio of industrial wastewater discharges as proxies for the stringency of environmental regulation. The higher the treatment ratio/compliance ratio is, the more stringent than the environmental regulation would be. All these data related to environmental performance are obtained from the *China City Statistical Yearbook (2004–2010)*.

Based on the data above, we construct a data panel for empirical analysis, consisting of 30 sectors and 261 prefectural-level cities during 2003–2009.

9.3 Spatial Division of Labour of Polluting Industries in China

9.3.1 Spatial Patterns of Regional Specialization in Polluting and Non-polluting Industries

Figure 9.1 uses the locational quotient to measure the regional division of labour of polluting industries in China. It conveys three significant findings. First of all, both polluting and non-polluting industries exhibit a clear coastal-inland division. Their spatial trends are the opposite. In the coastal regions, the east and southeast regions tend to specialize in non-polluting industries. This is especially the case for cities in the Pearl River Delta and the Yangtze River Delta. As for the inland regions, the specialized regions in non-polluting industries are confined to some core regions,

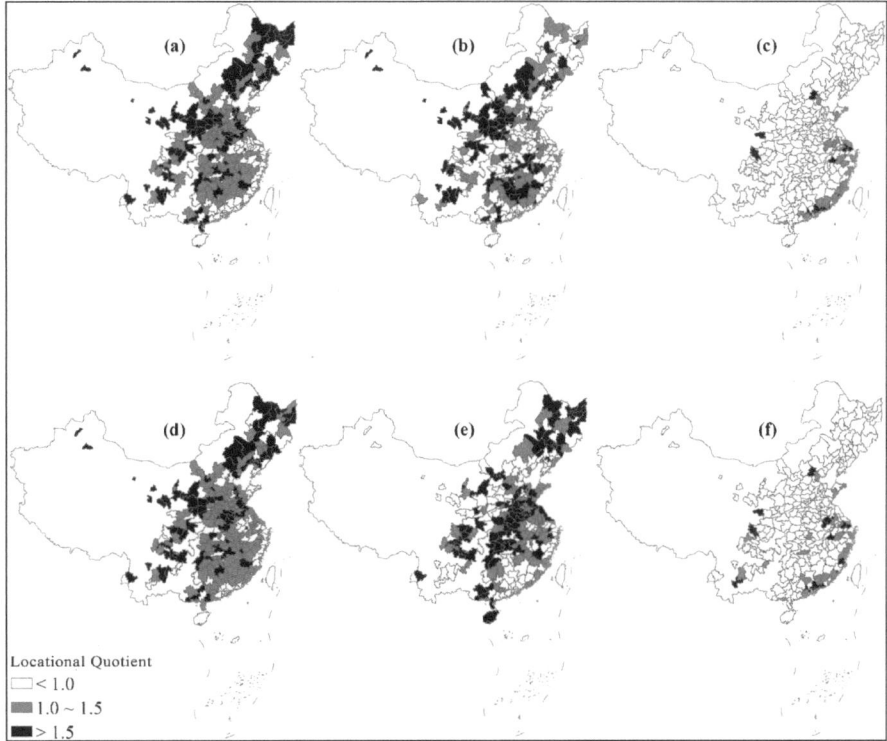

Fig. 9.1 Regional division of labour by air pollution-intensive sectors and water pollution-intensive sectors in China (2007). *Note* (a), (b), and (c) denote the air pollution, while (d), (e), and (f) denote the water pollution. (a) and (d) represent the polluting industries, including Q1 and Q2 industries. (b) and (e) represent the Q1 industries, while (c) and (f) represent the Q4 industries

such as the provincial capitals. In contrast, the inland regions overall exhibit a high level o f specialization in polluting industries.

Secondly, more interior regions tend to specialize in more polluting industries. As shown in Fig. 10.1b, e, regions specialized in Q1 industries are more interior than regions specialized in Q2 industries. The former primarily locates in the Liaoning, Hebei, Shaanxi, Shanxi, Jiangxi, Hunan, Guizhou, and Yunnan, which are identified as west provinces in China. On the contrary, the latter tends to be in Heilongjiang, Shandong, Henan, Hubei, and Anhui.

Last but not the least, the cleanest industries (the Q4 industries in this chapter) are highly concentrated on a few cities that lead the trends of national development, such as Beijing, Shanghai, Guangzhou, and Shenzhen. As for the relatively clean industries (the Q3 industries in this chapter), more cities get involved. The specialized regions cover the core cities at the national level as well as the regional level.

In sum, regions specialized in polluting industries are widely distributed across the nation. That is, the polluting industries lay the manufacturing foundation for most inland regions as well as some coastal regions in China. On the other hand, regions specialized in non-polluting industries are scattered, which are overlapped with those national and regional core cities. Such spatial pattern is subject to the feature of non-polluting industries. Non-polluting industries consist of various emerging industries, which are intensive in technology, capital, and knowledge. As such, non-polluting industries will gain more advantages in the core cities, which have stronger development bases as well as more agglomeration externalities.

9.3.2 Development Path of Regions Specialized in Polluting and Non-polluting Industries

Concerns about the trade-induced polarisation among regions are essentially built on the premise that foreign trade will continue to reinforce the specialization processes. As for environmental performance, foreign trade promotes the concentration of polluting industries in regions with locational advantages. This process is likely to crowd out the development of non-polluting industries and exerts more pressure on the local environment. However, the reinforcing of specialization is conditional.

On the one hand, studies on the agglomeration economies find that the growth of agglomeration has a limit (Dumais et al. 2002). The continuous development of agglomeration would result in a shift from scale economies to diseconomies (Zhou and Zhu 2013). Once the scale diseconomies emerge, the costs from negative impacts (such as increasing competition for factors and markets, environmental degradations, and so on) will exceed the benefit due to the reduction of spatial proximity. Hence, the growth of agglomeration comes to an end, and its scale reaches an equilibrium.

On the other hand, the evolutionary economic geography further reveals that the locked-in process of agglomeration growth requires a high level of similarity between incumbent industrial activities and new ones (Nooteboom 2000). Otherwise, the

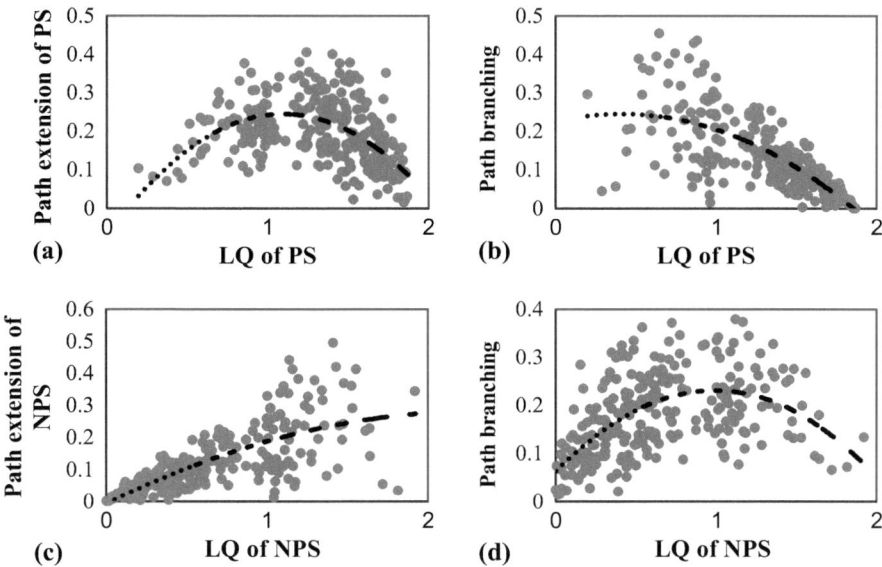

Fig. 9.2 Correlation between industrial specialization and development paths. *Note* PS represents the polluting sectors, and NPS represents the non-polluting sectors

introduction of new firms and activities will promote the incremental innovation in the agglomeration, and then trigger the industrial restructuring in the agglomeration by economic novelty (Boschma et al. 2017). Either way, the regional polarisation is less likely to emerge.

Figure 9.2 captures the inverted-U-shaped relationship between the level of specialization and the dynamics of specialization. As the level of specialization increases, the co-occurrence probability of different industries would also increase, but only to a point, beyond which increases in specialization would lead to a reduction in co-occurrence probability. This turning point indicates the shift from scale economies to diseconomies, suggesting the congestion effect or crowding-out effect due to excessive specialization. Such an inverted-U-shaped relationship points to the fact that regional specialization does not necessarily result in the development of lock-in.

Graphs in Fig. 9.2 convey three key findings. First, regarding those cities specialized in polluting industries, the higher the level of specialization is, the lower the co-occurrence probability of polluting industries would be. This fact suggests that excessive specialization will hamper the continuous development of similar industries. Second, the specialization of non-polluting industries does not reach the turning point yet. That is, across the whole nation, cities with higher levels of specialization in non-polluting industries will promote the co-occurrence of similar industries. This implies that the development of non-polluting industries in China is still fledgeling.

Third, the co-occurrence of polluting and non-polluting industries usually has a lower probability. This is especially the case in most cities specialized in polluting industries. This fact reveals that the development of polluting industries is likely to crowd out the development of non-polluting ones and results in the spatial separation of polluting and non-polluting industries. Polluting industries lay the foundations of most cities' development, while the specialization of non-polluting industries is confined to a few developed cities.

Overall, these findings capture the fact that the specialization in polluting industries does not necessarily lead to development lock-in. As the specialization level increases, the scale diseconomies will crowd out the co-location of similar industries, which would become a passive way to unlock the likely lock-ins. Moreover, considering the specialization of non-polluting industries does not reach the turning point, promoting the development of non-polluting industries will increase the co-occurrence probability between polluting and non-polluting industries. This allows cities to escape their reliance on polluting industries and become an active way to unlock the likely lock-ins. In this regard, the results of the descriptive analysis support the hypotheses preliminarily. Next, we present the results of empirical models to examine the hypotheses further.

9.4   Empirical Results

9.4.1   Path Extension and Industrial Pollution Intensity

Table 9.2 reports the coefficient estimations of the model regarding regional specialization and industrial pollution emissions. The results are in line with the theoretical prediction. Also, results based on sulphur dioxides emissions indicate the same trends as wastewater discharges. This fact supports the robustness of our results. Specifically, the results reveal that the path extension of polluting industries results in the increase of industrial pollution emissions. Meanwhile, the effects of path extension of non-polluting industries on industrial pollution emissions are insignificant. These two findings suggest that the development of polluting and non-polluting industries in China is inequal. The development of non-polluting industries is still fledgeling.

Regarding the path extension of polluting industries, the sources of congestion effects are primarily manifested by labour forces and firms. Their coefficients are positive, while the coefficients of their quadratic terms are negative. In theory, the concentration of homogeneous firms in one region will result in fierce competition for factors, and then raise the factor price as the concentration scale expands. Increasing costs will reduce the benefits stemming from geographical concentration, eventually generate scale diseconomies beyond a certain point. Hence, the crowding-out effect comes into being. In practice, looking into the regional development in China during the study period, the crowding-out effect first emerges in labour-intensive industries. In the 2000s, there is a significant relocation trend of labour-intensive industries, from the coastal to the inland, which is subject to the scale diseconomies (He and Wang

2010). By contrast, the capital exhibits a different role in affecting regional specialization dynamics. The coefficients of capital are negative, while the coefficients of its quadratic terms are insignificant. This represents that increasing capital inputs will hinder the path extension of polluting industries directly. The rationale behind this is that capital inputs can significantly promote firm growth and simultaneously accelerate the saturation of local markets. As a result, increasing capital inputs allows the growing incumbent firms to take up local markets and hinder the co-location of new homogenous firms. Regarding the specialization of non-polluting industries, the sources of congestion effects are also manifested by the firm competition. The crowding-out effect due to increasing capital inputs still holds, while the one due to increasing labour inputs is insignificant. This is partly because most non-polluting industries are not labour-intensive industries.

As for the role of industrial pollution emissions in path extension, the results reveal that increasing pollution emission is a significant factor hindering path extension. This forms a negative feedback loop between industrial pollution emission and path extension. Increasing specialization in polluting industries results in a high level of pollution emissions. However, more pollution emissions will hinder the growth of regional specialization. Therefore, this mechanism determines that the path extension of polluting industries will not result in development lock-in. Thus, it will not lead to the regional polarisation of environmental performance.

The results of trade expansion also echo the theoretical prediction. Export expansion in China significantly promotes the development of polluting industries in the previously specialized cities. However, the effects of export expansion on non-polluting industries are insignificant. This finding supports that China's participation in international trade has promoted its development of polluting industries (Sun et al. 2017). Internal regions exhibit different capacities in response to this trend. Regions with locational advantages tend to become specialized in polluting industries and then bear more environmental pressure. It is also notable that environmental regulation has significantly restricted the continuous concentration of polluting industries in those specialized cities, which supports the prediction of the second hypothesis.

9.4.2 Path Branching and Industrial Pollution Intensity

Table 9.3 reports the coefficients regarding the interactions between path branching on industrial pollution intensities. Concerning the effects of path branching on pollution intensities, the coefficients of *Dtr* are statistically significant and positive, which suggests that the current co-specialization in polluting and non-polluting industries still increase, rather than decrease, the pollution intensities. From another perspective, this finding supports that incumbent polluting industries restrict the development of their non-polluting counterparts, which impacts the improvement of regional environmental performance.

Table 9.2 Results of the effects of pollution intensity and path extension

	Industrial sulphur dioxides emissions						Industrial wastewater discharges					
	Pollution intensity		Path extension of polluting industries		Path extension of non-polluting industries		Pollution intensity		Path extension of polluting industries		Path extension of non-polluting industries	
	Dp	0.335***	X_{Lp}	0.452***	X_{Lnp}	0.113	Dp	0.284***	X_{Lp}	0.421**	X_{Lnp}	0.079
	Dnp	−0.002	$X_{Lp}{}^2$	−0.189**	$X_{Lnp}{}^2$	−0.034	Dnp	0.213	$X_{Lp}{}^2$	−0.187**	$X_{Lnp}{}^2$	−0.027
	lnP	−0.060	X_{Kp}	−0.312***	X_{Knp}	−0.323***	lnP	0.104*	X_{Kp}	−0.228***	X_{Knp}	−0.256***
	lnY	−0.660***	$X_{Kp}{}^2$	−0.003	$X_{Knp}{}^2$	0.101	lnY	−0.934***	$X_{Kp}{}^2$	−0.067	$X_{Knp}{}^2$	0.079
	lnT	−0.005	X_{Fp}	0.417***	X_{Fnp}	0.319***	lnT	0.021	X_{Fp}	0.415***	X_{Fnp}	0.360***
	G	−0.039***	$X_{Fp}{}^2$	−0.202**	$X_{Fnp}{}^2$	−0.123**	G	−0.043***	$X_{Fp}{}^2$	−0.199***	$X_{Fnp}{}^2$	−0.136***
	I	−0.114***	lnS	−0.317***	lnS	−0.268***	I	−0.042***	lnS	−0.158***	lnS	−0.173***
			G	0.049***	G	−0.009			G	0.049***	G	−0.005
			I	−0.082**	I	−0.044**			I	−0.040***	I	−0.008***
	No. obs	1564	No. obs	1564	No. obs	1564	No. obs	1564	No. obs	1564	No. obs	1564
	Adj-R²	0.897	Adj-R²	0.912	Adj-R²	0.972	Adj-R²	0.907	Adj-R²	0.924	Adj-R²	0.975

Note ***$p < 0.01$, **$p < 0.05$, *$p < 0.10$

Table 9.3 Results of the effects of pollution intensity and path branching

Industrial sulphur dioxides emissions				Industrial wastewater discharges			
Pollution intensity		Path branching		Pollution intensity		Path branching	
Dtr	0.464^{***}	X_{Lp}	0.028	Dtr	0.449^{***}	X_{Lp}	0.020
lnP	-0.053	$X_{Lp}{}^2$	-0.110	lnP	0.113^{**}	$X_{Lp}{}^2$	-0.032
lnY	-0.670^{***}	X_{Kp}	-0.117^{*}	lnY	-0.923^{***}	X_{Kp}	-0.025
lnT	-0.001	$X_{Kp}{}^2$	0.102^{*}	lnT	0.030	$X_{Kp}{}^2$	-0.008
G	-0.025^{*}	X_{Fp}	0.863^{***}	G	-0.035^{***}	X_{Fp}	0.967^{***}
I	-0.115^{***}	$X_{Fp}{}^2$	-0.186^{**}	I	-0.033^{**}	$X_{Fp}{}^2$	-0.264^{***}
		X_{Lnp}	0.373^{*}			X_{Lnp}	0.238
		$X_{Lnp}{}^2$	-0.157			$X_{Lnp}{}^2$	-0.142
		X_{Knp}	-0.448^{***}			X_{Knp}	-0.356^{***}
		$X_{Knp}{}^2$	0.086			$X_{Knp}{}^2$	0.037
		X_{Fnp}	-0.676^{***}			X_{Fnp}	-0.828^{***}
		$X_{Fnp}{}^2$	0.286^{***}			$X_{Fnp}{}^2$	0.384^{***}
		lnS	-0.295^{***}			lnS	-0.164^{***}
		G	-0.008			G	0.001
		I	-0.062^{***}			I	-0.002
No. obs	1564	No. obs	1564	No. obs	1564	No. obs	1564
Adj-R^2	0.894	Adj-R^2	0.952	Adj-R^2	0.908	Adj-R^2	0.956

Note $^{***}p < 0.01, ^{**}p < 0.05, ^{*}p < 0.10$

The results of the function regarding the effects of pollution intensities on path branching convey three key findings. First, the labour force does not exert significant influence in the co-specialization probability of polluting and non-polluting industries, since most non-polluting industries are not labour-intensive. Second, the increasing inputs of capital can promote firm growth and market saturation, and hamper the entrance of new firms. As a result, increasing capital will hinder the co-specialization in polluting and non-polluting industries. Meanwhile, increasing capital also allows non-polluting industries to develop regardless of the incumbent polluting industries.

Third, firm competition within either polluting industries or non-polluting industries will affect the probability that one region becomes co-specialized in them. In a region that has specialized in polluting industries, the development of non-polluting industries will benefit from the firm competition within polluting industries. However, this effect has its limits. Too many firms within polluting industries can still crowd out the development of non-polluting industries. On the other hand, if we look into the few regions that have specialized in non-polluting industries, the development of non-polluting industries avoids the co-location with polluting industries. This fact is manifested by the negative coefficients of X_{Fnp}. Moreover, the coefficients of $X_{Fnp}{}^2$ are significantly positive, suggesting that the path branching based on polluting industries into non-polluting industries requires a sustained development of non-polluting industries.

It is also noteworthy that the crowding-out effect of pollution emissions still hold for path branching, which hinders the co-location of polluting and non-polluting industries. Similarly, the environmental regulation hampers path branching as the development of polluting industries prevails at present. Export expansion does not exert significant influence in path branching, unlike the result of path extension.

Overall, empirical results reveal that export expansion has promoted path extension and exhibited little effects on path branching. However, the path extension does not lead to the development lock-in of polluting industries. The rationale behind is twofold. One is the crowding-out effect due to the emergence of scale diseconomies. When the agglomeration grows too large, costs is likely to increase as firm competition becomes fierce and factor price rises. Hence, the path extension of polluting industries is less likely to grow without limit. The other one is that increasing pollution emissions also serve as a centrifugal force for the concentration of polluting industries. Besides, environmental regulation also serves as an external force to restrict the continuous growth of polluting industries in specialized regions. All these findings point to the fact that although foreign trade promotes regional specialization, the dynamics of regional specialization does not lead to regional polarisation since regional specialization has its limit. There are both external and internal forces to increase the costs of production and curb the growth of specialization.

9.4.3 Robustness Check

This chapter applies the simultaneous equations model (SEM) to examine the interactions between regional environmental performance and regional specialization dynamic. Key variables in this model are pollution intensity, co-specialization probability in polluting/non-polluting industries, and co-specialization probability in polluting and non-polluting industries. In regard to the coefficient estimations of SEM, there are two common approaches, namely, the two-stage least square (2SLS) and the three-stage least square (3SLS). The 3SLS uses the residuals in 2SLS to do generalized least square estimation at the third stage so that it is a systematic approach. However, the third stage also determines that 3SLS is sensitive to variable selection. As such, in order to guarantee the robustness of the results of key coefficients, we use the results of 2SLS for analysis and report the results of 3SLS for the robustness test.

Tables 9.4 and 9.5 report the coefficient estimation results by the 3SLS. Compared with the results in Table 9.1 and Table 9.2, respectively, the coefficients and their significant levels are approximate. That is to say, empirical results in this chapter are not subject to a specific estimation approach. The results are robust in general.

Besides, we also regress the models with/without quadratic terms separately and compare their differences in the coefficients of labour, capital, and the number of

Table 9.4 Results of the effects of pollution intensity and path extension (3SLS)

Industrial sulphur dioxides emissions						Industrial wastewater discharges					
Pollution intensity		Path extension of polluting industries		Path extension of non-polluting industries		Pollution intensity		Path extension of polluting industries		Path extension of non-polluting industries	
Dp	0.335^{***}	X_{Lp}	0.508^{***}	X_{Lnp}	0.140	Dp	0.421^{***}	X_{Lp}	0.432^{***}	X_{Lnp}	0.131
Dnp	-0.335	X_{Lp}^2	-0.208^{***}	X_{Lnp}^2	-0.028	Dnp	0.067	X_{Lp}^2	-0.165^{***}	X_{Lnp}^2	-0.021
lnP	-0.043	X_{Kp}	-0.191^{***}	X_{Knp}	-0.190^{***}	lnP	0.112	X_{Kp}	-0.176^{***}	X_{Knp}	-0.202^{***}
lnY	-0.611^{***}	X_{Kp}^2	-0.045	X_{Knp}^2	0.038	lnY	-0.932^{***}	X_{Kp}^2	-0.054	X_{Knp}^2	0.040
lnT	0.001	X_{Fp}	0.108	X_{Fnp}	0.298^{***}	lnT	0.023	X_{Fp}	0.192	X_{Fnp}	0.323^{***}
G	-0.039^{***}	X_{Fp}^2	-0.049	X_{Fnp}^2	-0.120^{**}	G	-0.051^{***}	X_{Fp}^2	-0.082	X_{Fnp}^2	-0.125^{***}
I	-0.117^{***}	lnS	-0.321^{***}	lnS	-0.237^{***}	I	-0.049^{***}	lnS	-0.166^{***}	lnS	-0.156^{***}
		G	0.050^{***}	G	-0.003			G	0.052^{***}	G	-0.003
		I	-0.080^{***}	I	-0.043^{**}			I	-0.036^{***}	I	-0.003
No. obs	1564	No. obs	1564	No. obs	1564	No. obs	1564	No. obs	1564	No. obs	1564
Adj-R^2	0.900	Adj-R^2	0.911	Adj-R^2	0.973	Adj-R^2	0.902	Adj-R^2	0.919	Adj-R^2	0.976

Note $^{***}p < 0.01$, $^{**}p < 0.05$, $^{*}p < 0.10$

Table 9.5 Results of the effects of pollution intensity and path branching (3SLS)

Industrial sulphur dioxides emissions				Industrial wastewater discharges			
Pollution intensity		Path branching		Pollution intensity		Path branching	
Dtr	0.463^{***}	X_{Lp}	0.003	Dtr	0.457^{***}	X_{Lp}	-0.015
lnP	-0.063	X_{Lp}^2	-0.096	lnP	0.125^{**}	X_{Lp}^2	-0.070
lnY	-0.669^{***}	X_{Kp}	-0.116^{**}	lnY	-0.921^{***}	X_{Kp}	-0.095^*
lnT	0.001	X_{Kp}^2	0.100^{**}	lnT	0.022	X_{Kp}^2	0.091^*
G	-0.025^*	X_{Fp}	0.888^{***}	G	-0.033^{***}	X_{Fp}	0.942^{***}
I	-0.115^{***}	X_{Fp}^2	-0.190^{***}	I	-0.034^{**}	X_{Fp}^2	-0.226^{***}
		X_{Lnp}	0.373^{**}			X_{Lnp}	0.317^*
		X_{Lnp}^2	-0.161^*			X_{Lnp}^2	-0.149^*
		X_{Knp}	-0.434^{***}			X_{Knp}	-0.381^{***}
		X_{Knp}^2	0.078			X_{Knp}^2	0.046
		X_{Fnp}	-0.700^{***}			X_{Fnp}	-0.741^{***}
		X_{Fnp}^2	0.294^{***}			X_{Fnp}^2	0.340^{***}
		lnS	-0.293^{***}			lnS	-0.166^{***}
		G	-0.008	.		G	-0.003
		I	-0.063^{***}			I	-0.001
No. obs	1564	No. obs	1564	No. obs	1564	No. obs	1564
Adj-R^2	0.895	Adj-R^2	0.952	Adj-R^2	0.908	Adj-R^2	0.957

Note $***p < 0.01$, $**p < 0.05$, $*p < 0.10$

firms. The results do not exhibit significant differences, supporting the incorporation of quadratic terms into the models.

It is also notable that the results of key variables exhibit similar trends between industrial sulphur dioxides emissions and industrial wastewater discharges. This fact suggests that findings in this chapter are robust to types of pollutants to some extent.

9.5 Conclusion and Discussion

This chapter focuses on the concerns about the environmental effects of the spatial division of labour that is induced by foreign trade. Foreign trade reallocates resources across nations as well as domestic regions and such reallocation simultaneously reshapes the spatial division of labour within one nation. Regions are likely to become specialised in activities where they possess comparative advantages. These comparative advantages determine the diversification of regional development trajectories and result in the inequality of environmental performance. Some will be specialized in polluting industries, while others will be specialized in non-polluting ones. If things continue in this way, it would thereby raise an environmental-relevant question.

Whether trade-induced specialization in polluting/non-polluting industries results in the regional polarisation of environmental performance? If so, foreign trade will result in the development of some regions at the sacrifice of the others.

This chapter answers this question based on the observation of China's outward-oriented development during the last three decades. Foreign trade serves as an engine for China's regional development and reshapes its internal economic geography constantly. There is also evidence of the rising inequality associated with trade growth. However, the environmental performance among regions does not exhibit a trend towards polarization. Instead, the level of inequality of regional environmental performance decreases significantly, regardless of the fluctuated inequality of regional development. In this context, this chapter seeks to answer why the trade-induced spatial division of labour does not necessarily lead to the regional polarisation of environmental performance.

This chapter contributes to the literature by investigating the trade-induce spatial division of labour from an evolutionary perspective. Combining the theory of agglomeration economies and evolutionary economic geography (EEG), we propose that trade-induced spatial division of labour is essentially related to the process of path extension dynamically rather than the level of specialization statically. Path extension suggests the sustained growth of incumbent activities and foreign trade is one of the driving forces for it. In this regard, the occurrence of regional polarisation is built on the premise that once the path for regional development is created, it will extend endlessly and not change. Obviously, this is impossible. In fact, based on the theory of EEG, regional development has various models except for path extensions, such as path upgrade, path branching, and path diversification (Grillitsch et al. 2018). Regional development consists of the shifts from one path to another.

Thus, this chapter seeks to figure out the determinants that restrict path extension of polluting industries. Based on the theory of agglomeration economies, this chapter posits two forces that restrict path extension, namely, the external forces and internal forces. Internal forces stem from the sources of scale diseconomies as industrial agglomeration grows too large, such as the congestion effect and crowding-out effect. In contrast, external forces suggest the moderating effect of exogenous factors, such as institutions, policies, and nonlocal linkages. Empirical findings from China's regional development from 2003 to 2009 support the role of external and internal forces in restricting path extension of polluting industries, which explains the reason why regions specialized in polluting industries do not become locked-in polluting industries.

Empirical findings offer insights into the regional sustainable development in response to trade growth. In developing economies, traditional industries, which tend to be polluting industries, underlie the early development of regions. However, the boom of incumbent polluting industries is likely to constrain the development of non-polluting industries. The efforts on developing non-polluting industries should pay off in the long run. Nevertheless, there are both external and internal forces to shift regional development way from path extension in polluting industries. Previous efforts may focus on the role of external forces, such as environmental regulation. However, it is also beneficial to take advantage of the internal forces, particularly the role of investment.

References

Amiti, M. (1998). New trade theories and industrial location in the EU: A survey of evidence. *Oxford Review of Economic Policy, 14*(2), 45–53.

Bai, C., Du, Y., Tao, Z., et al. (2004). Local protectionism and regional specialisation: Evidence from China's industries. *Journal of International Economics, 63*(2), 397–417.

Boschma, R., Coenen, L., Frenken, K., et al. (2017). Towards a theory of regional diversification: Combining insights from Evolutionary Economic Geography and Transition Studies. *Regional Studies, 51*(1), 31–45.

Boschma, R., & Iammarino, S. (2009). Related variety, trade linkages, and regional growth in Italy. *Economic Geography, 85*(3), 289–311.

Cheng, Z. (2016). The spatial correlation and interaction between manufacturing agglomeration and environmental pollution. *Ecological Indicators, 61*(2), 1024–1032.

Coşar, A. K., & Fajgelbaum, P. D. (2016). Internal geography, international trade, and regional specialization. *American Economic Journal: Microeconomics, 8*(1), 24–56.

Cui, J., & Zhao, H. (2015). Spatial relocation of pollution-intensive industry and the mechanism in Yangtze River Delta. *Geographical Research, 34*(3), 504–512. (in Chinese).

Dietz, T., & Rosa, E. A. (1997). Effects of population and affluence on CO_2 emissions. *Proceedings of the National Academy of Sciences, 94*(1), 175–179.

Drut, M., & Mahieux, A. (2017). Correcting agglomeration economies: How air pollution matters. *Papers in Regional Science, 96*(2), 381–400.

Dumais, G., Ellison, G., & Glaeser, E. L. (2002). Geographic concentration as a dynamic process. *The Review of Economics and Statistics, 84*(2), 193–204.

Farole, T. (2013). Trade, location, and growth. In T. Farole (Ed.), *The Internal Geography of Trade: Lagging Regions and Global Markets* (pp. 15–31). Washington DC: World Bank Publications.

Fujita, M., & Hu, D. (2001). Regional disparity in China 1985–1994: The effects of globalisation and economic liberalisation. *The Annals of Regional Science, 35*(1), 3–37.

Ge, Y. (2009). Globalisation and industry agglomeration in China. *World Development, 37*(3), 550–559.

Grether, J.-M., & Mathys, N. A. (2013). The pollution terms of trade and its five components. *Journal of Development Economics, 100,* 19–31.

Grillitsch, M., Asheim, B., & Trippl, M. (2018). Unrelated knowledge combinations: The unexplored potential for regional industrial path development. *Cambridge Journal of Regions, Economy, and Society, 11,* 257–274.

Hassink, R., Isaksen, A., & Trippl, M. (2019). Towards a comprehensive understanding of new regional industrial path development. *Regional Studies, 53*(11), 1636–1645.

He, C. F., & Wang, J. S. (2010). Geographical agglomeration and co-agglomeration of foreign and domestic enterprises: A case study of Chinese manufacturing industries. *Post-Communist Economies, 22*(3), 323–343.

He, C. F., Yan, Y., & Rigby, D. (2018). Regional industrial evolution in China. *Papers in Regional Science, 97*(2), 173–198.

Huang, J. (2011). Foreign trade, interregional trade and regional specialisation. *South China Journal of Economics,* (6), 7–22. (in Chinese).

Isaksen, A. (2015). Industrial development in thin regions: Trapped in path extension? *Journal of Economic Geography, 15*(3), 585–600.

Jin, Y., Chen, Z., & Lu, M. (2006). Industry agglomeration in China: Economic geography, new economic geography and policy. *Economic Research Journal, 4,* 79–89.

Krugman, P. (1998). What's new about the new economic geography? *Oxford Review of Economic Policy, 14*(2), 7–17.

Krugman, P. (2011). The new economic geography, now middle-aged. *Regional Studies, 45*(1), 1–7.

Lin, H. L., Li, H. Y., & Yang, C. H. (2011). Agglomeration and productivity: Firm-level evidence from China's textile industry. *China Economic Review, 22*(3), 313–329.

Lin, H., & Lin, E. (2010). FDI, trade, and product innovation: Theory and evidence. *Southern Economic Journal, 77*(2), 434–464.

Lu, M., Chen, Z., & Yan, J. (2004). Increasing return, development strategy and regional economic segmentation. *Economic Research Journal, 1,* 54–63.

López, L. A., Arce, G., Kronenberg, T., et al. (2018). Trade from resource-rich countries avoids the existence of a global pollution haven hypothesis. *Journal of Cleaner Production, 175*(20), 599–611.

Mao, X., & He, C. (2019). Export expansion and regional diversification: Learning from the changing geography of China's exports. *The Professional Geographer, 71*(4), 692–702.

Marconi, D. (2012). Environmental regulation and revealed comparative advantages in Europe: Is China a pollution haven? *Review of International Economics, 20*(3), 616–635.

Markusen, J. R., & Venables, A. J. (1998). Multinational firms and the new trade theory. *Journal of International Economics, 46*(2), 183–203.

Mukim, M. (2013). Location and the determinants of exporting: Evidence from manufacturing firms in India. In T. Farole (Ed.), *The Internal Geography of Trade: Lagging Regions and Global Markets* (pp. 177–204). Washington DC: World Bank Publications.

Nooteboom, B. (2000). *Learning and Innovation in Organizations and Economies.* Oxford, UK: Oxford University Press.

Pindyck, R., & Rubinfeld, D. (1997). *Econometric Models and Economic Forecasts* (4th ed.). Singapore: McGraw-Hill/Irwin.

Poncet, S. (2005). A fragmented China: Measure and determinants of Chinese domestic market disintegration. *Review of International Economics, 13*(3), 409–430.

Ren, S., Yuan, B., Ma, X., et al. (2014). The impact of international trade on China's industrial carbon emissions since its entry into WTO. *Energy Policy, 69,* 624–634.

Rodríguez-Pose, A. (2012). Trade and regional inequality. *Economic Geography, 88*(2), 109–136.

Rodríguez-Pose, A. (2013). Trade openness and regional inequality. In T. Farole (Ed.), *The Internal Geography of Trade: Lagging Regions and Global Markets* (pp. 33–63). Washington DC: World Bank Publications.

Rodríguez-Pose, A., Tselios, V., & Winkler, D. (2013). Location and the determinants of exporting: Evidence from manufacturing firms in Indonesia. In T. Farole (Ed.), *The Internal Geography of Trade: Lagging Regions and Global Markets* (pp. 127–158). Washington DC: World Bank Publications.

Rosa, E. A., & Dietz, T. (2012). Human drivers of national greenhouse-gas emissions. *Nature Climate Change, 2*(8), 581–586.

Sanguinetti, P., & Martincus, C. V. (2009). Tariffs and manufacturing location in Argentina. *Regional Science and Urban Economics, 39*(2), 155–167.

Shen, J. (2008). Trade liberalisation and environmental degradation in China. *Applied Economics, 40*(8), 997–1004.

Shen, J., Xiang, C., & Liu, Y. (2012). The mechanism of pollution-intensive industry relocation in Guangdong Province, 2000–2009. *Geographical Research, 31*(2), 357–368. (in Chinese).

Sun, C., Zhang, F., & Xu, M. (2017). Investigation of pollution haven hypothesis for China: An ARDL approach with breakpoint unit root tests. *Journal of Cleaner Production, 161,* 153–164.

Tanner, A. N. (2014). Regional branching reconsidered: Emergence of the fuel cell industry in European regions. *Economic Geography, 90*(4), 403–427.

Villar, O. A. (1999). Spatial distribution of production and international trade: A note. *Regional Science and Urban Economics, 29*(3), 371–380.

Wei, H., & Bai, M. (2009). The characteristics, determinants and development trend of Chinese enterprise migration. *Development Research*, (10), 9–18. (in Chinese).

Wei, H., & Wang, C. (2011) Research on spatial agglomeration and its determinants of foreign trade in China. *Journal of Quantitative and Technical Economics* (11), 66–82. (in Chinese).

Woods, N. D. (2006). Interstate competition and environmental regulation: A test of the race-to-the-bottom thesis. *Social Science Quarterly, 87*(1), 174–189.

World Bank. (2009). *World development report 2009: Reshaping economic geography.* Washington DC: The World Bank.

Yang, D. (2005). International influences of China's becoming "World Factory". *China Industrial Economy, 25*(9), 42–49. (in Chinese).

Yao, P., & Sun, J. (2016). Trade openness and internal geography: Theoretical and empirical review. *Economist, 5,* 96–104.

York, R., Rosa, E. A., & Dietz, T. (2003). STIRAPT, IPAT, and ImPACT: Analytic tools for unpacking the driving forces of environmental impacts. *Ecological Economics, 46*(3), 351–365.

Zhang, X., & Zhang, K. H. (2003). How does globalisation affect regional inequality within a developing country? Evidence from China. *Journal of Development Studies, 39*(4), 47–67.

Zheng, D., & Shi, M. (2017). Multiple environmental policies and pollution haven hypothesis: Evidence from China's polluting industries. *Journal of Cleaner Production, 141,* 295–304.

Zhou, S., & Zhu, W. (2013) Must industrial agglomeration be able to bring about economic efficiency: Economies of scale and crowding effect. *Industrial Economics Research* (3), 12–22. (in Chinese).

Zhou, Y., Zhu, S., & He, C. (2017). How do environmental regulations affect industrial dynamics? Evidence from China's pollution-intensive industries. *Habitat International, 60,* 10–18.

Zhu, S. J., & Wang, C. (2018). Shifts in China's economic geography studies in an era of sindustrial restructuring. *Progress in Geography, 37*(7), 865–879. (in Chinese).

Chapter 10
Do Foreign Trade Contribute to Industrial Pollution?

Foreign trade can transfer pollution emissions spatially through relocating industrial production. It raises concerns about the likely pollution haven effect in lagging regions; that is, pollution-intensive production can move from the core to the periphery due to the changing comparative advantages (Copeland and Taylor 1994; Michida and Nishkimi 2007). Chapter 8 has confirmed these phenomena. Chapter 9 has probed further into the domestic changes when investigating the global pollution transfer. As shown in these two chapters, it is not satisfactory to ascribe the trade-induced pollution simply to the trading partners (Jakob and Marschinski 2013). After all, foreign trade is essentially a twin process of globalisation and localisation (Dicken 2003; Swyngedouw 2004). Besides the trading partners, foreign trade can also modify domestic industrial geography, such as changing the regional division of labour and promoting the industrial agglomeration (Farole 2013). Therefore, foreign trade can further relocate the industrial pollution emission across the country. However, in spite of the global and domestic shift of environmental burdens, one should also note that the local is more than a receptor of environmental burdens. Instead, the local place may react or even adjust to the shift of environmental burdens.

In such a case, this chapter investigates the trade-environment effect (TEE) at the sub-national level, based on a global-local perspective. We consider the TEE as a synergy of global transfer, domestic transfer, and local absorption. It is a cross-scale conceptualization. Firstly, at the global level, the global transfer captures the traditional pollution haven effect, identifying the environmental effect from the trading partners overseas. Secondly, at the national level, the domestic transfer concerns the responses of the country through spatial redistribution, which may offset or amplify the TEE. Thirdly, at the local level, the local absorption regards the relationship between industrial agglomeration and environmental performance, representing the trade-off between trade-induced growth and pollution.

We apply this framework to construct a dynamic panel model for empirical study, using the data covering 261 prefectural-level cities from 2003 to 2011. This period covers the first decade of China's membership in the World Trade Organisation. During this time, expansion into the overseas market has triggered a remarkable

© Springer Nature Singapore Pte Ltd. 2020
C. He and X. Mao, *Environmental Economic Geography in China*,
Economic Geography, https://doi.org/10.1007/978-981-15-8991-1_10

boom in China's economy. However, most researchers also insist that China reaps the benefits of globalization at the cost of the environment, since the low prices of primary factors underlie its advantage on foreign trade (Cole et al. 2011). China will inevitably become the pollution haven of developed economies in the end. In this chapter, our empirical findings support the validity of the pollution haven effect and the efficiency of local responses. What is more important is that our findings further caution the role of regional inequality of local responses. Leading regions can build stronger absorption capacity and crowd out the pollution-intensive production. In contrast, lagging regions with weaker absorption capacity have to face the pressure from both global and domestic transfer.

Theoretically, this chapter extends the line of research on the trade-environment effect by incorporating the global-local perspective. It contributes to the understanding of the controversial pollution haven effect. There is a debate in previous studies. Some studies advocate the pollution haven effect, while others highlight the active role of technological and institutional diffusion associated with foreign trade (Antweiler et al. 2001; Christmann and Taylor 2001; Wagner and Timmins 2009). Our findings demonstrate that spatial discrepancy can occur. That is, regions facing the trade-induced pollution emission are not necessarily with the capacity to absorb the technological diffusion. As such, investigating sub-national TEE is important, especially in a country with huge territories.

This chapter offers insights into the different opportunities and challenges for green development between leading and lagging regions in the context of foreign trade. The region-specific absorption capacities, combined with the domestic transfer, are likely to widen the gap of environmental performance between the leading and lagging regions. It requires that environmental regulation should develop its redistributive capacities to represent the environmental costs in the lagging regions.

The remainder of this chapter is as follows: Sect. 10.1 presents the literature review and the analytical framework. Section 10.2 shows the spatial pattern of TEE in China. Section 10.3 demonstrates the empirical model and reports the results. Section 10.4 concludes with a summary of the main findings and their policy implications.

10.1   A Research Framework Based on the Global-Local Linkage

The primary challenge in holding down TEE to the sub-national level is how to address its global-local linkage. Existing studies simply examine the validity of various hypotheses at the sub-national level, which are previously developed and tested at the supra-national level. However, TEE at the sub-national level is neither replacing the nation with the region nor simply incorporating international trade as trade openness (Zhang and Zhang 2003; Rodríguez-Pose 2013; Coşar and Fajgelbaum 2016). Previous studies overlook the fact that an open region will engage in multi-level connections, namely, the connection to the global market and other regions, and the internal connection between economic agents. In this chapter, we incorporate the global-local perspective in order to investigate the TEE at the sub-national level.

Specifically, we extend the line of the TEE research from the supra-national level to the sub-national level, which shifts the spatial unit from the country to the sub-national region. The region here refers to an open system, engaging in multi-level economic connections. As such, an open region has three basic attributes. Firstly, it connects to the global market via foreign trade. Secondly, it also participates in the regional division of labours, specializing in a particular production. Thirdly, its specialization will attract related firms to concentrate.

Building on these premises, we propose a framework to depict the global-local linkage during the foreign trade (Fig. 10.1). At the global level, this framework contains three types of countries: the home country, the developed country, and the developing country. At the local level, it divides the home country into the leading region and the lagging region. Hence, a region has three types of connections that affect its TEE.

First of all, both the leading and lagging regions will connect to the overseas markets. It is natural that the enlarging market confronts a region with incremental pressure on pollution emissions. In this framework, we argue that the geography of foreign trade also matters. We incorporate two major variables to explain the geography of foreign trade, namely, the development level of trading countries and the distance to them. According to the pollution haven effect, we can expect that trade with the developed country will bring the region more pollution emissions than with developing countries. It should apply to both the leading and the lagging regions.

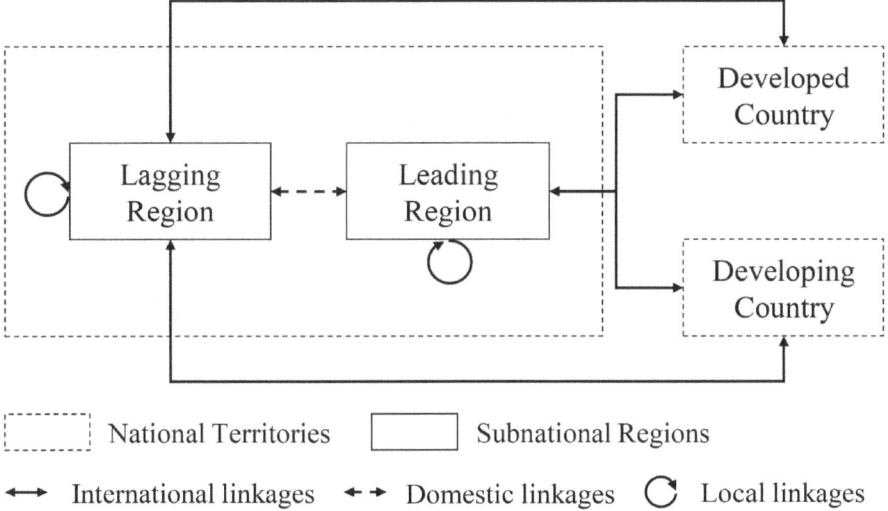

Fig. 10.1 Research framework for trade-environment effect at the sub-national level

With regard to the distance, we consider the effect of foreign trade on the pollution base rather than the emission stemming from transport. Considering the relationship between product quality and the distance of trade, the Washington Apple Effect suggests that the added value of trade goods increases together with the distance of trade (Baldwin and Harrigan 2011; Hummels and Skiba 2004). If this is the case, we can infer that trading with distant countries may achieve better environmental performance. There are two reasons. First, given the same trade good, higher quality (or higher added value) indicates fewer pollution emissions per unit of output. Second, the production of higher value-added goods requires upgrading the technology and improving the processing efficiency. Hence, we state the first hypothesis as follows,

Hypothesis 1: *The geography of foreign trade can alter TEE. Trade with developed countries and neighbouring countries may generate incremental pollution emissions.*

Secondly, foreign trade also fosters the regional division of labours, resulting in the trans-regional transfer of pollution-intensive production. In developed countries, such as the US, empirical studies find that the heavy industries transfer from the stringently regulated region to laxly regulated ones, constrained by the expensive costs of overseas shipment (Becker and Henderson 2000; Kahn 1997). In developing countries like China, previous studies report that polluting firms are transferring from the developed coast to the developing inland (Gereffi 2009; Wu et al. 2017).

Overall, foreign trade affects both the international division of labour and the regional division of labour simultaneously. Hence, foreign trade can also induce trans-regional connections, including the relocation of polluting firms from the core to the periphery. As such, the lagging regions also face domestic transfer from the leading regions. Hence, we propose the second hypothesis as follows,

Hypothesis 2: *Industrial relocation in the wake of foreign trade can amplify the TEE in the lagging regions.*

Last by not the least, the role of local linkage within the region is noteworthy. Local linkage refers to the interactions between economic agents, and between the agents and the environment, such as the industrial structure, industrial agglomeration, and knowledge diffusion. Theoretically, local linkages affect the comparative advantage of one region in foreign trade (Cole and Elliott 2003). They determine whether a region is engaged in international trade with "clean" or "dirty" production. Meanwhile, local linkages also affect the environmental performance of one region (Poon et al. 2006). Industrial agglomeration will generate a positive spill-over effect of improving the efficiency of production (Cheng 2016). A strong knowledge base underlies the capacity of absorbing green technology (Shi et al. 2013).

Local linkages in the framework represent the role of endogenous factors in altering the TEE. As foreign trade promotes the regional division of labour, the more related firms are likely to concentrate in the region. After controlling the scale effect due to the increasing number of firms, the agglomeration of firms is expected to improve environmental performance through enhancing efficiency. Moreover, given that there is trade-induced diffusion of green technology, leading regions are more likely to absorb and diffuse than the lagging regions, since a country's innovation and learning capacity can be highly concentrated in a few developed regions (Oinas and Malecki 2002). Hence, we propose the third hypothesis as follows,

Hypothesis 3: *Local linkage determines the regional capacity to offset the TEE, including the industrial agglomeration and the technology absorption.*

10.2   China's Foreign Trade and Industrial Pollution Emission

10.2.1   Data

We construct a "city-year" panel, covering 261 prefectural-level cities in China from 2003 to 2011. In this chapter, we choose sulphur dioxides emission (SO_2) as a proxy to indicate the industrial pollution emission, based on two properties of SO_2. Firstly, the emission of SO_2 almost comes from manufacturing activities and power generation. Meanwhile, transportations or household activities contribute little to its emission (Grether and Mathys 2013). Hence, SO_2 can indicate the environmental performance of industrial production directly. Second, the transboundary channel for SO_2 primarily depends on industrial mobility, which is embodied in trade. It is less likely to transport over a long distance physically. *The China City Statistical Yearbook* records the industrial SO_2 emissions of each prefectural-level city since 2003.

Information on firm agglomeration is compiled by the annual survey of industrial firms (ASIF) from 2003 to 2011. ASIF is a dataset from the National Bureau of Statistics of China, covering all state-owned firms and non-state-owned firms with annual sales of more than 5 million Chinese *yuan* (CNY). Sectors in this dataset follow the Standard Industrial Classification Codes of China (GB/T 4754-2002).

The data source of China's foreign trade is the same as the one in Chap. 9. In this chapter, we first eliminate the observations of trading agents to guarantee that the place of export is exactly the location of production. Secondly, we match the 6-digit HS code with the 2-digit Chinese coding system (GB/T 4754-2002) via the 4-digit standard industrial classification (SIC). Finally, the data includes 30 manufacturing sectors at the 2-digit level following the Chinese coding system.

Other socio-economic data are obtained from the *China Statistical Yearbook for Regional Economy* and *The China City Statistical Yearbook*.

10.2.2   Spatial Discrepancy Between Emission and Intensity of Industrial Pollution

We apply the Getis-Ord Gi* statistics to conduct the hot-spot analysis (Getis and Ord 1992; Ord and Getis 1995), and then show the spatial pattern of emission and intensity of industrial SO_2 pollution (Fig. 10.2). The results display that industrial pollution emission descents from the north to the south across the country. Many cities with a

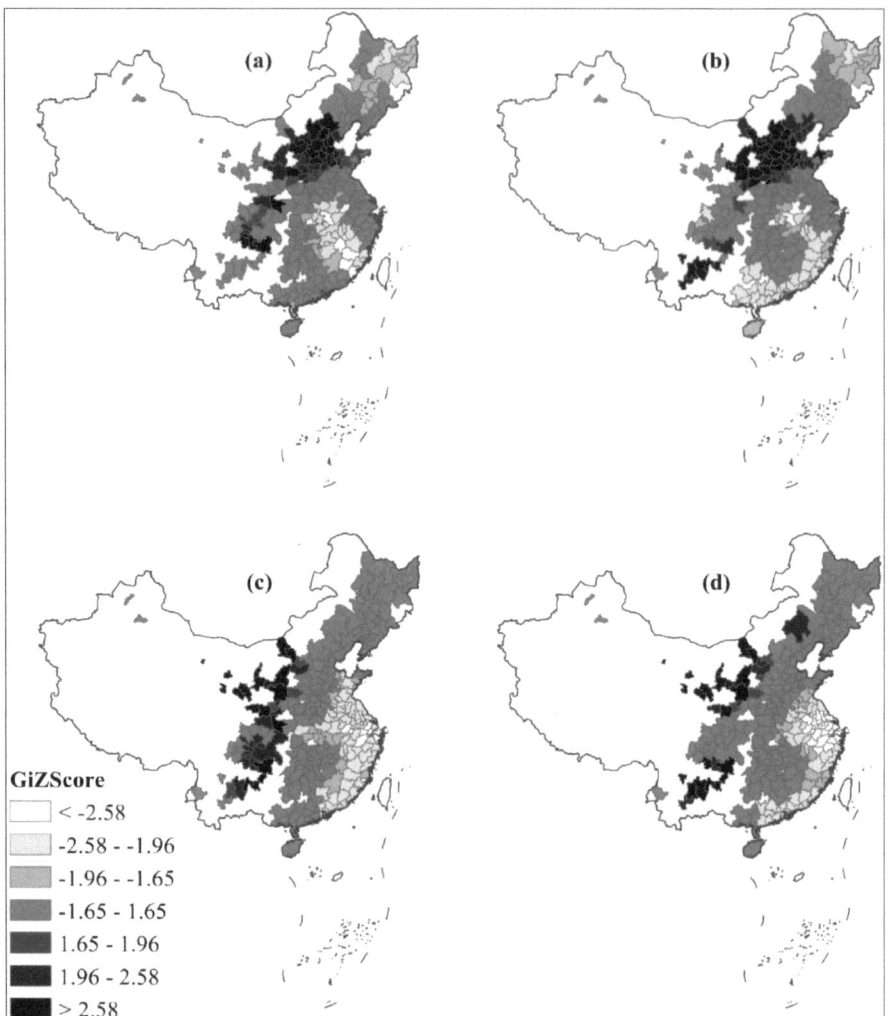

Fig. 10.2 Spatial pattern of pollution emission and intensity in China based on hot-spot analysis (**a** SO$_2$ emission in 2003; **b** SO$_2$ emission in 2011; **c** SO$_2$ intensity in 2003; **d** SO$_2$ intensity in 2011)

high level of pollution emission concentrate in the northern China. Northern China has long been specialized in the heavy and chemical industries. The entry rates of firms in heavy and chemical industries keep high there, especially the iron and steel firms.

Meanwhile, pollution intensity descends from the west to the east, differing from the pattern of pollution emission. Pollution intensity refers to the emissions per unit output. It depends on industrial structure and technological efficiency. Pollution intensity tends to be lower in the leading regions than the lagging regions. Cities

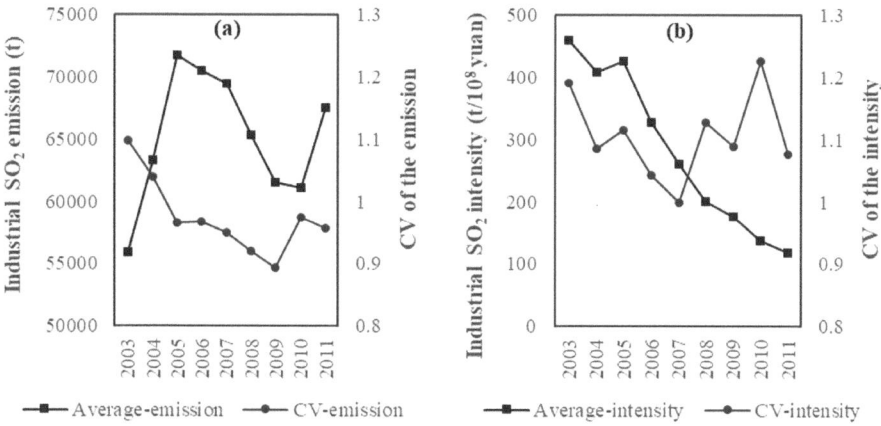

Fig. 10.3   Distribution of pollution emission (**a**) and intensity (**b**)

with lower pollution intensity highly concentrate on the Yangtze River Delta, which is the most developed and innovative region in China. By comparison, cities in the north-west overall have higher pollution intensity.

Spatial discrepancy between pollution emission and pollution intensity highlights the role of global-local linkage. On the one hand, it indicates that trading regions are facing different pressure on pollution emissions. On the other hand, the regions also respond to the pollution emission at various extents. Therefore, the region with a greater pressure of pollution is not necessarily equipped with capacities to absorb it.

Another evidence supports the importance of such spatial discrepancy. As in Fig. 10.3, industrial pollution emission fluctuates during the study period. However, the coefficient of variation (CV) decreases significantly, especially from 2003 to 2009. The gap of emission between leading and lagging regions shrinks during the foreign trade. By contrast, the pollution intensity shows a steady decline during 2003–2011. However, its CV fluctuates significantly, which is overall higher than the CV of emissions. These facts suggest that the development of the lagging regions generates more emissions, reducing the gap with leading regions. However, since their absorption capacities still lag behind, the gap in environmental performance keeps widening.

10.2.3   Foreign Trade and Industrial Pollution Emission

We further link the industrial pollution emission with foreign trade. Using the 30 sectors as a sample, Fig. 10.4 shows the relationship between the foreign trade and the industrial pollution. We first calculate the ratio of export to new countries in the total exports, indicating the expanding geography of exports. It is positively

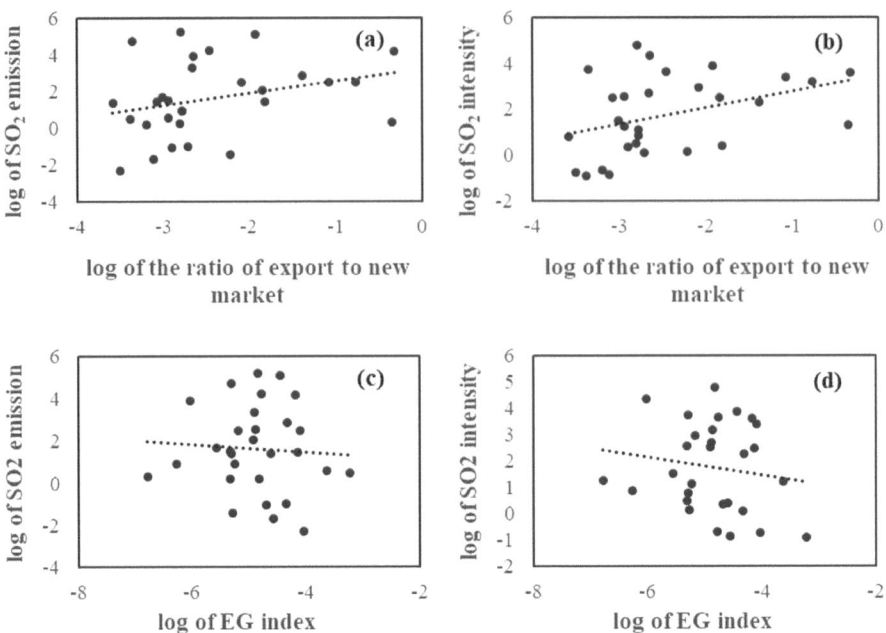

Fig. 10.4 Relationship between industrial SO$_2$ pollution and the global-local linkage of export (2007, 30 sectors as sample)

related to industrial pollution (both emission and intensity). Naturally, the enlarging markets will generate a higher demand for production, which would then increase the level of pollution. We then calculate the EG index to represent the industrial agglomeration (see Ellison and Glaeser 1997). The results show that the EG index is negatively related to industrial pollution intensity, while its linkage with pollution emission is insignificant. It supports that industrial agglomeration is likely to promote environmental performance by the spill-over effect.

Using the 261 cities as a sample, Fig. 10.5 shows different trends. The geographical expansion of exports is positively related to the pollution intensity, while its correlation with the emission is insignificant. It suggests that the enlarging market does not necessarily bring the trading region more pollution emission. By comparison, expansion into a new market holds a positive connection to the pollution intensity. That is, once considering the regional inequality of manufacturing output, the expansion of the trading network confronts one region with challenges of worsening environmental performance. With regard to the industrial agglomeration, we use the number of firms in one region to indicate the level of agglomeration. The results show that industrial agglomeration in one region will generate more pollution emissions and simultaneously help to reduce the pollution intensity. Taken together, the locals will alter the effect of the global. It supports the second and third hypotheses.

Charts in Fig. 10.6 further illustrate the linkage between local industrial pollution and the geography of exports. We use the ratio of exports to developed countries to

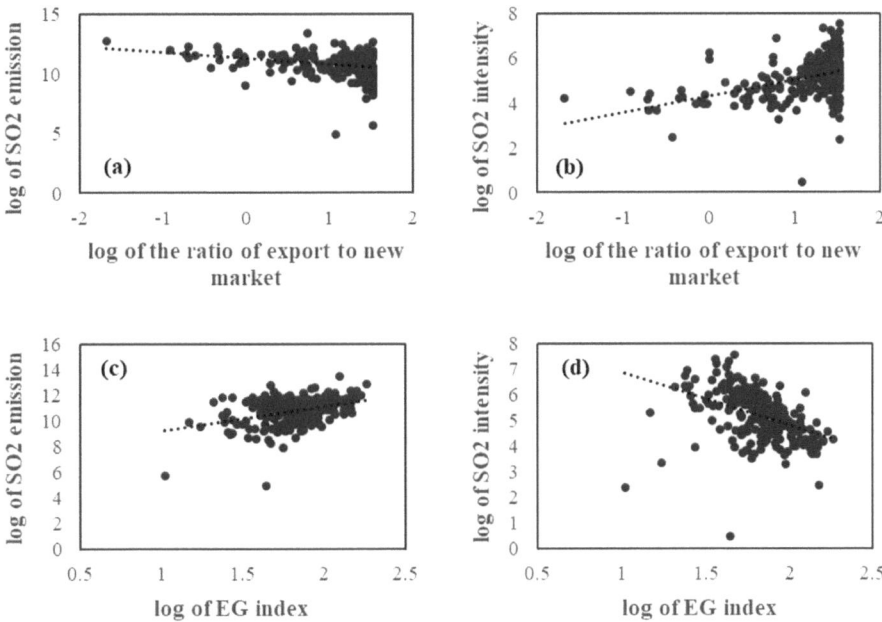

Fig. 10.5 Relationship between industrial SO$_2$ pollution and the global-local linkage of export (2007, 261 cities as sample)

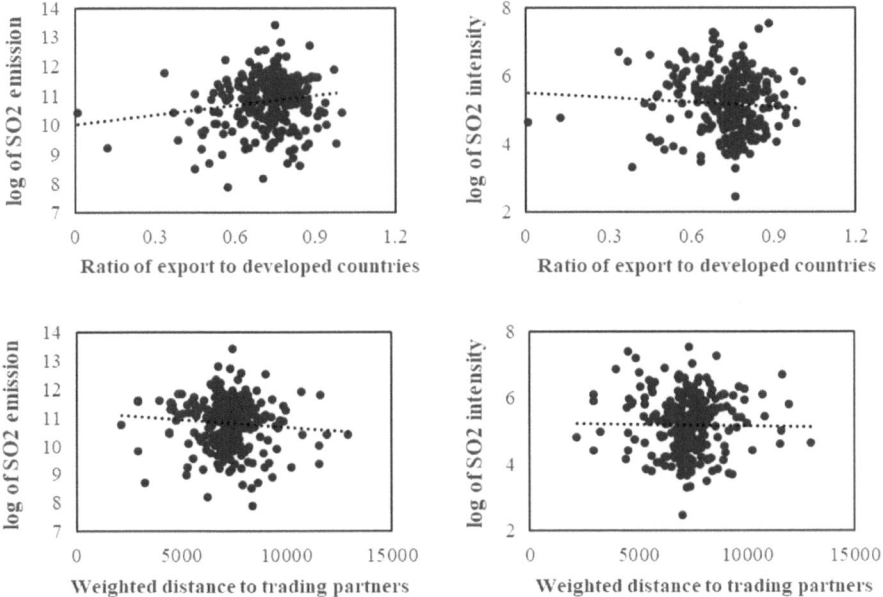

Fig. 10.6 Relationship between industrial SO$_2$ pollution and the geography of export (2007)

indicate the hypothesis 1. It holds a positive connection with the pollution emission, while its connection with the pollution intensity is insignificant, as predicted by the first hypothesis. Trading with developed countries is likely to generate more pollution emission. On the other hand, we also calculate the distance to trading partners, weighted by their values. The results show that the distance is negatively related to the pollution emission, while its linkage with pollution intensity is insignificant. Thus, cities trading with more distant markets tend to have a lower level of pollution emission.

Although the overall industrial pollution in China declines significantly, the regional inequality of pollution keeps growing. Particularly, the gap of pollution emission between the leading and lagging regions narrows, while that of pollution intensity widens. Hence, it urges the necessity to incorporate the global-local linkage in the TEE. The enlarging market, together with the industrial agglomeration, reshapes the spatial pattern of TEE. Moreover, the geography of the enlarging market also plays a role. In addition to the common pollution haven effect, it is also noteworthy that neighbouring market may become another major source of pollution transfer.

10.3   Empirical Analysis

10.3.1   Model Specification

The empirical analysis seeks to accomplish the major objective that incorporating the global-local linkage into the TEE. We start with a basic model regarding the driving forces of industrial pollution. Ehrlich and Holdren (1971) and Commoner et al. (1971) propose the IPAT (impact of pollution, affluence, and technology) model to capture the major anthropogenic determinants of environmental issues. It assumes that environmental problems originate from a set of factors comprehensively rather than a single factor independently (York et al. 2003). Hence, IPAT highlights three major factors: population, affluence, and technology. Affluence represents the comprehensive effect of economic development, including the role of resource endowment, income growth, and so on. However, the original version of the IPAT model fails to consider the possibility of non-monotonicity and non-proportionality. Dietz and Rosa (1997) revise the model and propose the STIRPAT (stochastic impacts by regression on population, affluence, and technology). This version is widely used in researches on pollution emission and greenhouse gas emission (Cramer 1998; Fan et al. 2006). As such, we develop a basic model based on STIRPAT (Eq. 10.1).

$$I_{ct} = \alpha P_{ct}^{\beta} A_{ct}^{\gamma} T_{ct}^{\delta} e_{ct} \qquad (10.1)$$

where for city c in year t, I represent the industrial pollution (emission or intensity), P is the population, T denotes technology, e is the error term.

Next, we extend the model to incorporate the changing geography of foreign trade, globally and locally. Firstly, we focus on the effect of trade-induced relocation of industry on the pollution. The size of the population in the original model is not an ideal proxy. Hence, we replace the size of the population with the number of employees in manufacturing sectors (P^*).

Secondly, we further specify the effect of affluence. Besides the conventional proxy of the gross domestic product per capita, we also consider the role of foreign trade. Foreign trade is a twin process of external and internal linkage. Regarding the external linkage, we incorporate the ratio of export to new countries (NT) as a proxy to denote the geographical expansion of China's foreign trade. Furthermore, in order to examine the first hypothesis, we employ the ratio of export to developed economies ($HIGH$) and the weighted distance to trading partners ($DIST$) as proxies. According to the first hypothesis, the coefficients of NT and $HIGH$ should be positive, whereas that of $DIST$ should be negative.

With regard to the internal linkage, we employ the number of firms in one city to indicate industrial agglomeration ($FIRM$). Also, we consider the interactions between economic agents and the environment, namely, the technological base and the environmental regulation. The number of employees in the sectors of advanced producer services is the proxy for the technological base (T^*). As for the environmental regulation, it is noteworthy that the hierarchically command-and-control mode prevails in China (Chen 2009). Its stringency represents the determination of the central government to improve the environment, whereas its enforcement depends on the green effort from the local government. Thus, the proxy for environmental regulation should denote the aspiration and capacity of the government. Therefore, we choose the percentage of domestic sewage treated in centralized sewage treatment (ER) as a proxy, since it is one of the most important public services regarding the local environment. Overall, the coefficients of $FIRM$, T^*, and ER should be negative, according to the second and third hypotheses.

As for control variables, in the context of transition and globalization, we control for the ownership of firms. The variables are the ratio of the output of foreign-owned enterprises (FOE) and the ratio of state-owned enterprises (SOE), respectively. Moreover, considering the likely path dependence of pollution emission temporally, we incorporate the one-year lag-term of industrial pollution in the model. Hence, the model becomes a dynamic panel data model.

The model including all the variables for estimation is as follows (Eq. 10.2).

$$
\begin{aligned}
lnS_{ct} = {} & \alpha_0 lnS_{ct-1} + \beta_0 lnP^*_{ct} + \gamma_1 lnNT_{ct} + \gamma_2 lnHIGH_{ct} \\
& + \gamma_3 lnWDIS_{ct} + \gamma_4 lnFIRM_{ct} + \gamma_5 lnY_{ct} + \gamma_6 lnFOE_{ct} \\
& + \gamma_7 lnSOE_{ct} + \delta_1 lnT^*_{ct} + \delta_2 lnER_{ct} + u_c + e_{ct}
\end{aligned}
\tag{10.2}
$$

where u represents the individual effect and e is the stochastic error term.

10.3.2   Estimation and Tests

The model in Eq. (10.2) has four properties. First of all, it is a "large N, small T" panel, suggesting that the number of individual observations (261 cities) is much larger than the period (9 years). Secondly, there is a lag-term of the dependent variable in the model, leading to the correlation between the independent variable and the stochastic term. Next, some independent variables are not strictly exogenous, especially the gross output per capita and environmental regulation. Finally, there are issues of heterogeneity and auto-correlations. In this regard, Arellano and Bond (1991) propose the system generalised method of moments (System GMM) to estimate the coefficients in the dynamic panel data model and to guarantee the efficiency of estimation.

Following the guideline from Roodman (2009), we control the two-way fixed effect to protect the premise of applying System GMM that auto-correlation only exists within the cross-section. Firstly, we apply the Hausman test to identify those variables which are not strictly exogenous. Next, we implement the two-stage system GMM for robust estimation. Then, we check the auto-correlation of the equation of first-order difference. The results show that the first-order correlation (AR (1)) exists, while the second-order correlation (AR (2)) does not. The insignificant of AR (2) is the premise of achieving consistent estimation.

Applying the system GMM may introduce many instrument variables. In this regard, we restrict the lagging period of lag-terms within two years. It turns out that the practical numbers of instrument variables are less than 35. Moreover, the results of the Hansen-test support the efficiency of these instrument variables, where the *p-values* range from 0.3 to 0.7. The Difference-in-Hansen-test fails to reject the null hypothesis, indicating that the system GMM is more efficient than difference GMM. Finally, we also compare the results of system GMM with those of pooled OLS and fixed effect models. The results suggest that system GMM can achieve a more robust estimation. Tables 10.1 and 10.2 report the results of estimation and tests.

10.3.3   Empirical Results

Tables 10.1 and 10.2 report the results of pollution emission and intensity, respectively. The results overall support the three hypotheses above and suggest the role of global-local linkage in shaping the spatial pattern of the TEE.

First of all, the global linkage serves as the channel of pollution transfer. It primarily increases the local pollution emissions, while its effect on the pollution intensity is insignificant. In Table 10.1, the coefficients of *NT* are significantly positive, representing that the enlarging market of foreign trade will generate more pollution emission locally.

Moreover, the effect of the global linkage also depends on the geography of foreign trade. The coefficients of *HIGH* are positive, whereas those of *WDIS* are negative.

Table 10.1 Results of the pollution emission

	Model 1	Model 2	Model 3	Model 4	Model 5
	Pooled OLS	Fixed effect	System GMM	System GMM	System GMM
$\ln S_{ct-1}$	0.917^{***}	0.550^{***}	0.565^{***}	0.580^{***}	0.580^{***}
$\ln NT_{ct}$	0.018	-0.009	0.049^{**}	0.039^{**}	0.039^{**}
$\ln HIGH_{ct}$	0.026^{*}	0.029^{*}	0.050^{**}	0.048^{**}	0.048^{**}
$\ln WDIS_{ct}$	-0.027^{*}	-0.020	-0.049^{**}	-0.045^{**}	-0.045^{**}
$\ln FIRM_{it}$	-0.003	-0.016	-0.010	0.003	0.004
$\ln P_{ct}^{*}$	0.062^{***}	-0.001	0.246^{***}	0.226^{***}	0.228^{***}
$\ln Y_{it}$	-0.004	-0.106^{***}	0.026	0.026	0.027
$\ln T_{ct}^{*}$	-0.009	-0.013	-0.024	-0.026	-0.026
$\ln FOE_{ct}$	-0.003	-0.004		-0.015	-0.015
$\ln SOE_{ct}$	-0.002	-0.005		0.031^{**}	0.031^{**}
$\ln ER_{ct}$	-0.020^{**}	-0.002			-0.005
Time fixed effect	Yes	Yes	Yes	Yes	Yes
Observations	2088	2088	2088	2088	2088
No. of Instrument variables	–	–	22	24	25
AR(1)	–	–	0.000	0.000	0.000
AR(2)	–	–	0.792	0.827	0.826
Hansen-test (p)	–	–	0.673	0.684	0.686
Difference-in-Hansen-test (p)	–	–	0.563	0.576	0.581

Note ***$p < 0.01$; **$p < 0.05$, *$p < 0.10$

On the one hand, it echoes the findings of previous studies that the pollution haven effect has occurred in China. Trading with developed countries will produce more pollution. On the other hand, it further highlights the role of distance. Trading with neighbouring countries will also generate a higher pressure of pollution emission. There are still debates on whether export firms tend to improve product quality or the production efficiency to offset the long-distance transport costs (Manova and Zhang 2009). However, either efficiency improvement or quality promotion will benefit the pollution emission reduction (Holladay 2010).

Regarding the pollution emission, it is also noteworthy that the coefficient of Y is insignificant. It suggests the findings in Sect. 10.2 that the gap of pollution emission between the leading and lagging regions has narrowed. The global linkage has become the most powerful force in reshaping the spatial pattern of pollution emission. Particularly, foreign trade exposes the lagging regions to larger markets as well as higher pressure of pollution emission. Hence, the global linkage is smoothing out the regional inequality of pollution emission.

Table 10.2 Results of the pollution intensity

	Model 1	Model 2	Model 3	Model 4	Model 5
	Pooled OLS	Fixed effect	System GMM	System GMM	System GMM
$\ln S_{ct-1}$	0.878^{***}	0.494^{***}	0.483^{***}	0.466^{***}	0.462^{***}
$\ln NT_{ct}$	-0.003	-0.016	-0.002	0.006	0.013
$\ln HIGH_{ct}$	0.014	0.019	0.029	0.030	0.031
$\ln WDIS_{ct}$	-0.010	-0.018	-0.014	-0.020	-0.023
$\ln FIRM_{it}$	-0.043^{***}	-0.040^{*}	-0.136^{***}	-0.122^{***}	-0.121^{***}
$\ln P_{ct}^{*}$	0.051^{***}	-0.020	0.179^{***}	0.160^{***}	0.153^{***}
$\ln Y_{it}$	-0.049^{***}	-0.461^{***}	-0.359^{***}	-0.329^{***}	-0.294^{***}
$\ln T_{ct}^{*}$	-0.026^{**}	-0.019	-0.067^{**}	-0.073^{*}	-0.073^{**}
$\ln FOE_{ct}$	0.004	-0.002		0.009	0.005
$\ln SOE_{ct}$	0.003	-0.004		0.044^{***}	0.044^{***}
$\ln ER_{ct}$	-0.024^{**}	-0.007			-0.023
Time fixed effect	Yes	Yes	Yes	Yes	Yes
Observations	2088	2088	2088	2088	2088
No. of instrument variables	–	–	28	30	31
AR(1)	–	–	0.001	0.001	0.001
AR(2)	–	–	0.397	0.364	0.346
Hansen-test (p)	–	–	0.336	0.359	0.380
Difference-in-Hansen-test (p)	–	–	0.346	0.321	0.329

Note $^{***}p < 0.01$; $^{**}p < 0.05$, $^{*}p < 0.10$

Secondly, the local linkage serves as the source of capacity-building in response to the pollution emission. Its effects on pollution intensity are overall significant, while those on pollution emission are not. On the one hand, the linkages between firms and the economic environment determine the regional capacity-building, such as the technological base. The coefficient of T^{*} is negative. A stronger technological base will effectively reduce pollution intensity. On the other hand, the linkages between firms help to absorb the increasing pressure of pollution emission. The coefficients of *FIRM* are significantly negative. Hence, industrial agglomeration will effectively reduce pollution intensity. It offers another evidence for the positive relationship between industrial agglomeration and pollution intensity (Cheng 2016).

The results of industrial agglomeration can also offer insight into the effect of industrial relocation. Given the relocation of pollution-intensive firms in the wake of foreign trade, a lumpy dispersion can occur (Sunley 2011). That is, firms may de-agglomerate in the existing core and re-agglomerate in the growing periphery (Dumais et al. 2004). Consequently, the relocation will increase the pollution intensity at both the leading and lagging regions at first. As the re-agglomeration keeps growing, the lagging regions are expected to promote their absorption capacity. Empirically, however, the coefficients of Y are significantly negative. Hence, there is still a gap in absorption capacity between the leading and lagging regions. When the lagging regions seek to catch up with the leading regions through expanding to the global market, its weak absorption capacity is likely to amplify the TEE. Moreover, the lagging regions may also face pollution transfer via industrial relocation (Zhou et al. 2017).

Thirdly, in the context of economic transition, the institutional factors can further alter the TEE at the sub-national level. The coefficients of *SOE* are significantly positive, both in the models of pollution emission and those of pollution intensity. As the legacy of the central planning system in China, previous studies argue that the state-owned firms will be less environmentally benign (Meyer and Pac 2013). There is strong evidence supporting that SOE tends to be less productive (Audretsch et al. 2016; Brandt et al. 2012). Institutional support may bridge the gap between their costs and benefits. As such, SOE can be reluctant to reduce undesirable output. Our findings suggest that the pollution emission and intensity of SOE tend to be higher. Therefore, the reform of state-owned firms will contribute to the capacity-building for pollution reduction.

Our findings suggest that the global-local linkage has reshaped the spatial pattern of the TEE at the sub-national level. They work in different ways. The global linkage generates more pollution emissions and narrows the gap between the leading and lagging regions. By comparison, the local linkage affects pollution intensity. The regional inequality of local linkage maintains the regional division of pollution intensity between the leading and lagging regions.

10.4 Conclusion

Foreign trade serves as an important engine for economic growth in developing countries like China. However, it also exposes the country to higher risks of environmental degradation. One of the main aims of the TEE studies is to answer whether foreign trade promotes economic growth at the expense of the environment. Conventional studies focus on the connections between countries, overlooking the domestic trade-induced changes. This chapter highlights the role of local capabilities and then provides a framework to consider the TEE at multiple levels comprehensively, which is one of the key perspectives of Environmental Economic Geography. It helps to explain the reason that the sub-national region reaping the benefit does not necessarily bear the environmental burden.

In the framework, the synergy of global transfer, domestic transfer, and local absorption determines TEE. Empirical findings confirm the validity of this model and contribute to understanding the spatial pattern of TEE in China.

The global transfer highlights that the enlarging market will generate a scale effect, so that produces more pollution. Moreover, the geography of foreign trade matters. Besides the pollution haven effect of developed countries, the neighbouring countries are likely to increase the level of emission due to lower transport costs. Because of the growing global linkage in the lagging regions, the regional division of pollution emission is fading.

On the contrary, the domestic transfer and the local absorption maintain the division of pollution intensity between leading and lagging regions. Nowadays, foreign trade in China, together with the foreign investment, fosters sectors with relatively lower added value to move out of the leading regions and relocate in the lagging ones. Therefore, the lagging region has to face the pollution transfer from both the foreign countries externally and the leading regions internally. Moreover, there remains a wide gap in absorption capacity between the leading and lagging regions, including the technological base and industrial agglomeration. Taken together, the local linkage may amplify the TEE in lagging regions.

Our findings in this chapter may offer insights into managing the trade-offs between foreign trade and the environment. Firstly, at the global level, a greening effort can be upgrading the product mix of export sectors, especially the exports to developed countries and neighbouring countries. Secondly, promoting the reform of state-owned firms will largely contribute to the local capacity-building for pollution reduction. It is useful to eliminate the institutional advantage and leave more room for productivity improvement. Thirdly, regional capacity-building for pollution reduction should also encourage the sharing and learning process stemming from industrial agglomeration. Last but not least, considering the regional inequality of absorption capacity, environmental regulation should be equipped with redistributive capacities, such as environmental zoning and making the region-specific negative list.

References

Antweiler, W., Copeland, B. R., & Taylor, M. S. (2001). Is free trade good for the environment? *American Economic Review, 91,* 877–908.

Arellano, M., & Bond, S. (1991). Some tests of specification for panel data: Monte Carlo evidence and application to employment equations. *Review Economic Studies, 58,* 277–297. https://doi.org/10.2307/2297968.

Audretsch, D., Guo, X. D., Hepfer, A., Menendez, H., & Xiao, X. Z. (2016). Ownership, productivity and firm survival in China. *Economia e Politica Industriale, 43*(1), 67–83. https://doi.org/10.1007/s40812-015-0021-6.

Baldwin, R., & Harrigan, J. (2011). Zeros, quality, and space: Trade theory and trade evidence. *American Economic Journal: Microeconomics, 3,* 60–88. https://doi.org/10.1257/mic.3.2.60.

Becker, R., & Henderson, V. (2000). Effects of air quality regulations on polluting industries. *Journal of Political Economy, 108,* 379–421. https://doi.org/10.1086/262123.

Brandt, L., Biesebroeck, J. V., & Zhang, Y. (2012). Creative accounting or creative destruction? Firm-level productivity growth in Chinese manufacturing. *Journal of Development Economics, 97,* 339–351. https://doi.org/10.1016/j.jdeveco.2011.02.002.

Chen, G. (2009). *Politics of China's environmental protection.* Singapore: The World Scientific Publishing.

Cheng, Z. (2016). The spatial correlation and interaction between manufacturing agglomeration and environmental pollution. *Ecological Indicators, 61,* 1024–1032. https://doi.org/10.1016/j.ecolind.2015.10.060.

Christmann, P., & Taylor, G. (2001). Globalization and the environment: determinants of firm self-regulation in China. *Journal of International Business Studies, 32,* 439–458. https://doi.org/10. 1057/palgrave.jibs.8490976.

Cole, M. A., & Elliott, R. J. R. (2003). Determining the trade-environment composition effect: The role of capital, labor and environmental regulations. *Journal of Environmental Economics and Management, 46,* 363–383. https://doi.org/10.1016/S0095-0696(03)00021-4.

Cole, M. A., & Elliott, R. J. R., Zhang, J. (2011). Growth, foreign direct investment, and the environment: Evidence from Chinese cities. *Journal of Regional Science, 51,* 121–138. https:// doi.org/10.1111/j.1467-9787.2010.00674.x.

Commoner, B., Corr, M., & Stamler, P. (1971). The causes of pollution. *Environment: Science and Policy for Sustainable Development, 13,* 2–19. http://dx.doi.org/10.1080/00139157.1971. 9930577.

Copeland, B. R., & Taylor, M. S. (1994). North-south trade and the environment. *Quarterly Journal of Economics, 109,* 755–787.

Coşar, A. K., & Fajgelbaum, P. D. (2016). Internal geography, international trade, and regional specialization. *American Economic Journal: Microeconomics, 8,* 24–56. https://doi.org/10.1257/ mic.20140145.

Cramer, J. C. (1998). Population growth and air quality in California. *Demography, 35,* 45–56. https://doi.org/10.2307/3004026.

Dicken, P. (2003). *Global shift: Reshaping the global economic map in the 21st century.* London, UK: Sage.

Dietz, T., & Rosa, E. A. (1997). Effects of population and affluence on CO_2 emissions. *Proceedings of the National Academy of Sciences, 94,* 175–179.

Dumais, G., Ellison, G., & Glaeser, E. L. (2004). Geographic concentration as a dynamic process. *Review of Economics and Statistics, 84,* 193–204. https://doi.org/10.1162/003465302317411479.

Ehrlich, P. R., & Holdren, J. P. (1971). Impact of population growth. *Science, 171,* 1212–1217.

Ellison, G., & Glaeser, E. L. (1997). Geographic concentration in U.S. manufacturing industries: A dartboard approach. *Journal of Political Economy, 105,* 889–927. https://doi.org/10.1086/262098.

Fan, Y., Liu, L. C., Wu, G., & Wei, Y. M. (2006). Analyzing impacts factors of CO_2 emissions using the STIRPAT model. *Environmental Impact Assessment Review, 26,* 377–395. https://doi.org/10. 1016/j.eiar.2005.11.007.

Farole, T. (2013). Trade, location, and growth. In T. Farole (Ed.), *The internal geography of trade: Lagging regions and global markets* (pp 15–31). Washington, DC: World Bank Publications.

Gereffi, G. (2009). Development models and industrial upgrading in China and Mexico. *European Sociological Review, 25,* 37–51. https://doi.org/10.1093/esr/jcn034.

Getis, A., & Ord, J. K. (1992). The analysis of spatial association by use of distance statistics. *Geographical Analysis, 24,* 189–206. https://doi.org/10.1111/j.1538-4632.1992.tb00261.x.

Grether, J.-M., & Mathys, N. A. (2013) The pollution terms of trade and its five components. *Journal of Development Economics, 100,* 19–31. https://doi.org/10.1016/j.jdeveco.2012.06.007.

Holladay, J. S. (2010). Are exporters mother nature's best friends? http://dx.doi.org//10.2139/ssrn. 1292885.

Hummels, D., & Skiba, A. (2004). Shipping the good apples out? An empirical confirmation of the Alchian-Allen conjecture. *Journal of Political Economy, 112,* 1384–1402. http://dx.doi.org/10. 1086/422562.

Jakob, M., & Marschinski, R. (2013). Interpreting trade-related CO_2 emission transfers. *Nature Climate Change, 3,* 19–23. https://doi.org/10.1038/nclimate1630.

Kahn, M. E. (1997) Particulate pollution trends in the United States. *Regional Science and Urban Economics, 27,* 87–107. https://doi.org/10.1016/S0166-0462(96)02144-8.

Manova, K., & Zhang, Z. W. (2009). *Export prices across firms and destinations.* NBER Working Paper No. 15342. Cambridge MA: NBER.

Meyer, A., & Pac, G. (2013). Environmental performance of state-owned and privatized eastern *European energy utilities. Energy Economics, 36,* 205–214. https://doi.org/10.1016/j.eneco.2012. 08.019.

Michida, E., & Nishikimi, K. (2007). North-south trade and industry-specific pollutants. *Journal of Environmental Economics and Management, 54*, 229–243. https://doi.org/10.1016/j.jeem.2007. 02.002.

Oinas, P., & Malecki, E. J. (2002). The evolution of technologies in time and space: From national and regional to spatial innovation systems. *International Regional Science Review, 25*, 102–131. https://doi.org/10.1177/016001702762039402.

Ord, J. K., & Getis, A. (1995). The analysis of spatial association by use of distance statistics. *Geographical Analysis, 27*, 286–306. https://doi.org/10.1111/j.1538-4632.1995.tb00912.x.

Poon, J. P. H., Casas, I., & He, C. F. (2006). The impact of energy, transport, and trade on air pollution in China. *Eurasian Geography and Economics, 47*, 568–584.

Rodríguez-Pose, A. (2013). Trade openness and regional inequality. In T. Farole (Ed.), *The internal geography of trade: Lagging regions and global markets* (pp. 33–63). Washington, DC: World Bank Publications.

Roodman, D. (2009). How to do xtabond2: An introduction to difference and system GMM in Stata. *The Stata Journal, 9*, 86–136.

Shi, Y. T., Sharma, K., Murphy, T., Hicks, J., & Arthur, L. (2013). Trade and environment in China: An input-output perspective on the pollution haven hypothesis. *International Journal of Economics & Business Research, 5*, 420–432. https://doi.org/10.1504/IJEBR.2013.054256.

Sunley, P. (2011). The consequences of economic globalization. In A. Leyshon, R. Lee, L. McDowell, & P. Sunley (Ed.), *The Sage handbook of economic geography*. Singapore: Sage.

Swyngedouw, E. (2004). Globalisation or 'glocalisation'? Networks, territories and rescaling. *Cambridge Review of International Affairs, 17*, 25–48.

Wagner, U. J., & Timmins, C. D. (2009). Agglomeration effects in foreign direct investment and the pollution haven hypothesis. *Environmental and Resource Economics, 24*, 245–262. https:// doi.org/10.1007/s10640-008-9236-6.

Wu, H. Y., Guo, H. X., Zhang, B., & Bu, M. L. (2017). Westward movement of new polluting firms in China: Pollution reduction mandates and location choice. *Journal of Comparative Economics, 45*, 119–138. https://doi.org/10.1016/j.jce.2016.01.001.

York, R., Rosa, E.A., & Dietz, T. (2003). STIRPAT, IPAT and ImPACT: Analytic tools for unpacking the driving forces of environmental impacts. *Ecological Economics, 46*(3), 351–365. https://doi. org/10.1016/S0921-8009(03)00188-5.

Zhang, X. B., & Zhang, K. H. (2003) How does globalisation affect regional inequality within a developing country? Evidence from China. *Journal of Development Studies, 39*, 47–67. http:// dx.doi.org/10.1080/713869425.

Zhou, Y., Zhu, S. J., & He, C. F. (2017). How do environmental regulations affect industrial dynamics? Evidence from China's pollution-intensive industries. *Habitat International, 60*, 10–18. http://dx.doi.org/10.1016/j.habitatint.2016.12.002.

Chapter 11
Does Export Upgrading Improve Urban Environment?

The previous three chapters have demonstrated that external and internal linkages work together to exposes one place to the global shift of environmental burdens, shaping the changing geography of industrial environmental performance in China. As a transitioning economy, foreign trade transfers environmental burdens to China from developed economies as well as developing economies. The imported environmental burdens are not evenly distributed across internal regions, since regions exhibit quite different absorption capacities in response to environmental burdens. Moreover, the interregional connections serve as other channels to shift environmental burdens domestically, usually from the leading regions to the lagging ones. Fortunately, the combined effects of external and internal linkages do not only build up channels for shifting environmental burdens. As discussed in last chapter, the interactions between external and internal linkages could block the transfer of environmental burdens by promoting structural changes of industries and reshaping the paths of industrial development.

However, in addition to the structural changes of industries, another crucial facet in response to environmental burdens is the so-called "technique effect", which has been mentioned frequently in previous chapters. Regarding the shift of environmental burdens, the "technique effect" implies the positive role of innovative knowledge embodied in external linkages as well as the pollution contents. On the other hand, the "technique effect" also indicates the local capacities in absorbing the new knowledge embodied in external linkages so as to improve the environmental performance. Taken together, the technique effect is subject to the interactions between external and internal linkages. As such, this chapter sets out to reveal how global-local linkages affect the environmental performance by technological improvements.

Export upgrading provides an ideal perspective to investigate this question. First, it allows us to continue our focus on the environmental performance of China's foreign trade based on the discussion in previous three chapters. This chapter will extend our understanding from the spatial trend of environmental burden shift to the spatial heterogeneity in response to environmental burden shift. Second, the process of export upgrading is a typical process of global-local interactions. At the global

© Springer Nature Singapore Pte Ltd. 2020
C. He and X. Mao, *Environmental Economic Geography in China*,
Economic Geography, https://doi.org/10.1007/978-981-15-8991-1_11

level, it relates to the way that one economy is inserted into the global value chain. At the local level, it is determined by the performance of local clusters (Navas-Alemán 2011; Pietrobelli and Rabellotti 2011). This feature matches our overall research interests in the Part II of this book, which delves into the role of global-local interactions in reshaping the geography of industrial environmental performance in China. Third, export upgrading takes a variety of forms, such as the improvement of product quality, the increase of production efficiencies, and the replacement of products with higher added value (Humphrey and Schmitz 2002). Apart from the heterogeneity of industries/firms, these various forms are also outcomes of the interactions between firms and places where they are located. Hence, understanding the environmental performance of export upgrading also requires geographical wisdom, leaving room for the Environmental Economic Geography to conduct investigation.

In this chapter, we start our discussion by summarising the drivers of export upgrading from the perspective of global-local interactions. Second, we discuss the environmental performance of export upgrading. The third section introduces the research design to assess the environmental performance of export upgrading. The fourth section reports the empirical results. The last section concludes.

11.1   Export Upgrading and Its Global-Local Drivers

Export upgrading refers to a process that one economy extends its export portfolios towards those goods with higher productivity levels. Trade theory argues that exporting serves as a powerful engine for economic growth of one economy, determined by not only how much it exports but also what it exports (Rodrik 2006). The strong empirical evidence from Hausmann et al. (2007) has supported that economies with more sophisticated exports tend to grow faster. Jarreau and Poncet (2012) echo this point with their study on the export sophistication of China's provinces. They report that provinces specializing in more sophisticated goods grow faster (Jarreau and Poncet 2012). Recent evidence argues that the positive effects of export upgrading on economic growth tend to be more significant in high-income and middle-income economies, while such effects are suspicious for low-income economies (Chrid et al. 2020). Overall, the capabilities of one economy to upgrade allow it to climb up the value chain and reap more benefits, thereby explaining why it grows (McMillan and Rodrik 2011; Hausmann and Hidalgo 2011).

What drives one economy to upgrade its exporting goods? A direct answer is the accumulation of knowledge capital (Zhu and Fu 2013). Knowledge capital is the immeasurable value of a firm/place's full body of knowledge and contributes to the improvement of efficiencies. A firm/place with a higher level of knowledge capital would be more productive than the one with lower level of knowledge capital (Scherngell et al. 2014). Knowledge capital can be accumulated by either internal knowledge creation or external knowledge diffusion. As such, economic activities that can contribute to the creation or diffusion of knowledge capital are expected to promote upgrading.

Regarding the external knowledge diffusion, the literature has examined the role of integration into global value chain in promoting upgrading, particularly the spill-overs from foreign investment and imports. A basic point is that foreign investment and imports are flowing with technological or knowledge contents, which may generate spill-over effects through their interactions with domestic firms. For example, firms that serve as suppliers for multinational enterprises are expected to improve their productivity by backward linkages (Javorcik 2004). As such, it is natural to assume that the gap of productivity between host and home economies determines the effects of technological spill-overs. A wider gap is likely to promote the upgrading in host economies (Pahl and Timmer 2019). For example, Harding and Javorcik (2012) report a positive effect of FDI in promoting export quality in developing economies based on the sample of 105 economies during 1984–2000. However, Amighini and Sanfilippo (2014) examine the role of South-South FDI and trade in the export upgrading of African economies. They report that both South-South FDI and trade have diversified the exports and simultaneously raised their average quality. Similarly, regarding the imports, Song et al. (2019) find that importing inputs with better quality allows domestic firms to improve product quality or diversify into higher value-added products. The pattern of processing trade typically reflects this mechanism, especially in the case of China. By utilizing foreign investment and importing intermediated inputs, China is able to anchor the sophisticated production by its comparative advantages in mass production (Zhu and Fu 2013; Poncet and de Waldemar 2013).

Regarding the internal knowledge creation, the upgrading-relevant studies are closely related to the agglomeration economies originated from Marshall's localization theory. From the classical view of agglomeration economies, firm interactions within the clusters play an important role in incremental upgrading. That is, geographical proximity increases the face-to-face communication and allows the diffusion and absorption of tacit knowledge, which contributes to local learning and innovation (Gertler 2003). Moreover, interactions within the agglomeration will produce the "collective efficiency", allowing co-located firms to maintain their competitive advantages by joint actions (Schmitz 1999). Therefore, the classical view tends to highlight the effects of firm interactions within the clusters.

Recent studies provide another perspective to understand the role of local agglomeration in upgrading by coupling the learning from interactions within local agglomeration and within the global value chain. The perspective looks into the co-evolution of local clusters and global value chain (Humphrey and Schmitz 2002). On the one hand, domestic firms have to upgrade to maintain their competitiveness in the face of increasing global competition. Their upgrading is largely subject to how resources would be allocated and flow within a global value chain, namely, the governance of value chains (Gereffi 2009). The governance of value chains also takes various forms, then exhibiting different upgrading routes (Humphrey and Schmitz 2002). Different upgrading routes further result in different dependencies on local agglomeration (Giuliani et al. 2005). On the other hand, the governance of global value chain would be adaptive to local agglomeration. A local agglomeration with strong learning capabilities may contribute to reduction of transaction complexity and then alter the governance of global value chain (Pietrobelli and Rabellotti 2011). Overall, the performance of upgrading is not solely underpinned by either local agglomeration or global value chain. Instead, it is determined by their co-evolution. For example, inefficient local absorption capacities are likely to disable the upgrading efforts of gaining access to global value chain (Crespo and Fontoura 2007; Poncet and de Waldemar 2013).

Apart from the knowledge capital, the institutional environment is also crucial to promote upgrading. At the global level, trade liberalization policies are expected to expand trade volume as well as promote product upgrading. Stimulus policies like tax rebate may encourage the diversification of exporting goods as well as increase product quality (Zhu and Fu 2013). In contrast, trade barriers may become a mixed blessing for export upgrading. Hu et al. (2019) report that technical barriers to trade will promote upgrading in twofold. On the one hand, it compels exporters to improve their product quality for compliance with technical standards. On the other hand, it also forces less productive exporters to exit, resulting in a sorting effect. Such a mixed blessing is the same with tariffs and environmental regulations (Gong et al. 2020).

At the local level, a supportive institutional environment contributes to the export upgrading in twofold. First, institutional environments provide a stable, clear, and efficient arrangement for exporters to make full use of the market mechanisms. Shimbov et al. (2019) investigate the export upgrading in Western Balkan countries and highlight the positive role of marketization in stimulating the upgrading of exports. Moreover, local institutional environment is an indispensable variable in determining the knowledge spill-overs and promoting local innovation (Grillitsch and Rekers 2016). Second, institutional environments play a positive role in cost-saving for exporters. In the case of China, Zhou et al. (2019) pointed to the fact that product upgrading has costs, confronted with the risk of failure. They further demonstrate that the coastal regions in China would be more supportive for upgrading than the hinterland due to their institutional contexts, which are more capable of reducing costs for upgrading. Song et al. (2019) find that institutional environment in China's regions contributes to export upgrading by mandatory information disclosure and technology adoption incentive. The former helps to reduce transaction costs, while the latter encourages the diffusion and application of new knowledge.

Overall, export upgrading is neither purely endogenous nor exogenous. It operates within multiple global value chains as well as local clusters. The global value chain and local clusters interact and then upgrade products in different ways. More importantly, different upgrading ways are not mutually exclusive within the same cluster (Navas-Alemán 2011). As a result, although export upgrading seemingly represents a positive factor in generating "technique effects", the co-evolution between global value chains and local clusters makes things more complicated. Implications of export upgrading vary across different sectors and local agglomerations.

11.2 Environmental Performance of Export Upgrading

Intuitively, export upgrading should improve environmental performance of industrial activities by increasing productivities. However, such a positive effect is conditional because export upgrading does not necessarily reflect a genuine advance in technology. The co-evolution of global value chain and local cluster determines that there are multiple ways to improve product sophistication, where the adoption of new knowledge is not indispensable. For example, processing trade allows most developing economies to participate in the global value chain by their advantages in the labour force but export sophisticated products. The technological contents in these products are less than expected, whose production still relies on resource-based and labour-intensive processes. In such a case, it is natural that export upgrading is less likely to improve environmental performance.

Based on the co-evolution of global value chain and local cluster, we propose four facets to examine the disjuncture between export upgrading and environmental improvement. The first facet is to distinguish between different ways to upgrade. Humphrey and Schmitz (2002) put forth four types of upgrading. They are processing upgrading, product upgrading, functional upgrading, and inter-sectoral upgrading. Processing upgrading refers to an improvement of input-output efficiencies by adopting new technology. Product upgrading means an improvement of product quality by adopting more sophisticated product lines. Functional upgrading indicates that increasing the sophistication of products by the replacement of product functions within the value chain. Similarly, inter-sectoral upgrading indicates that increasing the sophistication of products by the replacement of product types between different value chains.

It should be noted that processing upgrading and product upgrading suggest the changing technological contents of products themselves. It requires the genuine adoption of superior technology to increase both the productivity and quality of products. In contrast, functional upgrading and inter-sectoral upgrading are closely related to structural changes of products, which are more complicated and radical than previous two types. Marchi et al. (2013) posit that these upgrading types represent different environmental strategies for firms, then having different environmental implications. They assume that product upgrading and process upgrading refer to the strategy of "eco-efficiency" (cost-saving) and "eco-branding" (product differentiation), respectively. These strategies allow firms to achieve new competitive advantages. On the other hand, the functional upgrading and inter-sectoral upgrading refer to strategies of maintaining their competitive advantages by climbing up the value chain or moving into new fields. However, neither climbing up the value chain nor moving into new fields guarantees the improvement of environmental performance. This is especially the case for inter-sectoral upgrading.

Therefore, the environmental performance of export upgrading is closely related to the way to upgrade. Based on the discussion above, product- and process-upgrading tend to represent the genuine adoption of superior technology, generating a positive "technique effect" on the environment. In contrast, functional upgrading and inter-sectoral upgrading can be outcomes of structural changes more than technology adoption. As such, their environmental implications can be either positive or negative, depending on their initial position within the value chain and the fields where they newly entry.

The second facet is to distinguish between processing trade and ordinary trade. Exploring the environmental impacts of processing trade is particularly relevant for developing economies like China, since it usually accounts for a large portion of trade volume. Why does the difference between processing trade and ordinary trade matter? The answers can be twofold. On the one hand, it is the processing trade that makes developing economies having an income level that is incomparable with the sophistication of their exports (Schott 2008). In other words, processing trade is likely to make us overestimate the technological contents in exporting products. By utilizing foreign direct investment and importing sophisticated inputs, developing economies are able to manufacture and export sophisticated products. However, such productions still rest on their advantages on resource, energy, and labour rather than their knowledge bases (Xu 2010). If this is the case, upgrading of processing trade does not necessarily lead to better environmental performance. On the other hand, empirical studies also found that processing trade usually rests on a wide range of resource-based, energy-intensive, or even pollution-intensive industries (Hao and Liu 2015). Processing exporters are also found to be less productive and underperform non-processing exporters in many aspects (Dai et al. 2016). Hence, the environmental performance of export upgrading in processing trade should be less positive.

The third facet is to explore the role of export specialization. Export special-ization in an area with comparative advantages promotes local economic growth largely (Naudé et al. 2010). Given the spatial unevenness of comparative advan-tages, some exporting regions doom to specialize in polluting industries, while the others specialize in non-polluting ones. Regarding regions specialized in polluting industries, exporting first implies that these regions will serve a larger market by its resources and the environment. These regions are confronted with more signif-icant scale effects than their non-polluting counterparts. In such a case, export upgrading is expected to produce more significant environmental effects in these regions, particularly by upgrading ways relating to structural changes.

However, export specialization is not static. Export upgrading is closely related to the dynamic of export specialization, which implies the change of comparative advantages (Guerrieri and Iammarino 2007). The change of export specialization does not occur at random. It is determined by both the global-local linkages within commodity structures and trade networks (Mao and He 2019). As a result, the change of export specialization holds connections to the pre-existing specializations, exhibiting a likely path-dependent feature. The effects of path dependence can be twofold. On the one hand, the path dependence may contribute to the incremental development of technology. Economic activities tend to improve their performance in an already-existing development base that they are familiar with, which are expected to promote technological diffusion and knowledge absorption. On the other hand, the path dependence may also lead to lock-in of development. If this is the case, export upgrading by introducing new products in regions specialized in polluting sectors is less likely to improve the environmental performance drastically. Instead, the path dependence may even weaken the greening efforts of export upgrading.

The fourth facet is to explore the role of local environmental regulation. Environmental regulation is a classical determinant of trade-environmental effects in terms of Environmental/Ecological Economics. It is also a controversial determinant in promoting environmental innovation. On the one hand, environmental regulation directly controls the development of polluting industries by command-and-control mechanisms or market mechanism based on compliance costs (Jaffe and Stavins 1995; Xie et al. 2017). Thus, the differences of environmental regulation stringency and enforcement among places make polluting industries relocate to places with lax environmental regulation (as already discussed in Chap. 2). There are viewpoints insisting that enhancing the stringency and enforcement of environmental regulation would impact exporting (Dechezleprêtre and Sato 2017). On the other hand, there are positive viewpoints assuming that exporters have the motivation to increase their productivity for compliance with environmental regulation or in response to the compliance costs, such as the famous Porter Hypothesis. Porter and van der Linde (1995) propose that a well-design environmental regulation may encourage the innovation and then induce the competitiveness of firms. There is no consensus on the Porter Hypothesis, theoretically or empirically (Ambec et al. 2013). Even so, it is still reasonable to assume that exporters in the places with stringent environmental regulation are more likely to improve environmental performance by upgrading, since upgrading is an efficient way in response to environmental regulation.

Overall, the first two facets consider the effects of global linkages, while the last two facets consider the local linkages. These facets point to the fact that export upgrading does not necessarily improve environmental performance, going against our intuition. The co-evolution of global linkages and local linkages helps to explain the disjuncture between export upgrading and environmental improvement.

11.3   Assessing the Environmental Performance of Export Upgrading

11.3.1   Measure Different Ways of Export Upgrading

As mentioned above, export upgrading is manifested by the increase of product sophistication. There are several indices to measure the level of product sophistication, including the widely used *PRODY* and *EXPY*. *PRODY* is "a weighted average of the per capita GDPs of countries exporting a given product, and thus represents the income level associated with that product" (Hausmann et al. 2007). The weights here in *PRODY* exactly represent the revealed comparative advantages for one country to export the given product. *EXPY* is an average of the *PRODY* which is weighted by the value share of the given product in one country's total exports (Hausmann et al. 2007). As such, *PRODY* indicates the sophistication (productivity level) of one particular exporting good, while *EXPY* indicates the sophistication (productivity level) of one country's exporting portfolio. However, these indices do not take the quality of product into consideration, which may either overestimate the sophistication of developing countries or underestimate the sophistication of developed countries (Minondo 2010). Successive studies seek to address this issue by adjusting the *PRODY* and *EXPY* (Xu 2010; Kemeny 2011; Huber 2017).

In this section, we apply the novel measure, *TECH*, proposed by Kemeny (2011) to indicate the productivity level of a given product in one country's exporting basket. *TECH* considers both the average productivity level of one product (i.e. *PRODY*) and the relative quality level, which is formulated by Eq. (11.1).

$$TECH_{ij} = \sum_i \left(x_{ij} / \sum_i x_{ij} \right) Y_i \cdot \left(u_{ij} / \sum_i (s_{ij} \cdot u_{ij}) \right) \cdot x_{ij} \qquad (11.1)$$

where i denotes regions, j denotes exporting products, x_{ij} denotes the value share of product j in region j's exports, Y_i denotes the gross domestic production per capita of region I, u_{ij} denotes the unit price of product j in region I, s_{ij} denotes the region i's market share of product j in the whole country.

Export upgrading is manifested by the changes of sophistication levels. Various upgrading ways contribute to the sophistication changes to different extents. Hence, decomposition analysis can help to trace the channels of upgrading and measure their extents. By Eq. (11.2), *TECH* can be further decomposed into seven terms (Kemeny 2011).

$$\Delta TECH_i = \sum_{j \in C} x_{jt-1} p_{jt} \Delta q_{jt} + \sum_{j \in C} x_{jt-1} q_{jt-1} \Delta p_{jt} + \sum_{j \in C} (p_{jt-1} q_{jt-1}$$
$$- TECH_{jt-1}) \Delta x_{jt} + \sum_{j \in C} p_{jt} \Delta q_{jt} \Delta x_{jt} + \sum_{j \in C} q_{jt-1} \Delta p_{jt} \Delta x_{jt}$$
$$\sum_{j \in E} (p_{jt} q_{jt} - TECH_{jt-1}) x_{jt} - \sum_{j \in E} (p_{jt-1} q_{jt-1} - TECH_{jt-1}) x_{jt-1}$$
$$(11.2)$$

where $p_{jt} = \sum_i \left(x_{ijt} / \sum_i x_{ijt} \right) Y_{it}$, $q_t = u_{ijt} / \sum_i (s_{ijt} \cdot u_{ijt})$, $\Delta \cdot = \cdot_t - \cdot_{t-1}$.

Regarding the seven terms, the first, second, third, sixth, and seventh terms indicate different types of export upgrading. In addition, the fourth and fifth terms measure the covariance of the changing volume and quality.

The first term measures the change of unit price of products, indicating the change of product quality. Thus, the first term captures the product upgrading.

The second term measures the overall change of productivity, capturing the essence of process upgrading.

The third term measures the sophistication changes due to the shift from one product to another, exactly indicating the functional upgrading.

The sixth term measures the sophistication change due to the introduction of a new product. Likewise, the seventh term measures the change due to the exit of an incumbent product. These two terms together point to the process of inter-sectoral upgrading.

11.3.2 Examining the Environmental Performance of Export Upgrading

As different ways of export upgrading can be quantified, we can construct an empirical model to examine the environmental performance of different upgrading ways. Equation (11.3) shows the basic model.

$$\Delta PI_{ct} = \alpha_0 + UP_{ct}\alpha + \beta PI_{ct-1} + X_{ct-1}\gamma + \mu_c + \delta_t + e_{ct} \qquad (11.3)$$

where c denotes cities in China, t denotes the year. UP is a matrix consisting of five terms above that indicate different upgrading ways, indexed as Q (product upgrading), P (process upgrading), TR (functional upgrading), N (inter-sectoral upgrading by introducing new products), and E (inter-sectoral upgrading by abandoning incumbent products). PI denotes the pollution intensity measured by industrial SO_2 emissions per industrial output. Correspondingly, ΔPI_{ct} denotes the change of pollution intensity in city c between year t and year $t-1$. X is a matrix consisting of control variables. μ and δ control the time and city fixed effects, respectively. e denotes the error term.

Why does this chapter select industrial SO_2 emissions per industrial output as a proxy for environmental performance of industries? According to Grether and Mathys (2013), SO_2 is primarily manufacturing-driven, few of which will be generated by transportation or household activities. Moreover, it tends to a local pollutant rather than a transboundary or global one. Therefore, SO_2 emission intensity can be a reliable indicator to proxy industrial environmental performance. On the other hand, there are two ways to measure environmental performance. One is based on emission intensities, and the other is based on elimination costs. Considering the data availability and quality, we apply the former indicator.

The selection of control variables is in line with the co-evolution of global value chain and local clusters. Apart from the explanatory variables regarding the upgrading ways, control variables primarily regard the local clusters. First, the amount of labour (S) is used to capture the scale effect. Second, the amount of labour in advanced producer service sectors (T) serves as a proxy for the knowledge base of each city, capturing the local absorption capacity. Third, as mentioned in the fourth facet above, the dynamic of export specialization may exhibit path-dependent feature and then affect the environmental performance. We use the *density* index from Hidalgo et al. (2007) to capture the path-dependent feature, which has been frequently used in previous chapters. In this chapter, we calculate the *density* index of each polluting sectors and index it as C_{ct}. I t measures the average probability of city c to specialize in polluting sectors, given its specialization patters. Taken together, Eq. (11.4) shows the model for estimation.

$$\Delta PI_{ct} = \alpha_0 + \alpha_1 Q_{ct} + \alpha_2 P_{ct} + \alpha_3 T R_{ct} + \alpha_4 N_{ct} + \alpha_5 E_{ct}$$
$$+ \ \beta_1 PI_{ct-1} + \gamma_1 ln S_{ct-1} + \gamma_2 C_{ct-1} + \gamma_3 ln T_{ct-1} + \mu_c + \delta_t + e_{ct} \quad (11.4)$$

The estimation results of Eq. (11.4) provide a baseline. Corresponding to the four facets above in Sect. 11.2, it would be of great value to regress by different city groups. First, we will distinguish processing trade from ordinary trade. Second, we will further distinguish cities specialized in polluting sectors from ones in non-polluting sectors. Third, it is necessary to distinguish between cities with strict and lax environmental regulation.

In this chapter, the processing trade can be directly identified by a field in the dataset from China Customs. Regarding the identification of polluting sectors and non-polluting sectors, the strategy is the same as one in Chap. 9 so that we do not repeat it here. As for the proxy for environmental regulation, we use the ratio of domestic water centralized sewage treatment. The rationale behind this is the same as Chap. 7.

11.3.3   Data

For empirical studies in this chapter, we construct a city-year panel covering 261 prefectural-level cities in China from 2003 to 2011. Data for calculating the *TECH* index and its decomposition are taken from the Statistics of China Customs. This dataset also provides a field about trade patterns to distinguish between processing trade and ordinary trade. Data for calculating pollution intensity at the city level are taken from *The China City Statistical Yearbook*. Other socio-economic data at the city level are taken from *The China Statistical Yearbook for Regional Economy*.

11.4   Evidence from China's Foreign Trade

11.4.1   Disjuncture Between Export Upgrading and Environmental Improvement

Figure 11.1 depicts the simple correlation between export sophistication and pollution intensity. Charts on the left side and the right side distinguish between ordinary trade and processing trade. Charts on the top and charts at the bottom demonstrate the simple correlation from a static view and a dynamic view, respectively. From a static view, the simple correlations between product sophistication and pollution intensity are negative. This suggests that cities with higher levels of product sophistication tend to have less pollution intensity (i.e. better environmental performance). Such a

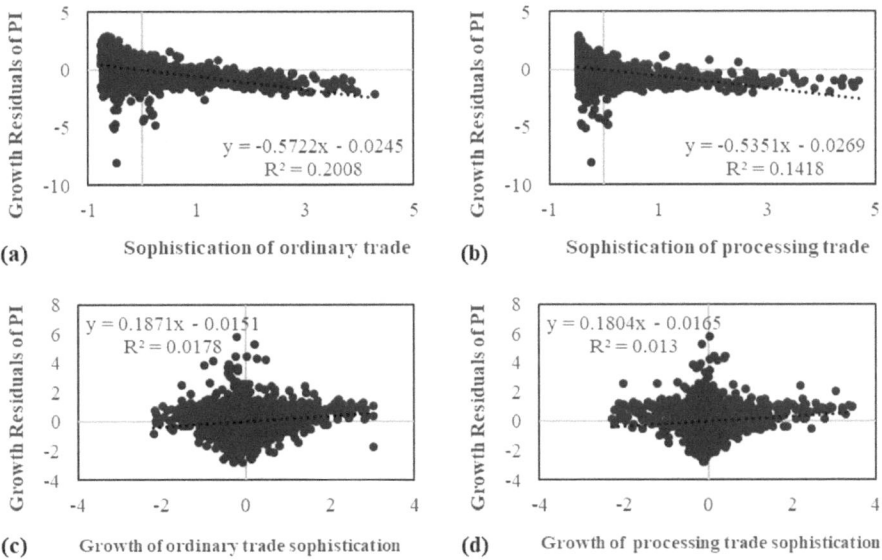

Fig. 11.1 Simple correlation between export sophistication and pollution intensity across Chinese cities during 2003–2011. *Note* Growth residuals of PI are predicted by regressing the growth of PI on its initial PI and simultaneously controlling the individual fixed effects as well as the time fixed effects

result helps to explain the spatial pattern in Chap. 3 that regions with a higher level of trade openness tend to have better environmental performance.

From a dynamic view, we regress the change of pollution intensity on its initial level and then predict the residuals, simultaneously controlling the individual and year fixed effects. The simple correlations between changes in product sophistication and growth residuals of pollution intensity become insignificant, which is against the spatial trend. This result points to the fact that export upgrading does not guarantee the improvement of environmental performance, although cities with higher productivity levels tend to perform better.

This disjuncture between export upgrading and environmental performance is related to the co-evolution of global value chain and local clusters. Regarding the global perspective, Fig. 11.2 demonstrates the simple correlation between different upgrading types and pollution intensity. Charts on the left side and the right side distinguish the processing trade and the ordinary trade. Charts in each row depict the simple correlation between pollution intensity and one particular type of upgrading. The results clearly reveal that the reduction of pollution intensity in China's cities is affected by various types of upgrading in different ways. Quality improvement, productivity improvement, and transformation of products, which tend to represent the genuine adoption of new technology, exert insignificant influence in the reduction of pollution intensity. Their simple correlation with the growth residuals of pollution intensity are rather weak. In contrast, the inter-sectoral upgrading (introducing

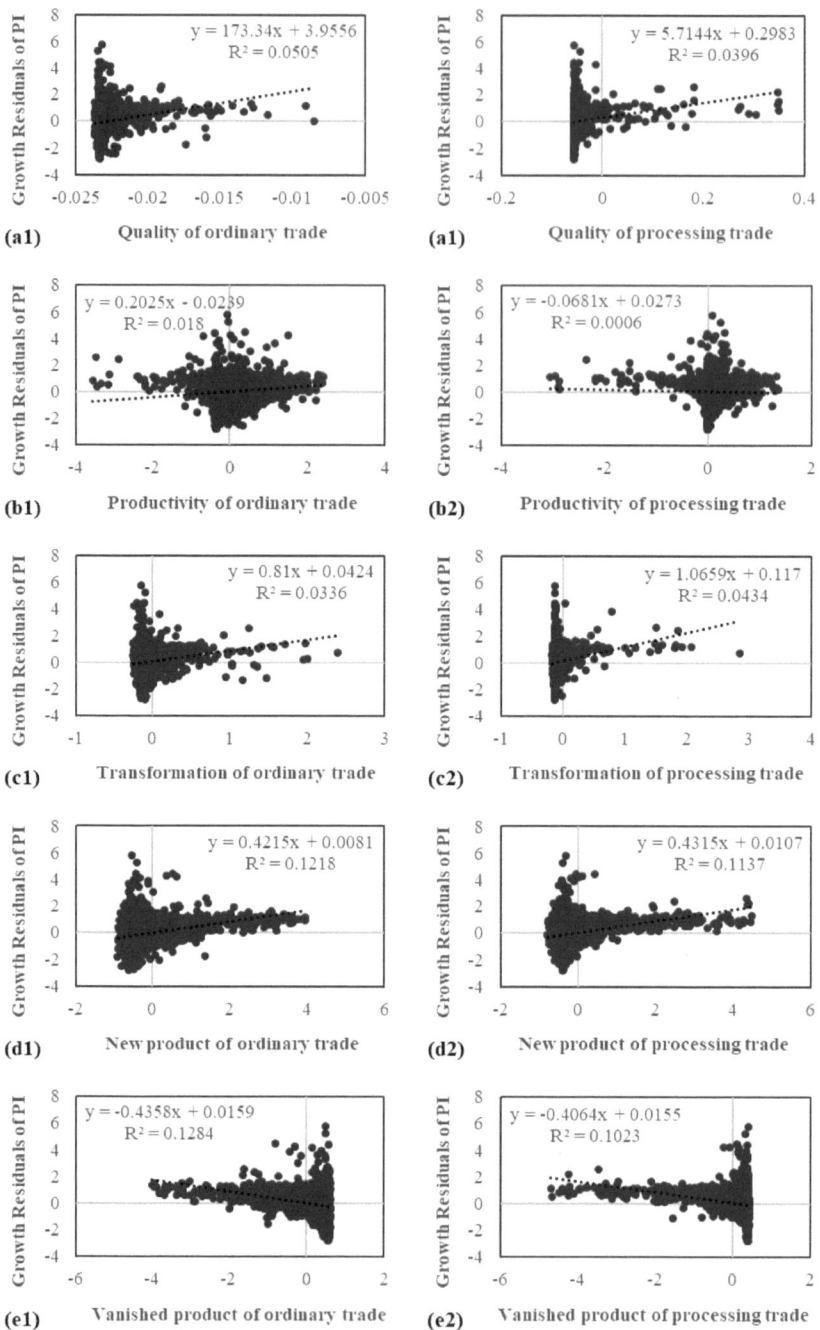

Fig. 11.2 Simple correlation between pollution intensity and various types of export upgrading

new products and abandoning existing ones) plays a role in the change of pollution intensity, producing a significant replacement effect. The introduction of new products contributes to the increase of pollution intensity, while the exit of incumbent products reduces pollution intensities and offsets the impacts of new products. These results mirror the development context of China since 2001, when exporting sectors underwent significant structural changes in response to the changing global business environment (Yang 2012).

Regarding the local perspective, we calculate the locational quotient of polluting sectors in one city to measure its specialization in polluting production. Then, we use t-statistics to test the difference in upgrading types between cities specialized in polluting and non-polluting productions. Table 11.1 reports the results, demonstrating a very significant difference between two city groups. Cities specialized in non-polluting production have an overall higher level of export upgrading than ones specialized in polluting production.

Apart from industrial specialisation, we also consider the difference between cities with strict and lax environmental regulation. Cities with environmental regulation stringency above the average are identified as "strict" and the rest are "lax". Table 11.2 reports the results. These results point to the fact that environmental regulation in China seems to be more capable of promoting structural changes than promoting the adoption of new technology. Regarding the functional upgrading and inter-sectoral upgrading, which are upgrading types related to structural changes, there are significant differences in upgrading levels between two city groups. Regarding the product upgrading and process upgrading, however, the differences in upgrading level between two city groups are less significant or even insignificant. These results provide another perspective to explain the positive effect of environmental regulation, as already examined in Chaps. 3 and 5. Based on the command-and-control regulation, environmental regulation can be more efficient in imposing restrictions on production and then lead the structural changes of products by compliance costs. The costs of adopting new technology or promoting innovation could be even higher than the costs of abandoning incumbent production. As such, environmental regulations tend to play a positive role by structural changes rather than technology adoption/innovation.

11.4.2 Environmental Performance of Various Upgrading Types and Trade Patterns

Table 11.3 reports the estimation results of Eq. (11.4), including both the results of pooled OLS and the fixed effect model. Considering the significant heterogeneity among cities in China, it is necessary to control the individual fixed effects. Thus, we suppose that results of the fixed effect models provide a baseline for the overall environmental performance of different upgrading types. The results of pooled OLS are reported for comparison. The modified Wald test for group-wise heteroscedasticity

Table 11.1 Differentiated components of sophistication growth between regions specialized in polluting and non-polluting production

	Ordinary trade					Processing trade				
	Q	P	TR	N	E	Q	P	TR	N	E
Mean (P)	−0.02	−0.03	−0.10	−0.27	−0.28	−0.05	0.06	−0.10	−0.26	0.27
Mean (NP)	0.07	0.09	0.28	0.80	0.81	0.14	−0.16	0.28	0.77	−0.78
p-value of *t*-test (two-tailed)	0.083	0.014	0.000	0.000	0.000	0.000	0.000	0.000	0.000	0.000
p-value of t-test (one-tailed)	0.041	0.007	0.000	0.000	0.000	0.000	0.000	0.000	0.000	0.000

Note Q indicates product upgrading, P indicates process upgrading, TR indicates functional upgrading, N indicates inter-sectoral upgrading by introducing new products, and E indicates inter-sectoral upgrading by abandoning incumbent products

Table 11.2 Differentiated components of sophistication growth between regions with stringent and lax environmental regulation

	General trade					Processing trade				
	Q	P	TR	N	E	Q	P	TR	N	E
Mean (stringent)	0.02	0.03	0.07	0.20	−0.21	0.04	−0.04	0.08	0.17	−0.18
Mean (lax)	−0.02	−0.04	−0.08	−0.25	0.26	−0.05	0.05	−0.11	−0.22	0.22
p-value of t-test (two-tailed)	0.378	0.119	0.001	0.000	0.000	0.058	0.066	0.000	0.000	0.000
p-value of t-test (one-tailed)	0.189	0.059	0.000	0.000	0.000	0.029	0.033	0.000	0.000	0.000

Table 11.3 Results of the baseline regression

	Ordinary trade		Processing trade	
	Pooled OLS	Fixed effect	Pooled OLS	Fixed Effect
Q_t	0.001	-0.014^{***}	0.004	0.002
P_t	-0.004	0.025^{*}	0.024^{**}	0.010
TR_t	0.002	-0.008	0.008	0.003
N_t	-0.026^{**}	0.073^{***}	0.017	0.056^{***}
E_t	-0.018	-0.147^{***}	-0.033^{***}	-0.096^{***}
PI_{t-1}	-0.566	-1.105^{***}	-0.572^{***}	-1.053^{***}
C_{t-1}	-0.063^{***}	-0.107^{**}	-0.026^{*}	-0.125^{**}
$\ln S_{t-1}$	-0.023	0.392^{***}	-0.004	0.323^{***}
$\ln T_{t-1}$	-0.020^{*}	-0.093^{*}	-0.008	-0.058
Observations	2063	2063	1950	1950
Groups	261	261	261	261
F	875.62^{***}	22.57^{***}	275.64^{***}	19.49^{***}
R-sq	0.328	0.368	0.343	0.378
Time fixed effect	Yes	Yes	Yes	Yes

Note $^{***}p < 0.01$; $^{**}p < 0.05$; $^{*}p < 0.10$

in the fixed effect regression has concluded the heteroscedasticity. Hence, a robust regression has been applied.

Results in Table 11.3 support that various types of export upgrading contribute to China's improvement of environmental performance to different extents. Intersectoral upgrading is the most important type of upgrading for the reduction of pollution intensity during the study period. Although the introduction of new products increases the pollution intensity, it can be offset by the even larger effects of the exit of incumbent products on reducing pollution intensity. The structural changes of exporting goods overall produce a conducive effect on environmental performance, representing a replacement effect. Besides, another conducive effect comes from product upgrading. The improvement of product quality contributes to the reduction of pollution intensity, exhibiting a quality effect. However, this effect is still confined to the ordinary trade and within a limited extent, while the processing trade does not show a significant quality effect. Overall, the conducive replacement effects point to the fact that products exported by China have become less polluting than before (Dean and Lovely 2010), through either processing trade or ordinary trade. The major difference between ordinary trade and processing trade lies in the quality effect.

Results in Table 11.3 provide another perspective to understand the debate over environmental effects of processing trade in China. Like most developing countries with export-oriented growth model, processing trade contributes to over 50% of China's foreign trade boom (Yu and Tian 2012). In such a context, there are rising concerns about the environmental effects of processing trade. Unlike ordinary trade, processing trade refers to a trade pattern that domestic firms purchase raw materials or supplementary inputs overseas and process domestically into finished goods for re-exports. Some studies insist that the overseas purchase of materials and inter-mediated inputs are able to alleviate the environmental impacts of exports (Soura et al. 2015), since the raw materials may reduce the demand for local resources and the intermediated inputs are expected to improve productivity by their technological contents. On the contrary, some empirical evidence also demonstrates that processing trade largely concentrates in resource-based, energy-intensive, and even emission-intensive industries (Liu 2009). Although the overseas purchase may benefit the environment to some extent, the domestic manufacturing processes still impact the environment largely (Song et al. 2015). Briefly speaking, this debate focuses on whether the "processing" can result in a conducive technique effect by process upgrading. Empirical results in Table 11.3 show that process upgrading does not contribute to the reduction of pollution intensity, at least not yet. Efforts from the Chinese government have been largely made to prohibit those commodities with emission-intensive processing, instead of improving the processing (Song et al. 2015). This fact also explains why the processing trade exhibits replacement effects rather than quality or technique effects.

11.4.3   The Role of Specialization in Polluting Production

Tables 11.4 and 11.5 report the results of regressing Eq. (11.4) group by group with the sample of ordinary trade and processing trade, respectively. Cities are divided into four quartiles based on the pollution intensity of industries in which these cities are specialized. Thus, there are four groups of cities. Overall, the results in Tables 11.4 and 11.5 illustrate that export specialization will affect the environmental effects of export upgrading.

Regarding the ordinary trade, the replacement effects are larger in cities specialized polluting industries, although the replacement effects are overall significant across four groups of cities. Besides, the scale effects are only confined to cities specialized in polluting industries. These results support the theoretical prediction above that specialization in polluting industries confronts one city with more environmental pressures. Thus, export upgrading is expected to produce more significant environmental effects in these regions. Apart from the replacement effects, it should be noted that the quality effect is confined to cities specialized in non-polluting industries, especially the cleanest sectors (Q4). Quality effects are absent or even detrimental in cities specialized in polluting industries. The reasons behind may be related to the inherent heterogeneity of industries, requiring more in-depth investigations. For instance, a positive quality effect requires the accumulation of knowledge capitals. The agglomeration of polluting sectors is less likely to provide a desired environment for the accumulation of knowledge capitals, particularly human capitals. Overall, considering the net effects of export upgrading and export specialization, export specialization is likely to disable the greening efforts of export upgrading. The

Table 11.4 Results of city groups specialized in polluting industries (based on ordinary trade)

| | Ordinary trade | | | |
	Q1	Q2	Q3	Q4
Q_t	-1.485	0.253^{***}	-0.010^{*}	-0.003^{**}
P_t	0.035	0.059^{*}	-0.015	0.023^{**}
TR_t	0.007	-0.052	0.014	-0.002
N_t	0.074	0.133^{***}	0.044	0.002
E_t	-0.310^{***}	-0.269^{***}	-0.110^{*}	-0.030^{***}
PI_{t-1}	-1.126^{***}	-1.083^{***}	-1.201^{***}	-0.987^{***}
C_{t-1}	-0.080	-0.090^{*}	-0.127	-0.029
$\ln S_{t-1}$	0.439^{**}	0.277^{**}	0.281	-0.094
$\ln T_{t-1}$	-0.120	-0.180^{**}	-0.048	0.043
Observations	1169	1171	665	321
Groups	187	171	115	65
F	16.02^{***}	14.96^{***}	11.33^{***}	198.57^{***}
R-sq	0.397	0.346	0.251	0.363
Time fixed effect	Yes	Yes	Yes	Yes

Note $^{***}p < 0.01$; $^{**}p < 0.05$; $^{*}p < 0.10$

Table 11.5 Results of city groups specialized in polluting industries (based on processing trade)

| | Processing trade | | | |
	Q1	Q2	Q3	Q4
Q_t	-0.520	0.151	0.004	0.003
P_t	-0.072	-0.013	0.001	0.009
TR_t	-0.155	-0.015	-0.001	0.003
N_t	0.049^{**}	0.098^{***}	0.029	0.004
E_t	-0.169^{**}	-0.204^{***}	-0.064^{**}	0.020
PI_{t-1}	-1.111^{***}	-1.081^{***}	-1.014^{***}	-0.985^{***}
C_{t-1}	-0.149^{**}	-0.107^{**}	-0.113	-0.011
$\ln S_{t-1}$	0.308^{*}	0.302^{**}	0.333	-0.064
$\ln T_{t-1}$	-0.059	-0.176^{**}	-0.067	0.032
Observations	1081	1118	650	318
Groups	187	170	114	65
F	17.98^{***}	13.09^{***}	11.23^{***}	15.70^{***}
R-sq	0.395	0.339	0.327	0.344
Time fixed effect	Yes	Yes	Yes	Yes

Note $^{***}p < 0.01$; $^{**}p < 0.05$; $^{*}p < 0.10$

replacement effects have to counteract the even larger scale effects and the quality effects may be absent in a place that specialized in polluting industries.

Regarding the processing trade, the effects of inter-sectoral upgrading (replacement effects) are confined to cities specialized in polluting industries, while they are insignificant in cities specialized in non-polluting industries. Also of note, the effects of product upgrading, process upgrading, and functional upgrading are insignificant. Regarding the scale effects, they are statistically significant only in cities specialized in polluting industries. Moreover, the effects of path dependence are also confined to cities specialized in polluting industries, highlighting the positive role of pre-existing base in reducing pollution intensity. Overall, the results are consistent with the theoretical prediction. From these results, we further recognize that the upgrading of processing trade would be more crucial for cities specialized in polluting industries. One of the primary reasons can be that processing trade is more common in polluting industries. Regarding the upgrading of processing trade, structural changes have played an essential role in improving the environmental performance of China's processing trade. The structural changes are manifested by the inter-sectoral upgrading globally as well as the dynamics of specialization locally.

11.4.4 The Role of Environmental Regulation

Table 11.6 reports the results of regressing Eq. (11.4) by two city groups with the sample of ordinary trade and processing trade, respectively. Two city groups here are categorized by their stringency of environmental regulation. One group consists of cities with environmental regulation stringency above the average, and the rest belongs to the other group. The results in Table 11.6 clearly point to the fact that environmental regulations contribute to the environmental performance of export upgrading. By comparing the coefficients between the strict group and the lax group, the impacts of introducing new products are insignificant in the strict group but significant in the lax group. On the contrary, the desirable effects of product upgrading are significant in strict group but insignificant in the lax group. Overall, a strict environmental regulation is expected to keep the desirable effects of export upgrading and simultaneously keep export upgrading from undesirable effects.

Also of note, the coefficients of *density* index (C) are statistically significant only in strict group. This finding suggests that a strict environmental regulation is capable of leading the dynamic of specialization towards a more "green" way, which is in line with the previous findings in Chaps. 5 and 6.

Table 11.6 Results of city groups with strict and lax environmental regulation

	Ordinary trade		Processing trade	
	Strict group	Lax group	Strict group	Lax group
Q_t	-0.012^{***}	0.158^*	0.002	-0.290
P_t	0.030	0.073^*	0.004	0.059^{**}
TR_t	-0.007	0.015	-0.001	0.219^{***}
N_t	0.058	0.193^{***}	0.038^*	0.138^{***}
E_t	-0.125^{***}	-0.364^{**}	-0.079^{**}	-0.333^{***}
PI_{t-1}	-1.296^{***}	-1.176^{***}	-1.195^{***}	-1.208^{***}
C_{t-1}	-0.176^{***}	-0.023	-0.191^{***}	-0.085
$\ln S_{t-1}$	0.471^{**}	0.324^{**}	0.341^{**}	0.413^{***}
$\ln T_{t-1}$	-0.053	-0.118	-0.003	-0.135
Observations	1156	907	1109	841
Groups	224	205	222	201
F	24.12^{***}	15.38^{***}	9.87^{***}	13.89^{***}
R-sq	0.331	0.450	0.294	0.535
Time fixed effect	Yes	Yes	Yes	Yes

Note $^{***}p < 0.01$; $^{**}p < 0.05$; $^*p < 0.10$

11.5 Conclusion

As the last chapter regarding the global-local interactions in reshaping the geography of environmental performance, this chapter discusses how the global-local interactions determine what to export and then affect the environment. The basic viewpoint of this chapter is that export upgrading is an outcome of the co-evolution of global value chain and local clusters. Building on this premise, what determines the environmental performance of export upgrading should be explored in four facets, namely, the upgrading types, the trade patterns, the local specialization, and the institutions. The former two facets represent the global linkages, while the latter two facets represent the local linkages. These four facets comprise the basic research framework in this chapter, which further enrichs the framework in Chap. 10.

Empirical findings in this chapter change the perceptions that export upgrading can benefit the environment by superior technology and new knowledge, which is more or less taken for granted previously. However, if export upgrading is seen as an outcome of the co-evolution between global value chain and local clusters, export upgrading does not necessarily represent the genuine adoption of new knowledge or superior technology. It can be achieved by a combination of overseas purchase of sophisticated inputs and domestic processing by energy-intensive, emission-intensive productions. It can also be achieved by the suppression of incumbent products and the introduction of new ones. These alternatives do not guarantee an environmentally benign outcome, although the overall technological contents in exports augment superficially.

At the other end of the global-local interactions, local clusters are different from one to another. Some of them are specialized in polluting industries and then confronted with more severe environmental pressure. It leaves more rooms for the effects of export upgrading. However, export upgrading in these clusters largely relies on the replacement of product types. The improvement of products themselves (i.e. the quality of products and the efficiency of processing) is rather weak or even inexistent. On the contrary, the quality improvement tends to manifest itself in places specialized in non-polluting sectors. Such a spatial difference also serves as a force to prevent the spatial polarisation in terms of environmental performance, as already discussed in Chap. 9, since there are various types of upgrading for different places in response to environmental pressures.

Environmental regulation is also different spatially and has long been identified as a primary determinant of the shifts of environmental burdens. From the perspective of upgrading, this chapter highlights the role of environmental regulation in promoting desirable effects of upgrading and simultaneously suppressing undesirable effects. More importantly, empirical findings in this chapter further indicate that the current environmental regulation in China is largely dependent on imposing restriction so that the effects of export upgrading are manifested by inter-sectoral upgrading. Since its environmental regulation fails to consider costs and risks of firms' adopting new technology and knowledge, its effects on product upgrading and process upgrading are still fledgling.

From a perspective of environmental economic geography, the four chapters in Part II of this book successively demonstrate how the geography of environmental performance in China will be affected by the global shift of environmental burdens. It delves into the shift of environmental burdens from overseas to China, the shift of environmental burdens between regions in China, and the responses of different places in China to the shift of environmental burdens. These chapters demonstrate how various theories in Economic Geography are needed to explain the global shift of environmental burdens across nations and places. Empirical findings in these chapters also indicate that the incorporation of context-dependent, border-crossing, multi-scalar, and co-evolutionary views allows us to find new inspirations for ongoing debates.

References

Ambec, S., Cohen, M. A., Elgie, S., et al. (2013). The Porter Hypothesis at 20: Can environmental regulation enhance innovation and competitiveness? *Review of Environmental Economics and Policy, 7*(1), 2–22.

Amighini, A., & Sanfilippo, M. (2014). Impact of South-South FDI and trade on the export upgrading of African economies. *World Development, 64,* 1–17.

Chrid, N,, Saafi, S., & Chakroun, M. (2020). Export upgrading and economic growth: A panel cointegration and causality analysis. *Journal of the Knowledge Economy.* https://doi.org/10.1007/s13132-020-00640-6.

Crespo, N., & Fontoura, M. P. (2007). Determinant factors of FDI spillovers—What do we really know? *World Development, 35*(3), 410–425.

Dai, M., Maitra, M., & Yu, M. (2016). Unexceptional exporter performance in China? The role of processing trade. *Journal of Development Economics, 121,* 177–189.

Dean, J. M., & Lovely, M. E. (2010). Trade growth, production fragmentation, and China's environment. In R. Feenstra & S. J. Wei (Eds.), *China's growing role in world trade* (pp. 429–469). Chicago IL: University of Chicago Press.

Dechezleprêtre, A., & Sato, M. (2017). The impacts of environmental regulations on competitiveness. *Review of Environmental Economics and Policy, 11*(2), 183–206.

Gereffi, G. (2009). Development models and industrial upgrading in China and Mexico. *European Sociological Review, 25*(1), 37–51.

Gertler, M. S. (2003). Tacit knowledge and the economic geography of context, or the undefinable tacitness of being (there). *Journal of Economic Geography, 3*(1), 75–99.

Giuliani, E., Pietrobelli, C., & Rabellotti, R. (2005). Upgrading in global value chain: Lessons from Latin American clusters. *World Development, 33*(4), 549–573.

Gong, M., Zhe, Y., Wang, L., et al. (2020). Environmental regulation, trade comparative advantage, and the manufacturing industry's green transformation and upgrading. *International Journal of Environmental Research and Public Health, 17*(8), 2823.

Grether, J.-M., & Mathys, N. A. (2013). The pollution terms of trade and its five components. *Journal of Development Economics, 100*(1), 19–31.

Grillitsch, M., & Rekers, J. V. (2016). How does multi-scalar institutional change affect localised learning processes? A case study of the med-tech sector in Southern Sweden. *Environment and planning A: Economy and space, 48*(1), 154–171.

Guerrieri, P., & Iammarino, S. (2007). Dynamics of export specialisation in the regions of the Italian Mezzogiorno: Persistence and change. *Regional Studies, 41*(7), 933–948.

Hao, Y., & Liu, Y.-M. (2015). Has the development of FDI and foreign trade contributed to China's CO_2 emissions? An empirical study with provincial panel data. *Natural Hazards, 76,* 1079–1091.

Harding, T., & Javorcik, B. S. (2012). Foreign direct investment and export upgrading. *The Review of Economics and Statistics, 94*(4), 964–980.

Hausmann, R., & Hidalgo, C. A. (2011). The network structure of economic output. *Journal of Economic Growth, 16,* 309–342.

Hausmann, R., Hwang, J., & Rodrik, D. (2007). What you export matters. *Journal of Economic Growth, 12,* 1–25.

Hidalgo, C. A., Klinger, B., Barabási, A. L., et al. (2007). The product space conditions the development of nations. *Science, 317,* 482–487.

Hu, C., Lin, F., Tan, Y, et al. (2019). How exporting firms respond to technical barriers to trade? *The World Economy, 42*(5), 1400–1426.

Huber, S. (2017). Indicators of product sophistication and factor intensities: Measurement matters. *Journal of Economic and Social Measurement, 42*(1), 27–65.

Humphrey, J., & Schmitz, H. (2002). How does insertion in global value chains affect upgrading in industrial clusters? *Regional Studies, 36*(9), 1017–1027.

Jaffe, A. B., & Stavins, R. N. (1995). Dynamic incentives of environmental regulations: The effects of alternative policy instruments on technology diffusion. *Journal of Environmental Economics and Management, 29*(3), S43–S63.

Jarreau, J., & Poncet, S. (2012). Export sophistication and economic growth: Evidence from China. *Journal of Development Economics, 97,* 281–292.

Javorcik, B. S. (2004). Does foreign direct investment increase the productivity of domestic firms? In search of spillovers through backward linkages. *American Economic Review, 94*(3), 605–627.

Kemeny, T. (2011). Are international technology gaps growing or shrinking in the age of globalization? *Journal of Economic Geography, 11*(1), 1–35.

Liu, J. (2009). Comparative analysis on the influence of general trade and processing trade on environmental pollution of China. *World Economy Study, 6,* 44–48. (in Chinese).

Mao, X., & He, C. (2019). Product relatedness and export specialisation in China's regions: A perspective of global-local interactions. *Cambridge Journal of Regions Economy and Society, 12*(1), 105–126.

Marchi, V. D., Maria, E., & Micelli, S. (2013). Environmental strategies, upgrading and competitive advantage in global value chains. *Business Strategy and the Environment, 22,* 62–72.

McMillan, M. S., & Rodrik, D. (2011). *Globalisation, structural change and productivity growth.* NBER Working Paper No. 17143.

Minondo, A. (2010). Exports' quality-adjusted productivity and economic growth. *Journal of International Trade and Economic Development, 19*(2), 257–287.

Naudé, W., Bosker, M., & Matthee, M. (2010). Export specialisation and local economic growth. *World Economy, 33*(4), 552–572.

Navas-Alemán, L. (2011). The impact of operating in multiple value chains for upgrading: The case of the Brazilian furniture and footwear industries. *World Development, 39*(8), 1386–1397.

Pahl, S., & Timmer, M. P. (2019). Do global value chains enhance economic upgrading? A long view. *Journal of Development Studies.* https://doi.org/10.1080/00220388.2019.1702159.

Pietrobelli, C., & Rabellotti, R. (2011). Global value chains meet innovation systems: Are there learning opportunities for developing countries? *World Development, 39*(7), 1261–1269.

Poncet, S., & de Waldemar, F. S. (2013). Export upgrading and growth: The prerequisite of domestic embeddedness. *World Development, 51,* 104–118.

Porter, M. E., & van der Linde, C. (1995). Toward a new conception of the environment-competitiveness relationship. *Journal of Economic Perspectives, 9*(4), 97–118.

Rodrik, D. (2006). What is so special about China's exports? *China and the World Economy, 14*(5), 1–19.

Scherngell, T., Borowiecki, M., & Hu, Y. (2014). Effects of knowledge capital on total factor productivity in China: A spatial econometric perspective. *China Economic Review, 29,* 82–94.

Schmitz, H. (1999). Collective efficiency and increasing returns. *Cambridge Journal of Economics, 23*(4), 465–483.

Schott, P. K. (2008). The relative sophistication of Chinese exports. *Economic Policy, 23*(53), 5–49.

Shimbov, B., Alguacil, M., & Suarez, C. (2019). Export structure upgrading and economic growth in the Western Balkan countries. *Emerging Markets Finance and Trade, 55*(10), 2185–2210.

Song, P., Mao, X., & Corsetti, G. (2015). Adjusting export tax rebates to reduce the environmental impacts of trade: Lessons from China. *Journal of Environmental Management, 161,* 308–416.

Song, Y., Wu., Y., & Deng, G., et al. (2019). Intermediate imports, institutional environment, and export product quality upgrading: Evidence from Chinese micro-level enterprises. *Emerging Markets Finance and Trade.* https://doi.org/10.1080/1540496x.2019.1668765.

Soura, J. D., Hering, L., & Poncet, S. (2015). *Has trade openness reduced pollution in China?* CEPII Working Paper No. 2015-11, Paris: CEPII.

Xie, R., Yuan, Y., & Huang, J. (2017). Different types of environmental regulations and heterogeneous influence on "green" productivity: Evidence from China. *Ecological Economics, 132,* 104–112.

Xu, B. (2010). The sophistication of exports: Is China special? *China Economic Review, 21*(3), 482–493.

Yang, C. (2012). Restructuring the export-oriented industrialization in the Pearl River Delta, China: Institutional evolution and emerging tension. *Applied Geography, 32*(1), 143–157.

Yu, M. J., & Tian, W. (2012). China's processing trade. In H. McKay & L. G. Song (Eds.), *Rebalancing and sustaining growth in China* (pp. 111–148). Canberra: Australian National University E Press.

Zhou, Y., Zhu, S., & He, C. (2019). Learning from yourself or learning from neighbours: Knowledge spillovers, institutional context and firm upgrading. *Regional Studies, 53*(10), 1397–1409.

Zhu, S., & Fu, X. (2013). Drivers of export upgrading. *World Development, 51,* 221–233.

Chapter 12
Summary and Implications

12.1 The Value of Environmental Economic Geography

In this book, we have demonstrated some attempts to practice Environmental Economic Geography (EEG) with the case of China's industrial pollution intensities. As discussed in Chap. 2, we are intended to consider EEG as a way to understand environmental implications of the spatial configuration of production, exchange, and consumption as well as to figure out spatial conditions for the development of environmental-benign economic activities. In other words, EEG should never be simply considered as neither mapping environmental issues led by economic activities nor exploring environmental issues with the sample of geographical units. On this basis, we briefly distinguish EEG from other relevant fields in Fig. 12.1, which helps to reveal that EEG is still expected to fill in a blank among the seemingly overwhelming amount of studies on the environment.

In comparison with non-geographical studies, EEG operates within the geographical space rather than a uniform space or a stylized space. This means EEG should look into the spatial configuration of production, exchange, and consumption more than these activities themselves. This determines that EEG seeks to find the environmental implication of the economy by economic actors' dependence on differentiated external conditions to gain advantages and their spatial interconnections to make full use of the advantages. Based on such thinking, EEG is able to change some perceptions that are taken for granted as well as contribute to ongoing debates in non-geographical studies.

Empirical studies in this book investigate the spatial configuration of production, exchange, and consumption by looking into the interactions spanning from the local to the global. It starts at the local level. Chapter 3 unravels that the unique development conditions of one place determine the environmental performance of its hosted economic activities. Even the same industrial activities can perform differently across spaces. This determines that environmental issues are essentially a term with relativity since industrial activities are likely to trigger an environmental problem in one place but not in others. As such, from the perspective of EEG, environmental

C. He and X. Mao, *Environmental Economic Geography in China*,
Economic Geography, https://doi.org/10.1007/978-981-15-8991-1_12

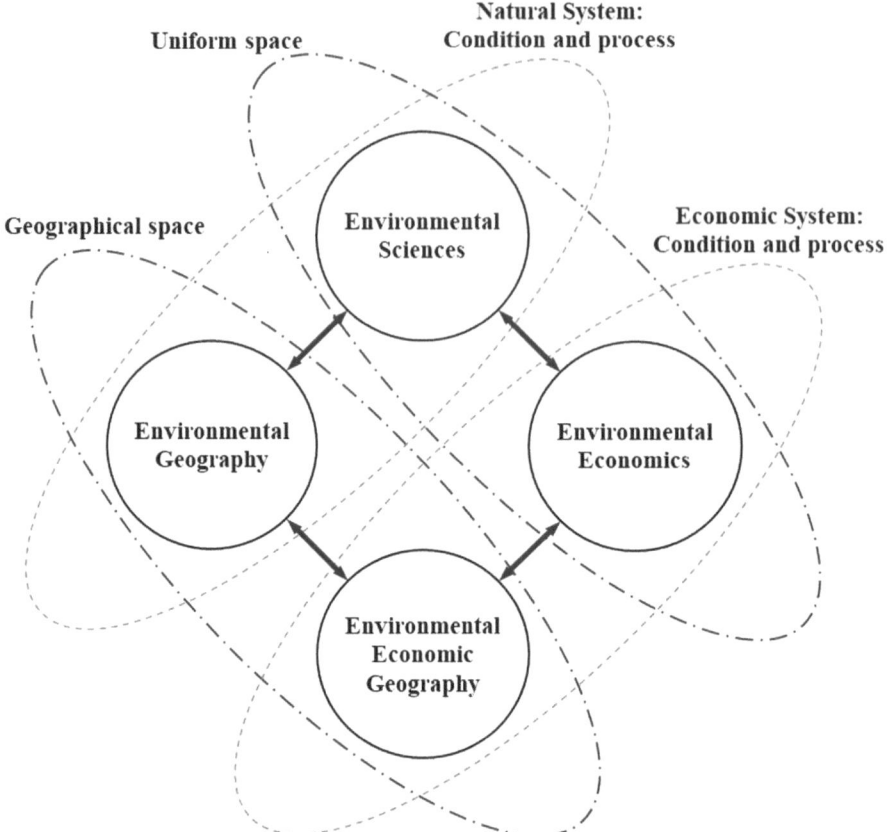

Fig. 12.1 Environmental economic geography in the environmental-related fields

conservation is not only a matter of eliminating "dirty" activities. More importantly, it is a matter of finding the right place for economic activities.

Empirical evidence from Chap. 4 challenges a seemingly geographical question in non-geographical studies, the "pollution-thy-neighbour" issues. It is natural to posit that polluting firms tend to locate at the border regions so that they can reduce local emissions by taking advantages of geographical conditions. From the perspective of economic geography, this is partly true but not always the case. Economic activities are unevenly distributed across spaces, and various core-periphery patterns manifest economic geography at multiple levels. The division between the core and the periphery thereby confronts economic actors with significant opportunity costs for location choice. Admittedly, a peripherical location allows polluting firms to become free riders and reduce their costs in compliance with environmental regulation, as highlighted in most previous studies. However, the opportunity costs of locating at the core and benefiting from the agglomeration economies there have been somehow overlooked. Empirical findings from Chap. 4 exactly evidence the value of the core location for polluting firms.

Both Chaps. 3 and 4 point to the value of EEG by changing the stereotypes of "polluting activities" in non-geographical studies. "Polluting activities" is a relative term which is contingent on the spatial-temporal contexts. Conventionally, studies get used to identifying them as undesirable activities with negative externalities. As a result, studies tend to assume that polluting firms would like to be free riders. In this regard, EEG points to the basic fact that sustainability is not always a case of identifying and eliminating polluting activities. Instead, it is crucial to find the right place. In this regard, EEG will demonstrate the value of location theory.

Besides the likely stereotype of "polluting activities", another term frequently used in previous studies is environmental regulation. There are also ongoing debates over the role of environmental regulation in improving environmental performance and impacting economic competitiveness. To some extent, this strand of the literature becomes centred on environmental regulation but somehow loses a big picture of the institution. Empirical studies in Chaps. 5 and 6 demonstrate that environmental regulation does not always stand at the centre of various determinants of environmental performance. During the economic transition in China, a set of institutional changes has occurred at the same time, such as the fiscal decentralization, the reform of state-owned enterprises, the market access for foreign investment, and a wide array of pilot policy for economic reform.

These institutional changes will never be considered as environmental regulation. However, these changes have generated far-reaching effects on China's environmental performance. Institutional changes are intended to be place-specific or even place-biased, while places also react or respond to these changes in different ways. Places with policy supports will gain comparative advantages in development so that there is also increasing inter-region competition for policy supports from the central government. Evidence from Chaps. 5 and 6 reveals the mixed environmental effects of China's decentralization and marketization efforts.

From Chaps. 5 and 6, we seek to highlight the value of EEG from the institutional perspective. Just in line with the theoretical development of economic geography, the institution lays the foundation for economic growth and simultaneously determines the spatial differences of economic performance. As mentioned in Chap. 2, recent advances point to the fact that economic activities operate in the institutional environment at multiple levels. Empirical cases are particularly interested in environmental industries. Thus, EEG provides an integrated perspective to see the environmental performance through institutional arrangement at multiple levels, going beyond the conventional perspective centred on environmental regulation.

Empirical studies in Chaps. 7 and 9 follow the evolutionary perspective of EEG, corresponding to the "evolutionary turn" in Economic Geography. In non-geographical studies, the structural changes in production have always been considered as a panacea for improving environmental performance. However, they seldom consider the feasibility to achieve the so-called "optimized structures" or the trajectory of structural changes. As a result, the "structural changes" in these studies remain still static, consisting of several structures. The use of "structural changes" essentially compares the difference between structures rather than investigate the linkages between structures. From an evolutionary view, EEG provides an opportunity to internalize the structural changes into the process of regional development, which are given as external factors.

Empirical evidence in Chaps. 7 and 9 support that structural changes of industries do not occur at random but hold connections to pre-existing bases to different extents. Regarding environmental performance, new industries can benefit from such connections, exhibiting an incremental improvement. However, too many connections may also hinder the improvement of environmental performance due to the diminishing increment of knowledge. Such a connection can exist either between local industries or with nonlocal industries. As a result, the environmental performance of structural changes varies across places and their development phases. It is subject to not only local conditions but also their nonlocal connections. In this regard, EEG can help to change the perception that structural change is a panacea for sustainability. Moreover, EEG can unravel the conditions for a trajectory of structural change towards sustainability.

Finally, empirical studies in Chaps. 8–11 demonstrate the relational perspective of EEG. They look into the industrial environmental performance of China from a broader context—the global shift of environmental burdens due to the spatial separation between production and consumption and the international division of labour. These chapters exactly show how the spatial configuration of production will reallocate industrial pollution emissions globally and then affect environmental performance locally. In non-geographical studies, a stylized space consisting of developed and developing economies lays the foundation for most classical hypotheses in Environmental Economics. However, in a geographical space, developing economies are still different from one to another. The economic relations between economies are not solely underpinned by their income levels. Moreover, a stylized space has also overlooked the crucial role of distance costs in determining interregional connections.

By incorporating more geographical features into the conventional duality of developed and developing economies, empirical studies in Chaps. 8–11 have changed some perceptions that are taken for granted previously. Evidence from Chap. 8 challenges the pollution haven hypothesis, which should be an outcome in a particular development period of the world economy rather than a rule for the world economy. That is, when the catching-up economies become rich, they are not necessarily able to shift their environmental burdens to other developing economies. Evidence from Chap. 9 challenges the concerns of spatial inequality (or even spatial polarisation) in response to the global shift of environmental burdens. Conventional thinking concerns the utilization of nonlocal resources allow leading regions' development to be at the costs of lagging regions' environment and resources. However, this is not the case, at least in China. Lagging regions can also benefit from the utilization of nonlocal resources to developing new paths.

On the other hand, the path extension in leading regions will not continue without limit. Evidence from Chaps. 10 and 11 further reveals that the local does not only a receptor of the global shift of environmental burdens. On the contrary, they are active in response to the global shift. From the evidence above, through investigating the spatial configuration of the global economy, EEG can explore the sources and causes of local environmental issues at multiple geographical scales as well as from a global perspective. As such, EEG can better explain why there are no one-size-fits-all problems and solutions regarding environmental issues.

12.2 Empirical Findings from China's Case Based on EEG

Environmental challenges are locally unique and globally interactive. EEG provides a panoramic view to understanding the formation of and feedbacks to environmental challenges. Empirical findings in this book unravel the changing nature-economy interactions in China during the past two decades. There are several marked transition processes interwoven in this period, including the shift from a passive involvement in the globalization to an active one, the strengthening of regional integration and interregional competition, the deepening of marketization reform, the starting of sustainable development, and the adjustment of spatial governance. These processes impact the environment interactively but work in different manners.

12.2.1 The Core-Periphery Interactions

The reshaping core-periphery patterns generate mixed environmental effects. The core regions in economic terms attract resources and factors to concentrate. The spatial effects of economic concentration on the environment are twofold. On the one hand, economic concentration increases the intensity and then increasingly generates pressure on the local environment. On the other hand, due to agglomeration externalities, economic concentration enhances efficiency and then improves environmental performance. Taken together, the core regions tend to exhibit better environmental performance but still concern environmental capacities. In contrast, the periphery regions in economic terms suffer from their lack of locational advantages and economic efficiencies. The lack of locational advantages usually suggests that the local environment would be more fragile and sensitive to economic activities. The lack of efficiencies further amplifies the impacts of economic activities on the environment.

At the national level, the coastal-inland division is assumed to be a core-periphery pattern. The coastal regions tend to be environmentally benign and have a majority of economic activities concentrated there. The economic concentration generates significant environmental pressures locally but also improves environmental performance globally. However, as the economic density continues to grow, a congestion effect comes into being, which would then trigger the spatial relocation of some economic activities from the coastal to the inland. The inland regions tend to be more ecologically fragile and environmentally sensitive. Simultaneously, economic development here strives for catching-up. Spatially, economic activities are overall scattered across the region but highly concentrated in a few places. The resilience in terms of both economic development and environmental conservation is particularly relevant here. Hence, the spatial relocation of economic activities increases the difficulties in balancing the nature-society interactions. Noteworthily, empirical findings show that the rapid development of coastal regions is not at the sacrifice of the inland regions' environment.

At the regional level, the core-periphery patterns of conventional economic activities take a variety of forms. Typical types include the core city and border cities within the same province, the core city and its connected cities within the same city agglomerations, and the core city and its surrounding places within a physio geographic unit. Various core-periphery patterns in China thereby raise two noteworthy questions. One is the changing location of polluting activities. Conventionally, theoretical thinking posits that along with the regional development, polluting activities will relocate from the core to the periphery. On the one hand, the polluting activities in the core regions will be replaced by non-polluting ones. On the other hand, the relocation to the periphery can make use of the border effects. However, empirical findings from China reveal that neither the replacement effects or the border effects can uniquely determine the location of polluting activities. The location of polluting activities is subject to the trade-off between the agglomeration externalities in core regions and the border effects in periphery regions. There is not always a pollution-thy-neighbour phenomenon for polluting activities at the regional level.

The other one is the environmental effects of accelerating regional integration. Empirical findings in this book demonstrate that the spatial division of labour does not result in the polarisation of environmental performance among regions. On the contrary, it is expected to narrow the gap of environmental performance between the core and the periphery. Recent regional development in China has witnessed an accelerating process of integration. A bunch of city agglomerations and strategic regions emerge, seeking to break up conventional administrative confines and strengthen inter-local cooperation. Regional integration is likely to extend the scale economies along with the cooperation network, and thereby improve local environmental performance. Besides, regional integration will also reduce the risk of environmental race to the bottom phenomena by promoting inter-local cooperation.

12.2.2 The Global-Local Interactions

The global-local interactions promote environmental upgrading but simultaneously introduce environmental risks. Integration with the global market is one of the most important engines that spark China's economic miracle. By interacting with the already-existing local linkages, the establishment of global linkages further reshapes the internal geography of China. Empirical studies in this book investigate several channels through which global-local interactions affect the environment. At the national level, empirical findings reveal that China faces the risk of being stuck in a "pollution trap". The intra-industry trade with developed economies worsens Chinese pollution terms of trade. Regarding the inter-industry trade with developing economies, the diluting advantages of the labour force, and the accumulating advantages of the capital stocks keep China specializing in polluting activities. Hence, continuous economic growth is less likely to make China relocate polluting activities to other developing economies. Taken together, integration with the global market is likely to make China become a convergence of embodied pollution unless an efficient economic transition can be accomplished.

At the regional level, empirical findings suggest that the environmental effects of establishing global linkages in a certain region depend on the synergy of global transfer, domestic transfer, and local absorption. As for the global transfer, regions have different trading partners overseas, which would generate different environmental impacts. Trade with developed economies or surrounding economies tends to expose one region to more pollution emissions. The domestic transfer is subject to the industrial mobility of low-value added sectors. Overall, there is a spatial trend from developed regions to lagging ones. Local absorption represents the positive effects of agglomeration externalities, which can offset the adverse nonlocal impacts by local capabilities. Our findings show that lagging regions in China have witnessed a rapid increase in both global and domestic transfer, while their local absorption is rather weak. By contrast, the leading regions offset the risks of global transfer through the channel of domestic transfer and local absorption.

Besides the spatial relocation of environmental burdens, the global-local interactions also work to promote economic upgrading, which addresses the environmental issues in an endogenous manner. In this book, we posit that foreign investment and foreign trade interact with the configurations of local clusters and promote the firm dynamics there. Their combination promotes the local economic upgrading in different ways. Empirical findings indicate that global-local interactions during the past two decades accelerate the firm dynamics and product diversification, generating significant displacement effects on the environment. That is to say, in China's case, the global-local interactions improve environmental performance by industrial restructuring to a larger extent instead of efficiency promotion.

From an evolutionary perspective, the global-local interactions serve as an important force to avoid polarised environmental performance among regions as well as the overdependence on polluting activities. Empirical findings reveal that industrial restructuring in China does not always replace dirty industries with relatively clean ones. The already-existing development base exhibits strong power in maintaining the incumbent development path and resists significant deviation. However, such a path dependence does not always lead to poor environmental performance, which is also related to the interactions within local clusters. On the other hand, empirical evidence also indicates that global linkages tend to introduce new firms with better environmental performance, contributing to environmental improvement as well as path breaking. Taken together, the environmental-benign path creation in China is an outcome of the global-local interactions.

12.2.3   The Government-Market Interactions

The government-market interactions determine the efficiencies of environmental-relevant governance. The interactions between governments and markets are especially relevant in China's cases. Empirical findings in this book echo this point from three different perspectives, namely, the environmental regulation, the state-owned enterprises, and the interregional competition.

The effects of China's environmental regulation on environmental improvement are overall positive. At the national level, although the levels of stringency and enforcement of China's environmental regulation are relatively lower than some developed economies, they are the highest among most developing economies. This fact indicates that environmental improvement in China does not rely on further shifting the environmental burden to other developing economies. Instead, environmental improvement is an outcome of economic restructuring and environmental regulation. At the regional level, environmental regulation overall contributes to pollution reduction and the improvement of environmental performance. Particularly, empirical evidence suggests that environmental regulation has supported intersectoral upgrading by promoting firm dynamics. However, it is also noteworthy that the stringency and enforcement of environmental regulation vary from place to place, which would thereby result in the pollution haven phenomenon among domestic regions.

The government-market interactions are also manifested by the modes of environmental regulation, including the command-and-control mode and the market-oriented mode. Empirical results suggest that these two modes are efficient in China's context, largely reducing pollution emissions in Chinese regions. However, the market-oriented mode has more sustained effects. More importantly, the market-oriented mode of environmental regulation has exhibited its potential in simultaneously promoting economic performance and environmental performance. Therefore, we argue that the market-oriented mode represents the direction of environmental governance reform.

As for the state-owned enterprises (SOEs), the marketization reform in China has significantly changed the ownership structures of firms. The share of private firms keeps increasing, and most of the low-efficient SOEs have been transferred or closed. As a result, empirical results indicate that SOEs are generally larger in size and older in age. Most SOEs belong to the heavy and chemical industries, which rank among the most pollution-intensive sectors. Therefore, empirical findings show that the environmental effects of SOEs are overall negative.

Interestingly, empirical results further reveal that SOEs affect the environment of the coastal cities but improve the environment of the inland cities. It is noteworthy that the coastal cities in China are specialized in export-oriented sectors, while heavy and chemical industries tend to concentrate in the inland cities. In this context, we can't simply criticize that SOEs impact the environment due to the lack of efficiencies and productivities. According to the empirical results in this book, if we control the fixed effects of industries, the adverse effects of SOEs on the environment will be less significant, or even insignificant. These findings point to the fact that although not always sensitive to profits and costs, SOEs are sensitive to regulations and social responsibilities. Hence, the environmental effects of SOEs are quite complicated, depending on their balance between governmental regulations and market rules.

Last but not least, the decentralization reform allows local governments to have primary responsibility and great autonomy for economic development in their jurisdictions. Local governments have to self-finance budgets and development. Hence, this reform increases the gaming between central and local governments and the inter-local competition for resources and opportunities. In environmental terms, there is a high risk that local governments seek to compete for resources and opportunities by lowering environmental standards. Empirical results in this book support that these processes lead to the environmental race to the bottom phenomenon to some extent. Along with the industrial transfer processes, the adverse effects of decentralization on the environment tend to be observed in the central cities rather than the western ones.

12.3 Is EEG a Fresh Start?

In this book, we have argued that economic geographers should not keep silent in response to the growing environmental challenges. On the one hand, the twin processes of global environmental change and globalization are increasingly interwoven. The environment is no longer an external factor of human activities. As the techno-economic paradigm shifts towards green-oriented innovation, there is no reason for economic geographers to keep overlooking the environmental issues. On the other hand, responding to environmental challenges has become an urgent task for human development, and attracted interdisciplinary efforts from both physical sciences and social sciences. It is safe to say that responding to environmental challenges provides a good opportunity for various scientific fields to communicate with each other and support decision-making. Also, as a discipline concerning space-economy interactions with a long history, there is no reason for economic geography to retreat from environmental issues.

Advances in economic geography provide rich and deep theoretical thinking about regional development, globalization, innovation, institution, and transition. These topics are closely related to the key themes of environmental challenges, such as the sustainable development, resource management, energy transition, industrial transformation, cleaner production, responsible consumption, global environmental governance, green innovation, and sustainability transition. Economic geography can make distinctive contributions by unravelling the rationales of why economic activities can/cannot occur in one place in particular manners and the rules of how economic activities in one place interact with those in the other places. These rationales and rules are usually overlooked by conventional studies, which lead to unpractical policy implications for responding to environmental challenges.

Admittedly, there are sustained efforts on developing EEG, although they do not yet influence the mainstream of economic geography. Also, incumbent studies on EEG are fragmented and polyvocal without a coherent research agenda. However, we argue that the current practices of EEG share some similarities, which may capture the essence. Firstly, compared with other environmental studies, EEG focuses more on the context-dependence of environmental problems. Even for the same environmental problems, the sources, causes, state, effects, and feedbacks are different among places. Therefore, EEG seeks to answer the question of why economic activities challenge the environment in one place but not the others.

Secondly, EEG also delves into the border-crossing nature of environmental issues, based on its theoretical thinking of spatial interdependencies. The spatial separation between production and consumption has created various types of channels to transfer environmental burdens across places. As such, EEG can answer how spatial interdependences will extend the environmental effects of one place's economic activities to others.

Thirdly, EEG points to the fact that economic activities operate within the environmental conditions and institutional contexts at multi-scales. Since the determinants of economic activities are not necessarily local, it is natural that the environmental effects of economic activities are subject to nonlocal forces. This is particularly the case in the environmental governance at the global and national levels. In this regard, EEG can respond to the question of how the national and supra-national institutional environment will moderate the environmental effects of economic activities.

Last but not least, EEG highlights the co-evolution between the environment and the economy. Environmental evolution represents the shift from one equilibrium state to another, while economic evolution looks for novelty. Economic novelty may either maintain or disturb the environmental equilibrium. Likewise, the shift in environmental equilibrium will also benefit or impact economic novelty. In this context, EEG can answer the question of how to coordinate the interactions between economic novelty and environmental equilibrium, which are essential to sustainable transition/sustainable development.

These four facets are closely related to advances in economic geography. The context-dependence is based on both the neoclassical locational theory and the cultural/institutional turns of economic geography. The border-crossing and multiple scales rely on the relational turn of economic geography. The border-crossing represents the horizontal linkages, while the multiple scales indicate the vertical linkages. The co-evolution between environment and economy holds strict connections with the evolutionary turn of economic geography. Therefore, we conclude that developing EEG does not have to create a new intellectual territory. Instead, EEG can start with linking the advances in economic geography with environmental issues.

12.4 Implications for Moving Forward

The practice of EEG with the Chinese case in this book tries to highlight the importance of empirical works in developing EEG at this stage. As we review in Chap. 2, we appreciate the continuous efforts in establishing a consistent research agenda for EEG. But theoretical thinking needs to be examined in practice, which may, in turn, provide lessons to improve the thinking. In this regard, this book tries to take one step further. Nevertheless, environmental issues take various forms. This book is less likely to provide a panorama for the full potential of EEG. At last, we discuss the likely prospects for the future as follows.

First of all, we notice that the incorporation of economic geography theory can enrich our understanding of controversial theoretical thinking. For example, empirical findings in this book provide several new insights into the pollution haven hypothesis phenomenon, the environmental race to the bottom phenomenon, the beggar-thy-neighbours phenomenon, and so on. From the spatial perspectives, findings reveal that the debates over these phenomena are not always ascribed to the technical deficiencies. In essence, these phenomena are problems with the location selection of polluting activities. Conventional approaches centre on the role of environmental regulations. However, the spatially various environmental regulations can not uniquely pin down the location of economic activities. In this regard, advances in economic geography provide a variety of theories and approaches, which span from the neoclassical framing to the new economic geographies. There are still other similar cases that EEG can raise geographical questions from the ongoing debates, then making a distinct contribution.

Secondly, we find that the geographical perspective is under the development trend of global scientific projects on environmental issues. Also, we see the value of geographical wisdom in addressing environmental concerns, such as transboundary pollution, multi-level governance, and sustainability transition. For example, empirical findings in this book indicate the value of spatial interdependencies, trans-scale interactions, and path dependence in responding to environmental challenges. They point to the fact that the spatial patterns, configurations, and dynamics of economic activities have their own rules, which can not be modified by assumption. In this regard, EEG points out that it requires a broad picture to understand better why environmental challenges occur in one certain place, thereby delving into its socio-economic linkages with surrounding or even distant regions. On this basis, EEG should further focus on whether one place can replicate other places' successful experience in responding to environmental challenges.

Third, environmental studies are currently an arena with considerable policy relevance. Conventional studies tend to make various recommendations based on controlling undesirable environmental outcomes. However, they seldom ask why and how these economic activities come into being. They usually overlook the rationales behind these economic activities. More importantly, they merely concern the feasibility of their recommendations in socio-economic terms. In this regard, EEG does not have to follow the studies on tracing the sources of environmental issues. Instead,

EEG can trace the reason why environment-unfriendly activities come into being and answer the question of how these activities can be removed.

Last but not least, besides the industrial pollution reduction, which is used as examples in this book, there are a wide array of hot topics including greenhouse gas emissions and climate changes, depletion of natural resources, energy issues, and sustainability transition. In the related literature, although the voice from economic geographers is still weak, studies from economic geographers have already exhibited how theoretical thinking based on economic geography will make a difference. As such, the development of EEG requires more practices in various topics of environmental issues as well as adaptive learning from these practices. The current literature on EEG follows a top-down approach, devoting to a consistent research agenda for EEG. However, considering the diversity and complexity of environmental issues, we would like to propose a bottom-up approach that is a learn-by-doing process with the practice of theoretical thinking in economic geography.